FIXING SEX

FIXING SEX

Intersex, Medical Authority,

and Lived Experience

KATRINA KARKAZIS

DUKE UNIVERSITY PRESS
Durham and London
2008

Printed in the United States
of America on acid-free paper ∞
Designed by Amy Ruth Buchanan
Typeset in Quadraat by Keystone
Typesetting, Inc.
Library of Congress Cataloging-in-
Publication Data appear on the last
printed page of this book.

To my family, for everything

And to Carole, Suzanne, and Bo,
without whom . . .

Do we *truly* need a *true* sex? With a persistence that borders on stubbornness, modern Western societies have answered in the affirmative. . . . For a long time, however, such a demand was not made. . . . Indeed, it was a very long time before the postulate that a hermaphrodite must have a sex—a single, a true sex—was formulated.

—**MICHEL FOUCAULT,** introduction to *Herculine Barbin: Being the Recently Discovered Memoirs of a Nineteenth-Century French Hermaphrodite*, 1980

CONTENTS

ACKNOWLEDGMENTS

My greatest debt of gratitude goes to the adults and parents who agreed to be interviewed for this project. For their willingness to share details about deeply personal and often difficult and painful experiences I am profoundly grateful. It goes without saying that there would be no book without them. It is my hope that this book both captures and conveys their experiences in a meaningful way.

I am exceedingly grateful to the physicians who agreed to be interviewed for the open and sincere way in which they reflected on their clinical practice especially during such a contentious period. I have also benefited greatly from the expertise of many clinicians and researchers who have been very generous with their time and expertise. They include: Ian Aaronson, Sheri Berenbaum, Bill Byne, Bill Cromie, Mel Grumbach, Phil Gruppuso, Elizabeth McCauley, Heino Meyer-Bahlburg, Melissa Parisi, John Park, Bill Reiner, David Sandberg, Justine Schober, Nina Williams, and Bruce Wilson.

I owe special thanks to Cheryl Chase who has dedicated her life to improving medical care for those with gender-atypical bodies and to formulating a new treatment paradigm driven by extraordinary compassion and concern for the lives of those treated. Regardless of whether one agrees with her, it is undeniable that without her initial efforts, there would be no debates of which to speak and much slower progress on improvements in care.

Several outstanding scholars have helped me with my work. I owe a tremendous debt to Carole S. Vance, whose keen insight and intellectual rigor inspire me. I have benefited enormously from our friendship, and I only hope to give back some of what I have received over the years. Sherry Ortner, Shirley Lindenbaum, and Lesley Sharp have served as role models who have inspired me through their thinking and research. Their insightful comments have helped me clarify my thinking, and their enthusiasm is always appreciated. This book would not have been possible without the pioneering work of Suzanne Kessler. It is my hope that this research complements her trailblazing work. I thank her and Rebecca Young for their knowledgeable, detailed, and thoughtful readings of the manuscript and their critical insights.

Numerous other colleagues and friends have provided wonderfully provocative ideas, intellectual insight, or encouragement during the course of this project. They are: Marcus Arana, Anthony Asciutto, Arlene Baratz, Karl Bryant, David Cameron, Adele Clarke, Alice Dreger, Betsy Driver, Steve Epstein, Ellen Feder, Jennifer Fishman, Jane Goto, Janet Green, Thea Hillman, Emi Koyama, Geertje Mak, Ilan Meyer, Ali Miller, Iain Morland, Sharon Preves, Alison Redick, Anne Tamar-Mattis, and David Valentine.

Several institutions have supported this research through its many stages. I could not have completed it without the generous funding of an Individual Research Grant and the Richard Carley Hunt Publication Grant, both from the Wenner-Gren Foundation for Anthropological Research. I am indebted to the Sexuality Research Fellowship Program (SRFP) of the Social Science Research Council (with funds provided by the Ford Foundation) for supporting my work through a Sexuality Research Fellowship and a Sexuality and Policy Fellowship. For ten years SRFP provided funding to projects that otherwise might never have been completed, and I am sorry to have seen it end. Many thanks are due to Diane DiMauro for shepherding the program during its existence. I also received excellent research assistance from Dana Greenfield at Barnard College and from Candace Frazier through the Program in Feminist Studies at Stanford University.

Stanford University and especially the Center for Biomedical Ethics have provided a valuable incubation period for me to reflect on this work. I am extraordinarily grateful for the support and intellectual engagement from everyone there, but especially from Barbara Koenig, David Magnus, Judy Illes, Anne Footer, Sandra Lee, Holly Tabor, and Chris Scott.

I thank Ken Wissoker for seeing value in what I had to say and Courtney Berger, Molly Balikov, and everyone at Duke for bringing the book into being. I am indebted to Erika Büky for her editorial wisdom and Petra Dreiser for her copyediting. Kate Ouderkirk diligently transcribed the hundreds of hours of interviews on which this book is based, and I am very thankful for her keen ear and able fingers. Chris Stave at Lane Library at Stanford University provided excellent assistance locating references. Margie Towery proved herself to be the champion of indexers.

Many have encouraged and supported me over the years of research and writing. This support has sustained me and was especially critical as I was finishing this book and was diagnosed with breast cancer. I have never known such bodily vulnerability, such fear, and such love, and my deepest appreciation goes to those who comforted and cared for me during the work's completion and above all during the year in which I underwent treatment: Jackie Karkazis, Sharon Lazaneo, Mildred Cuyler, Dennis Cuyler, Agnes Cuyler, Howard Cuyler, Devin Cuyler, Reese Pecot, Bee Davies, Justine and Darryl Saussey, Rodger and Margaret Davies, Edna Marsh (Little G), Joy Leahy, Betty Cronin, Roxanne Varzi, Elena Muldoon, Rosalind Wiseman, Isolde Brielmaier, Craig Handy, Svati Shah, Deb Pelto, Elizabeth Kolsky, Monica De Leon, Ng'ethe Maina, Carole Vance, Beck Young, Sal Cooper, Marysol Asencio, Penelope Saunders, Robin Mathias, Penny Huang, Gillian Hadfield, Marian Chapman, Dan Ryan, Jennet Robbins, Holly Tabor, David Magnus, Rob Wallace, Melanie Suchet, Ruth Goldman, Margaret Harding, Jennifer Fishman, Nicholas King, Murat Akalin, and Bill Barnes and Charlene Clark (and their rowdy clan). Most of all I thank Jon Davies—for his love, patience, computing know-how, and his willingness to adopt a new homeland. I am so very grateful.

INTRODUCTION

It is the evening of May 17, 2000, and I am seated in a packed auditorium at the New York Gay and Lesbian Center on Little West Twelfth Street in Manhattan for the presentation of the Felipa de Souza Award. The prize, given by the International Gay and Lesbian Human Rights Commission (IGLHRC), is awarded to individuals and organizations that have made significant contributions toward securing human rights and freedom for sexual minorities. Tonight, Cheryl Chase, the founder of the Intersex Society of North America (ISNA), is being honored for her efforts to change the medical treatment for people with intersex diagnoses, work she began in earnest in 1993. The audience is standing room only, filled with human rights and gay activists, intersex adults and their supporters, a world renowned specialist on intersexuality, and John Colapinto, who had just published his book *As Nature Made Him: The Boy Who Was Raised as a Girl*. The book, about the tragic story of David Reimer, whose penis was burned off during a circumcision accident and who was subsequently raised as a girl, would hit the *New York Times* bestseller list several months later. Although Reimer did not have an intersex condition, he was treated by the psychologist John Money, the man largely responsible for developing the current treatment paradigm for intersexuality, making Reimer's heartbreaking story important for intersex activists fighting to change medical care.

The executive director of IGLHRC, Surina Khan, excitedly introduces Chase, noting that before ISNA, "very few of us knew about intersexuality." Arguing that there is no justification for early genital surgery other than "doctors' quest for normalcy," she likens the procedures to torture, calling it intersex genital mutilation. "This is wrong. It's torture. These children are subjected to involuntary surgery. Intersex people are not sick, they are not in need of care, but so-called rational medicine is coming after these kids with knives in their hands." The audience responds with a huge round of applause.

Although I am deeply sympathetic to intersex adults' criticisms of their medical care, Khan's comments paint a disturbing image of half-crazed doctors running down hospital corridors wielding knives—one that clashes with my knowledge of clinicians working in the field of intersexuality, whose intentions are more benevolent. I wonder what the clinical specialist in the audience makes of the claim that he and his colleagues are mutilating children. I furtively glance over to gauge his response; his face is expressionless, unreadable. It is hard not to get caught up in the heartfelt emotion of the moment and to be excited by the radical rethinking of intersexuality implied by the director's critique: Why are gender-atypical bodies construed as a medical problem? Should the medical profession have the right to make treatment decisions at all for people born with intersex conditions? But it is equally hard to believe that the issue is as simple as doctors torturing children. I am brought back to the moment by a huge round of applause as Chase takes the stage.

Chase tells a deeply moving story about her parents' experience at her birth. She explains that when she was born her mother knew something was wrong with her baby; and that doctors sedated her mother for three days while they figured out what to tell her. Eventually the doctors told her parents that they had a "deformed boy" and sent them on their way. At age one and a half, Chase's parents took her to be evaluated by specialists who concluded that their son was actually a girl with a large clitoris. As she tells it tonight, "the doctors decided to remove my clitoris, told my family to leave town and not tell anyone what had happened, and to destroy all the old photos of me as a boy," recommendations that sent a loud message that their daughter's condition was shameful and something to be hidden at all costs. "They felt that by doing this I would become a well-adjusted girl, marry a man, and have children," Chase says with a wry smile, adding: "Well, their prediction did not work out." The last point elicits uproarious laughter—Chase is a lesbian—and then she pointedly adds, "They picked the wrong person to do this to."

Although her language and demeanor is more tempered, Chase's analysis of what is problematic about the treatment of intersexuality is not far removed from Khan's. Doctors, she argues, do not understand female sexuality, think homosexuality is a failure of treatment, refuse to refer families to therapists and social workers, and encourage parents never to discuss the diagnosis with others or the child, thus instilling extraordinary shame in parents (and hence the child) about the condition. Focused on normalizing infants, she notes, doctors have failed to ask what intersex individuals themselves want. Early genital surgery, she says, is intersex genital mutilation, and "the number of people targeted is one in two thousand or five a day," recasting what doctors understand as medical treatment as, instead, a gender battle waged on the bodies of small children.

Fast forward to October 2005 when fifty international experts in such fields as pediatric endocrinology, pediatric urology, genetics, and gender-identity development are gathered in Chicago to revise treatment guidelines for infants born with what are broadly called intersex diagnoses.[1] It is the first time researchers and clinicians will so thoroughly revisit the medical standard of care for these diagnoses since Money and his associates first proposed treatment standards in the 1950s. Chase and another patient advocate have also been invited to participate, a kind of collaboration unimaginable a few years earlier. The meeting would not have happened without intersex activists' growing chorus of demands, beginning in the early 1990s, for changes in medical treatment practices that had driven the field into a deep and divisive crisis by the year 2000.

The earliest challenges to the traditional treatment paradigm came from a number of adults who had been treated as children and who felt that this treatment paradigm, with its focus on rapid gender assignment and genital surgery, had caused extraordinary and irrevocable harm, even though it had been designed to ease their psychological and social adjustment. They pointed to the lack of complete and honest disclosure to parents about the child's anatomy and condition and to the child about her or his treatment history, to the rush to normalize atypical genitals by performing surgery, and to the desire to erase gender atypicality in the name of care. In addition, some of these individuals have asked why bodies challenging traditional beliefs about gender difference have been construed as problematic and forced to conform to male and female ideals through hormonal and surgical shoehorns (Fausto-Sterling 1993). While doctors decided how to deal with bodies that transgressed naturalized ideas about gender difference, intersex activists and others argued that it was the rigid ideas of gender difference that were transgressing the nature of their bodies.

At the Chicago meeting, participants agree to recommend several important changes to care that demonstrate a significant shift in thinking. Providing recognition and advising caution, the guidelines state that intersex conditions are not shameful and suggest that psychological care should be integral to medical treatment. Given that patients and parents (and even clinicians) often find the terminology and labels surrounding intersex diagnoses confusing, misleading, stigmatizing, and distressing, participants agree to change the medical nomenclature. In the proposed system, the term *intersex* would be replaced by the more general descriptor "disorders of sex development" (DSD), referring to congenital conditions in which chromosomal, gonadal, or anatomical sex development is atypical. Other terms such as *hermaphroditism*, *sex reversal*, and gender-based diagnostic labels were to be replaced by a system based on clinically descriptive terms (e.g., *androgen insensitivity syndrome*).[2]

Suggesting a willingness to think more expansively about culture-bound assumptions about gender and its relationship to sexuality, the guidelines note that homosexuality should not be construed as an indication of incorrect gender assignment, a point that Chase had made five years earlier at an annual meeting of pediatric endocrinologists, the specialists primarily responsible for treating children with DSD. Participants also recommend that the potential for fertility—originally emphasized for female gender assignment only—be an important consideration for male gender assignment as well.

Despite the unquestionably positive developments they encompass, the guidelines also encapsulate the more entrenched obstacles to medical understandings of and treatment practices for intersexuality. From the standpoint of treatment, the new guidelines fail to resolve an issue that lies at the center of current controversies—early genital surgery. The guidelines acknowledge that there are minimal systematic outcome data about genital surgery, that orgasmic capability may be harmed or even destroyed by such surgery, and that there is little documentation to support the widely held belief that early surgery relieves parental distress about atypical genitals. The statement nevertheless says that surgery can be considered for young girls with "severe" genital virilization, which would include procedures to reduce a "too-large" clitoris or to create, open, or elongate a vagina on babies and toddlers.[3]

Although the guidelines incorporate important changes based on intersex adults' and others' criticisms of care and will foster increased awareness of these criticisms among physicians, they demonstrate an unwillingness (or inability) to think about intersexuality in terms other than biomedical (and pathological). From the physician's point of view, gender assignment or surgical techniques are controversial, but the existence of intersex bodies and the need to treat them are not. From a medical standpoint, what is at issue in debates over intersexuality is not the category intersex per se, but what theories and technologies are most appropriate to treat individuals with intersex diagnoses. It is rare for clinicians to view the line separating intersex from non-intersex as culturally determined because most physicians see their taxonomies as apart from culture, not as reproducing culture (Kessler 1998). But the whole reason intersex even exists as a category is because these bodies violate *cultural* rules about gender. These rules assume an agreement among a series of somatic characteristics (chromosomes, gonads, genitals, and secondary sex characteristics) and more phenomenological processes such as gender identity, gender role, and sexuality. The debates over when to perform surgery and how best to decide gender assignment obscures the fact that in trying to make infants with intersex diagnoses "normal" boys and girls, physicians and parents are necessarily drawing on cultural ideas about what constitutes male and female.

By avoiding these broader questions and issues, including whether gender atypical bodies require treatment, the guidelines sustain the assumptions that physicians should intervene in embodied processes to control the "sex" of the body, that treatment might be a wholly unambiguous good, and that good intentions result in good care. These assumptions stem from the fact that biomedicine, characterized by pragmatic thinking and preoccupied with materiality and physicality, often atomizes or individualizes issues and pinpoints the body and its parts as appropriate sites of intervention. Science and medicine normalize the view that adjusting the material world to human aspirations is a positive goal (Rapp 1999).

My reason for lingering on these guidelines is not because I (or anyone else) expected them to solve all aspects of these debates once and for all, but because they illustrate that questions still linger of exactly which aspects of intersexuality are open to questioning, who belongs "inside" to evaluate medicoscientific problems, and to what extent they are allowed to participate (Harding 1991). Perhaps unsurprisingly, given their narrow, medicalized view of intersexuality, the guidelines fail to address the widespread social fears and assumptions that drive responses to intersexuality: namely, a vision of gender fraught with overdetermined investments, desires, and anxieties that creates the need for consistent and unified gendered ways of looking and being. From this perspective, the guidelines appear as little more than a new biomedical technology for the management of intersexuality (see Rapp 1999: 45).

Far from existing outside culture, biomedicine is a cultural entity that not only has unparalleled discursive and practical powers to define and determine what it is to be normatively human but also to withstand alternative constructions and challenges to its version of normativity (Rapp 1999: 13–14). Indeed, the guidelines assume that what constitutes normal bodies and ways of being is uncontested, unambiguously understood, and that the impact of treatment might be universally assessed. They also embody the presumptive universality of terms and phrases such as *good sexual function*, as if this were self-evident, understood and experienced similarly by all persons. The guidelines are consistent with assumptions that biomedical experts alone can and should judge what would constitute a "good" outcome to treatment. They provide a sealed and self-confident narrative of the important issues in the treatment of intersexuality.

Eschewing any wider social and cultural context for understanding intersexuality, the guidelines instead enclose it in a discourse of medical management. Their vocabulary is predominately medical, with no language available to frame alternative descriptions or understandings. Nor does the document provide much critique of the assumptions that drive treatment. The lack of sustained attention to gender as a way of marking difference, or to how ideas about gender variance are rooted in our cultural practices of thinking about the body, circumscribes how we understand intersexuality and what we, as a society, believe should be done for those born with gender-atypical anatomies. Human sexual difference is seemingly obvious and certainly real on many levels, but in another sense it is a carefully crafted story about the social relations of a particular historical time and place, mapped onto available bodies (Fausto-Sterling 1995: 21).

Nevertheless, the meeting itself and the gradual shift in recommendations would not have happened without a decade of efforts to change medical protocol by intersex adults who felt harmed by their treatment. The inclusion of ISNA in particular marked a shift in the aims and the reputation of a group that began as a collection of outsiders toward whom clinicians felt deep suspicion and antagonism; instead, it now emerged as a (limited) partner and resource for ideas about how to improve care. Early activism attempted to make intersexuality visible and to convince clinicians of existing problems with the standard of care. A decade later, the topic has received extensive coverage in documentaries, newspapers, magazines, on television, and even in novels.

At birth, the sex of every infant is determined based on an inspection of the external genitalia and the understanding of the newborn as a "girl" or a "boy." The process of gender assignment at birth is usually uneventful, but each day, somewhere in the United States, an infant is born for whom gender assignment is not obvious. These infants may have any one of numerous diagnoses,[4] but their common feature is gender-atypical anatomy—a combination of what are typically considered male and female chromosomal, gonadal, and genital characteristics—which is often signaled by the presence of what clinicians call ambiguous genitalia.

Prior to the middle of the twentieth century, medical intervention in intersexuality was not routine in the United States and in some cases occurred only in adolescence or adulthood at the request of the individual.[5] Although physicians had long been assigning gender at birth based on the predominant characteristics of the external genitalia, and medicine at this time was developing increasing authority in matters of the body and thus of sex determination, its involvement in theses cases was ad hoc. This is in part because medicine had little to offer in terms of treatment, successful or otherwise. By the middle of the twentieth century, however, intersex births had come to be labeled a medical and social emergency.[6] Raising a child with a gender-atypical anatomy (read as gender ambiguity) is almost universally seen as untenable in North America: anguished parents and physicians have considered it essential to assign the infant definitively as male or female and to minimize any discordance between somatic traits and gender assignment. Since the 1950s, clinicians have used a treatment protocol developed by John Money and his colleagues at Johns Hopkins University to assign a gender. Emphasizing thorough but swift clinical workups to determine the etiology, clinicians determine a sex for these infants, and surgeons then modify the infant's body, especially the genitals, to conform to the assigned sex.

The shift to interventionist treatments for intersexuality resulted from developments in areas such as plastic surgery, urology, biology, and endocrinology. The most important development, however, was the work on gender identity by Money and his associates. Gender identity and role, they argued, was not something individuals were born with, but something built up cumulatively over time, much as the acquisition of a language. It was also malleable to a certain extent: Money suggested a small window of gender flexibility (until eighteen months of age) before which gender should be assigned. These authors stressed the importance of thorough assessments of infants born with so-called ambiguous genitalia to identify the etiology, assess the intervention possibilities, and determine the intervention most congruent with anticipated physical developments in puberty and adulthood. This treatment model was quickly and broadly adopted.

For roughly four decades, Money's recommended treatment was widely accepted by clinicians and virtually unchallenged by those treated. This situation changed dramatically in the 1990s. As some children treated according to the protocol reached adulthood, they began reexamining what had happened to them. Traumatized and angry, they started to look for sources of information about their situation and for people who had undergone similar experiences. The rapid expansion of the Internet, with the development of relatively confidential, anonymous online discussion forums, facilitated these efforts. Simply finding ways to live with the consequences of their diagnoses and treatments did not satisfy some affected individuals; they have instead sought to change the context in which intersexuality is understood and treated to, they hope, save others from experiencing similar pain and distress.

Drawing energy from social movements for women's rights, civil rights, and gay liberation and building on recent challenges to and shifts in medical authority, intersex adults and their supporters have increasingly claimed knowledge and authority about the meaning and appropriate medical response to intersexuality. On one level, those in the intersex advocacy movement have variously objected to the timing and necessity of genital surgeries, to the biomedical notion that genitals are naturally dimorphic, and to the presumption of heterosexuality implicit in the treatment recommendations. Some have also questioned the right of the medical profession to make treatment decisions at all. On another level, this movement has argued for the acceptance, dignity, and humane treatment for those with gender-atypical bodies in an effort to challenge ideology, practices, and consciousness.

These developments emerged in a context of broader social changes in attitudes toward the body, gender, and sexuality, and toward the authority of science and medicine. At about the same time, feminist academics, parents of children with intersex diagnoses, and eventually clinicians, ethicists, and legal scholars began offering alternative views on the meaning and construction of intersexuality and appropriate medical responses to it. These views provoked turbulent controversies in the popular media, propelling intersexuality from the shadows to popular consciousness. What was once known only to physicians, researchers, and those affected suddenly became showcased in national newspapers and on network television programs. In academia, intersexuality emerged as a staple in many women's studies courses for what it revealed about the social construction of gender.

What made these criticisms possible and perhaps even inevitable? The naming and treating of intersexuality represents one aspect of the increasing tendency to turn social issues into biomedical problems that can be solved by clinical intervention. The current treatment protocol provides a structure and a method for addressing violations of gender rules that individualize, privatize, and depoliticize the meaning of those transgressions. In this view, gender problems exist within the body of the individual, not in social and medical understandings of what is gender-typical or even healthy. Intersex is a catchall category that encompasses dozens of medical diagnoses, but the defining feature is that intersex bodies in some way violate the commonly understood biological differences between males and females. Intersexuality does not represent a point of pure liminality between sexes. The category intersex relies on the very categories of the medicalization of sex, and it is meant to cover a range of disparate diagnoses and biologically diverse individuals. But the breadth of human physical variance is more complex than the category allows for.

Bodies, Medicine, and Gender

Although this book explores a seemingly exotic or rare issue, intersex is unique only because it makes explicit the cultural rules of gender. Put another way, because the treatment for intersexuality exemplifies attempts to codify normality and abnormality, the frequency of intersexuality is less important than ideas about how to make these infants "normal" boys or girls. As clinicians and parents try to make sense of a situation in which the expected bodily concordances have not occurred, they are forced to answer complicated questions: Can a girl have XY chromosomes? Are some penises too small for a male gender assignment? The controversies about the adequacy and consequences of these decisions bring into sharp focus mainstream cultural rules about the proper relationships among bodies, gender, and sexuality that apply to all persons and raise fundamental questions about all bodies by forcing a reassessment of what is understood as natural and normal in connection with the human body (Lindenbaum and Lock 1993: xi). What bodily parts, experiences, and capabilities are necessary for an individual to feel he or she is a man or a woman? And, ultimately, how are bodies, gender, and genitals involved in the creation of men and women? The discussions about whether and how to treat intersexuality expose our cultural anxieties about sex, gender, and their relationship to sexual desire and behavior.

I eavesdrop on these discussions and debates in relatively unguarded moments, when doctors and parents discuss what to do about infants with gender atypical bodies. Their opinions about what constitutes a good outcome reveal the underlying logic of how different bodily parts and functions (e.g., chromosomes, gonads, external genitals, and intercourse) and ways of being normally go together, and what different cultural logics can be used to bring about the supposedly next best outcome.

Bodies whose appearance departs from social and cultural expectations are often subject to various forms of medicoscientific disciplining. Surgical normalization has been one method of reconfiguring such "deviant" bodies. Although surgery can sometimes ameliorate otherwise unsustainable lives—as in some cases of cleft palate—such procedures performed with the intention of normalizing the body often fail to offer that which is hoped for or promised. Whether cast as corrective, reconstructive, or cosmetic, such a surgical reshaping of atypically embodied persons has the effect of limiting human variation and expressing a disdain for atypical bodies. Intersex embodiments are congenital variations that are disabling not so much in that they present functional limitations—which all embodiment does to one degree or another—but rather in that they are corporeal configurations that violate cultural standards. Bodies occupy terrain at the boundary between self and society; they are both subject and object. As a consequence, they are highly politicized and have played significant roles in social crises. Biomedical knowledge and practices have proven instrumental in regulating bodies and populations through their ability to delve deeper into the body, producing highly technical knowledge, facts, concepts, taxonomies, and categories. Medicine has established itself as the sole decoder of the body's many signs. Indeed, medicine has become a dominant discourse in all aspects of life—birth, growth, death, health, and sexuality. Biomedicine's cultural and material authority—the source of its power—is constantly produced and reproduced, creating a cycle critical to its continued existence. But its power, as this book will show, is not fixed or all encompassing.

Difference in the body has long been used as the basis for supporting social projects (Terry and Urla 1995). The body becomes a primary way to locate and mark difference because of its materiality; its realness as it were makes difference appear concrete and unassailable. Situating differences in the body through scientific endeavors reinforces their natural status. The body, far from being a self-evident organic whole, is at best a nominal construct and phantasmatic space imagined very differently over time and across various cultural contexts (Martin 1987). The understanding, knowledge, and representation of the body constitute means to structure complex social relations and establish flows of power. One central articulation of bodily difference is gender. Gender difference has been assumed to extend not simply to the social but to all aspects of the body and biology, and both aspects are assumed to stand in fixed relationship to one another.

Until recently, intersexuality was framed exclusively in medical terms, in large part because over a century ago biomedicine assumed authority over its classification and treatment. The privileging of biomedical understandings has naturalized categories such as those of normal males and females—and their corollary, supposedly abnormal males and females—at the expense of exploring how these understandings are produced and reproduced. Whatever intersexuality may be physiologically (and it is many things), intersexuality as a category of person (requiring medical treatment) is not natural (Holmes 2002). That is, although all bodies are natural in the broadest sense, medicine's formulation of some as intersex is not natural, creating instead a category accomplished in culturally distinctive ways.

Indeed, the power of medicine and science lies in their ability to define what is natural, to name nature and human nature, and in their claim or hope to return individuals to a more natural state or way of being. Medicine and science are grounded in the taken-for-granted status accorded to biological "facts." The distinction between nature and culture relies on a model of nature that is eminently cultural—that is, on a specific concept of the natural that can stand for itself as a domain of immutable and fixed properties. Fierce debate about what the category of natural comprises, as well as about how society deals with that which does not fit into the concepts of the natural or normal, hardly seems surprising. The demarcations of some bodies, behaviors, and ways of being as normatively gendered—and others as not—is a biopolitical project that raises fundamental questions about how to understand the role of medicine in shaping and governing humans and human experience (Adams and Pigg 2005).

Sex, Gender, and Sexuality

The categorization of individuals as male or female is woven into the fabric of daily life, often in ways that elude our awareness. Cultural understandings of categories such as male and female and ideas about appropriately gendered subjects drive treatment decisions for intersexuality. Male-female gender dichotomies permeate discussions of intersex at numerous levels—from debates over where sexual difference is located in the body, and which organs or genitals properly sexed subjects can or should have, to the sexual orientation and sexual behavior of appropriately gendered subjects. This book examines debates over intersexuality for what they reveal about cultural understandings of gender difference. It exists on numerous levels: anatomical difference (males and females), behavioral and psychological difference (masculine and feminine behavior and gender identity), and erotic difference (masculine and feminine sexuality).

Prior to the 1970s many took for granted that biological sex determined one's identity and behavior. Unlike sex, a biological concept, gender was imagined as a social construct specifying the socially and culturally prescribed roles that men and women should follow; it either grew out of biology or was mapped onto it. Several sequelae derive from this distinction. First, biology and culture are constructed as distinct realms. Second, biology is assumed as stable, immutable, and internally consistent for males and females. Third, biology is the substance on which culture works, rather than the other way around.

In the 1970s feminists theorized a distinction between biological sex and gender, highlighting that reproduction did not cause gender difference in any natural or obvious way (Rubin 1975: 159). Arguing that the relationship between the two is far from natural but instead a system by which sex is fashioned into gender, they sought to weaken biological essentialist arguments that ascribed women's inferior status to innate biological differences. In this early work, the material body remained essentially male or female; it was the system of gendered social behaviors attached to these bodies that was open to critique.

In this critique the sex-gender system not only required identification with one sex but sexual desire also was directed toward the other sex (Rubin 1975: 122). Although sexuality and gender are interwoven in complex and varying ways, and cultures tend to experience these linkages as natural and seamless, the specific configurations and points of connection vary historically and across cultures (Vance 1991). Cultures in modernity have assumed an intimate connection between being male or female—a distinction based on a set of physical organs, traits, and characteristics, as well as on reproductive capacities—and the "correct" form of erotic behavior, namely, penile-vaginal intercourse between a man and a woman (Weeks 1986: 13). The construct of difference between males and females and the belief that reproduction is the natural and desired outcome of sexual activity undergird this view.

Our insistence on a so-called true sex is tied to a deep and abiding social interest that individuals engage in "correct" (i.e., socially sanctioned) forms of sexual behavior (Foucault 1980). This *moral* interest in limiting licentious behavior (largely, but certainly not exclusively, focused on same-sex behaviors) has driven the social interest in the *medical* determination of a single true sex. The assumed concordance among bodies, gender, and sexuality provides an ideological framework for the treatment of intersexuality. Clinicians and parents aim to provide a gendered picture that they recognize as cohesive and consistent, in turn producing norm-abiding gendered subjects.

In a theoretical move that represented a departure from other feminist theory of the time and that still resonates today, scholars sought to dismantle two particularly pernicious and taken-for-granted beliefs: that only two genders are given in nature and that sex exists as a biological fact independent of time and place (Kessler and McKenna 1978). Dichotomous gender, far from being natural or innate, or based in our being, is accomplished, constructed, and reproduced in interactions and interpretive processes. Most radically, this critique asserts that the biological is as much a construction as the social; through an understanding of the social construction of gender, biological sex becomes completely destabilized as a separate and coherent category. From this perspective there is little need to distinguish between gender-dichotomous characteristics defined as biological (e.g., chromosomes) and all other aspects of the male-female dichotomy (McKenna and Kessler 2000). When the word *gender* is used to include what we commonly understand as sex (i.e., biological differences), female and male are no more objective or real than the socially constructed categories of woman and man. What was thought to be the base or root of gender difference is actually an *effect* of gender (Kessler and McKenna 1978; Butler 1990; Laqueur 1990; Kessler 1998).

Despite more than twenty years of thinking on this subject both within and outside academe, these formulations have largely not taken hold. We often do not treat the category of gender as problematic, do not view biology as constructed, and even continue to believe that our genitals make us who we are (McKenna and Kessler 2000: 70). Researchers in many disciplines have learned, and continue to teach, that whereas gender is cultural, sex is biological. For those who have grown accustomed to this conceptual distinction, the case for sex itself as a social construction remains to be made.

In this book, I draw heavily on Kessler and McKenna's formulation and on Kessler's later work on intersexuality that derives so beautifully from her earlier work with McKenna. Following their lead I ask, if one postulates bodies (including genitals, gonads, chromosomes, and hormones), what more does the word *sex* buy us? By saying that sex is really about gender, however, I run the risk of failing to problematize the body as a culturally formed dimension, suggesting that the body has no significance in the process of gender construction or minimizing the body as a material fact. The body as a material fact is given, but sex is not. By exploring the interweaving among gender's constructions and its embodiment and lived experience, I hope to show how the lens of gender literally shapes the body, and what this means for individuals who undergo treatment procedures and interventions for intersexuality.

It has been said that feminist academics concern themselves more often with meaning than with flesh or real people (Tuana 1996). To say this is not to diminish the importance of meaning but rather to argue that it is inadequate by itself. We need to understand and interrogate what these meanings imply for lived experience. In the same way that bodies are co-constituted by biological and cultural concepts, they are also both theoretical and lived. Intersexuality raises interesting theoretical questions, but it is not only a theoretical question.

Current clashes over taxonomies and treatments of intersexuality are not merely questions of semantics. Because intersex management rests on cultural understandings of masculinity, femininity, sexuality, and the body, the very process of defining these categories and their relationships has significant consequences not only for those directly involved (children, parents, and doctors) but also for larger contemporary debates about how to understand these relationships. Debates over intersexuality raise basic questions about how bodies, gender, and genitals are involved in the creation of men and women. Consequently, any redrawing of the boundaries of intersex has tremendous potential to redefine how we understand commonsense categories such as male and female.

Intersex, then, is a core location at which, to borrow a phrase from Andrew Lakoff and Stephen Collier (2004: 427), "how to live is at stake"—a location both made problematic and resolved by a binary and discrete conceptualization of gender as a primary alignment, as a regime of living that guides medical practice and holds binary gender together. With so much at stake, the intensity of the debates comes as no small wonder.

Methods

Research on intersexuality from other than biomedical perspectives is relatively recent and constitutes a small, though growing, body of work. Scholars working in disciplines such as history, sociology, social psychology, philosophy, and comparative literature have made important contributions to understandings of intersexuality in a historical and cultural context (Foucault 1980; Epstein 1990; Fausto-Sterling 1993, 2000; Daston and Park 1995; Dreger 1998a, 1999; Kenen 1998; Kessler 1998; Hausman 2000; Feder 2002; Holmes 2002; Hester 2003, 2004; Preves 2003, 2004; Redick 2004). Much of this work is excellent. My aim is not to provide a corrective to these other accounts, but to build on and extend them by closely examining contemporary controversies.

This building on and extending of previous work largely results from my methodology, which I turn to shortly, and from the fact that I conducted my research at a high point in the controversies. The medical management of intersexuality has moved into scholarly and public consciousness in an unprecedented and significant way over the past decade. Much of the previous social scientific work on intersexuality has tended to focus on one group of actors (e.g., clinicians or intersex adults), to utilize one methodology (e.g., textual analysis), or to address one time period (usually prior to the current debates). A large portion of the previous analytical work has looked exclusively to texts (Foucault 1980; Epstein 1990; Fausto-Sterling 1993, 2000; Daston and Park 1995; Dreger 1998a; Kenen 1998; Hausman 2000; Hester 2003, 2004; Redick 2004). Fausto-Sterling, for example, provided important conceptual ideas about the medical perspective on intersexuality, much of which were based on her close readings of the medical literature (1993, 2000).

Textual analysis, however, cannot tell us how clinicians conceive of their work, interpret theories and guidelines, and make decisions. Interviews with clinicians, a rich source of data, have remained largely unexplored since Kessler conducted her groundbreaking interviews with six clinicians in 1985. Drawing on these interviews, and using gender as a primary analytical frame, Kessler provided a cogent and still resonant analysis of how clinicians think about treatment for intersexuality (Kessler 1990). My interviews with clinicians extend those by Kessler by interviewing a larger, contemporary sample of clinicians *after* intersex activism and the ensuing controversies began. This enabled me to ask pointed questions about clinical decision making in the context of controversies as they were taking place, to explore how Money's paradigm is applied in clinical settings, and to examine areas of disagreement among clinicians. I was also able to examine how medical paradigms of intersexuality change in response to internal professional developments and challenges by nonexpert medical consumers and the public. How do clinicians react to internal and external challenges to the dominant medical paradigm? Do practices change in response to these criticisms and, if so, in what ways?

Clearly the views of those individuals treated by clinicians provide another important perspective, as do the views of parents. Preves, the first to interview intersex adults about their treatment experiences, examined how individuals cope with the stigma of being labeled as gender deviant, arguing that the medical intervention into intersexuality often creates, rather than mitigates, this stigma (Preves 2003). Feder conducted interviews with parents about their experiences during the same period I did, arguing that medicine has not only failed to examine parents' experiences but that parents' feelings of confusion and isolation are built into the treatment processes themselves: often they receive neither full information abut the child's condition nor referral to psychosocial resources (Feder 2002).

Both Preves and Feder have provided rich analyses of the viewpoints and experiences of individuals with intersex diagnoses and their parents. Where my work departs from theirs is in the range of subject matter covered in the interviews and in my triangulation of both viewpoints with that of clinicians. This triangulation afforded me the opportunity to contrast the main conceptualizations of intersexuality for the participants, as well as the explicit or implicit understandings of gender and sexuality among the three groups. It also allowed me to ask: Who has the authority to determine what constitutes a good result in medical treatment? What are the legitimate boundaries of medical intervention, especially regarding treatments meant to address social, rather than medical, difficulties? What constitutes good data in evaluating outcomes—like sexual function, pleasure, or satisfaction with treatment—that are always subjective?

Building on this work, my aim was to capture, clarify, and contextualize the current controversies over a treatment protocol and treatment practices that have dominated intersexuality care since the 1950s. Despite heated debates, scarcely any data are available on the practices of treatment for intersexuality. How do physicians arrive at a gender assignment for an infant? Although the typical wisdom is that gender assignment is largely based on phallus size, my work shows that medical decision making proves much more complex and contradictory. Phallic length matters, but so, too, do chromosomes and hormones. We do not know why parents may choose surgery or not, what factors contribute to these decisions, and how they ultimately feel about their choices. What fears and hopes shape surgical decision making?

The lack of data about medical treatment for intersexuality has serious consequences. Without some understanding of what is actually happening and why, physicians, parents, and others cannot thoroughly evaluate current practices and devise ways to improve treatment. Without these data, questions will remain about treatment practices, treatment's efficacy, and the long-term impact on the children who are treated. My hope is thus to provide a snapshot of the current treatment practices for intersexuality in the United States and, more important, to explore the rationale behind it.

As with any such study, it is important that the reader understand how I came to know what I claim to know (Rapp 1999: 11). I employed a range of ethnographic methods, including participant observation and semistructured open-ended interviews, to investigate the spheres of activity under scrutiny here. I supplemented this ethnographic work with primary and secondary literature research in post–World War II reproductive science, biomedical ethics, U.S. political theory, and feminist and other writing on intersexuality itself. I also researched media accounts of controversies about intersex, starting with the earliest protests at medical conferences.

Often one hears of going to the field in anthropological studies. For this study, the field encompasses hospitals, Web sites, conferences, family discussions, protests, the *Oprah Winfrey Show*, and *Dateline*—anywhere discussions, negotiations, and contestations of intersexuality take place. Although debates about intersexuality have exploded into so many venues in the United States that it would be impossible to trace all of the discussions, I attempt to trace and present a broad array of voices.

I use several methodologies to assess the varied and often competing understandings of intersexuality. Those involved—primarily clinicians and researchers, parents, and intersex adults—interact in varied and complicated ways, often neither directly nor publicly. I followed the actors to the loci of these discussions using a wide-ranging, itinerant approach aimed at tracing connections (Heath et al. 1999: 452).

My data include over fifty-three in-depth interviews with clinicians and researchers, intersex adults, and the parents of children with intersex conditions, most of which were conducted from 2000 through 2002.[7] I did the majority of these interviews in person; however, when costs made face-to-face interviews prohibitive, I conducted interviews over the phone. Interviews lasted one to four hours each, with clinician interviews often being the shortest and those with adults with intersex conditions the longest. The length of the interviews depended on the time available to the person, the rapport we developed, and the interviewees' willingness to share their experiences and thoughts.

I interviewed nineteen clinicians or researchers of various specialties who treat intersex conditions: eight surgeons (either pediatric urologists or pediatric surgeons), nine pediatric endocrinologists, one child psychiatrist, and one research psychologist. Many of these clinicians are leaders in their field. I identified them through reading the medical literature on intersexuality and following the debates, as well as through consultation with medical insiders, participant observation at professional meetings, and recommendations by other clinicians. I tried to capture a range of opinions by interviewing clinicians with different specialties, years in practice, and geographic locales of training and practice. I sought to interview a gender-diverse group of clinicians; however, men dominate the surgical fields that treat intersexuality and outnumber women in the other specialties. The youngest clinician I interviewed had been in practice one year; the most experienced clinician had more than twenty years of clinical experience treating intersexuality. Finally, I sought out clinicians trained at diverse institutions who are now practicing at hospitals serving varied populations. Of course, given the relatively short period in which pediatric urology and endocrinology have been specialties and the rather narrow opportunities for apprenticeship, diversity was more of an ideal than a reality.

In addition to the clinicians, I interviewed fifteen intersex adults and fifteen parents of children with intersex diagnoses. Before explaining how I selected these individuals, I want to say a word about terminology. The term *intersex* is used by all clinicians, but not by all parents or affected persons. Part of the debate concerns precisely what conditions or types of body count as intersex. Some people consider the label *intersex* as central to their sense of self; for others it holds no personal relevance, and yet others see it as incorrect and even deeply offensive (see chapter 8). My aim is not to serve as an advocate or detractor for any particular term but to note that much is at stake in contestations over terminology. That this issue is debated so fiercely highlights the importance of the power to name things and to self-identify. Few terms define the boundaries of this varied and diverse group. Although some, as I noted earlier, have recently begun using the phrase *disorders of sex development* (DSD) in an effort to lessen the stigma tied to *intersex*, my sense is that this term, though in some ways less culturally loaded than *intersex*, still leaves intersexuality fully medicalized and construes gender difference as a disorder requiring treatment—a position with which I do not agree. With reservations, then, I use the terms *intersex* and *persons with intersex conditions*.[8] I do not mean to impose a label or an identity that the individuals concerned do not employ. Moreover, I am deeply aware that this usage may privilege clinical interpretation and understanding over personal experience, or vice versa, and that terms such as *intersex person* emphasize intersexuality before personhood. It is an imperfect solution. I will later discuss the objections to this term raised by participants, which I hope will provide a context for the limits of its descriptive usefulness.

More important and problematical, the category *intersex* blurs as soon as one attempts to draw its borders, and this uncertainty itself makes for a component of and a complication in the debates. Any assumptions about what conditions are intersex are refigured in conversations with clinicians, affected adults, and parents. Clinicians may refer to females with congenital adrenal hyperplasia (CAH), for example, as intersex; but many women with CAH do not consider themselves intersex, and some vehemently oppose the inclusion of CAH under the rubric of intersexuality. Parents also express varying levels of willingness to consider their child's diagnosis (and thus their child) as intersex. Had I interviewed only people who identified as intersex or who identified their children's condition as intersex, I would have missed an important and emotionally charged aspect of the current debates.

Studies of people with intersex diagnoses or their parents necessarily work with samples of convenience, rather than with sampling frames. There is no national census of individuals with intersex conditions to utilize, no directory, no registry. Consequently, I was concerned about what researchers call selection bias, particularly for a population that is partially hidden, lacks a consensus for membership, and is dealing with stigma. Indeed, disputes over representativeness frequently recur in these debates (see chapter 8). As a result, I was acutely aware of arguments already circulating; namely, that intersex activists were the "disgruntled minority" and that there was a silent and essentially happy majority from which the public did not hear because they were doing well. My sensitivity to these arguments led me to exercise great care selecting my sample.[9]

To capture the heterogeneity of adults with intersex diagnoses, it would have been inappropriate to employ only one method of recruiting participants in the study. To avoid capturing only activist perspectives, I purposely did not rely on ISNA to recruit my sample. Instead, I contacted numerous support groups for both adults with intersex diagnoses and for parents and visited Web sites geared toward affected adults and parents. Some may rightly argue that those belonging to groups somehow differ from those who do not (it is an empirical question whether they are better or worse off than those who do not participate). For this reason I also pursued referrals by clinicians, parents, and intersex adults (see below). As in any research, however, if someone does not agree to participate, it is hard to know how they are different, if at all, from those who did.[10]

I mostly limited myself to contact with individuals and groups oriented toward CAH and androgen insensitivity syndrome (AIS), or intersex diagnoses in general, for two reasons. First, CAH accounts for roughly 60 percent of all infants born with ambiguous genitalia, meaning there are simply more individuals with this diagnosis (Grumbach and Conte 1998: 1360). Although AIS is not as prevalent, AIS support groups and resources are relatively well organized. A focus on these two diagnoses allowed me to trace common themes in the concerns, treatments, and lives of those with each diagnosis (and thus the variability between diagnoses) that might not have been possible had I only interviewed one or two people with each diagnosis. As a consequence, all of the adults I interviewed were women, save one man with partial AIS (PAIS) who was raised female and began living as a male in early adulthood. (Another woman with PAIS was raised male and began living as a woman in adulthood.) Also, the majority of genital surgeries that have sparked so much controversy are performed on individuals with these diagnoses, making these groups an important source of information.

Unquestionably that I interviewed only one person who identifies as male is a limitation of the present study. Yet I had little trouble locating adults with intersex diagnoses willing to be interviewed; many affected adults referred others to me. Parents are rightfully protective of their children, and concerns about how their children's condition may be represented often make them cautious and reserved; nonetheless, I also found more parents than I could interview. In three instances, adults I interviewed told their parents about my work, so that I later interviewed the parents as well. A number of clinicians I interviewed referred me to their patients, and a few individuals made efforts to contact me after hearing about this project.

My status as an outsider, though a caring one, made it possible for me to interview adults and parents with diverse backgrounds and treatment experiences. Although some were "out" to friends and others, a surprising number had never shared their stories with anyone. As an outsider, I was not the parent, primary caregiver, or clinical caregiver, and I had no vested interest in what they said. I think this status helped enormously in ensuring the honesty and integrity of my conversations.

I do not claim my data to be representative of the viewpoints or experiences of all individuals with intersex diagnoses and their parents. It bears repeating that this is not the aim of qualitative research. Nevertheless, my interviewees represented a broad cross-section of individuals who were diverse on numerous levels. Participants were located in more than twenty states; the intersex adults I spoke with ranged in age from twenty-one to fifty-five. The youngest parent I interviewed was twenty-four and the oldest fifty-nine. Their children ranged in age from five months to thirty-eight years.

The participants were also socioeconomically diverse, almost equally divided between the working and the middle class. More than half had completed college. All families except one had health insurance that covered the treatment of their children. Fewer than half of the participants identify as or are comfortable with the label *intersex* for themselves or (to a lesser extent) their children. Half object to the term and do not use it. Roughly half of the adults with intersex diagnoses that I interviewed have had same-sex sexual experiences; however, they do not necessarily identify as lesbian. Some were uncertain about their sexual orientation. Some have had multiple sexual partners; others have had none. Most have had genital surgery, but several have not. Most parents chose genital surgery for their infants, but some did not. Of those parents who chose surgery, some regretted the decision, while others did not.

I found some common themes striking. For example, adults with intersex diagnoses had similar experiences regarding clinical visits, genital exams, secrecy about their diagnosis and treatment, and shame about their bodies. Many parents spoke of very similar birth and treatment decision-making experiences. In addition, some of the arguments for performing early genital surgery, for example, are frequently echoed by both parents and clinicians. In contrast, arguments for not performing genital surgery were similar among women who have had genital surgery, irrespective of other demographic characteristics.

The current book is not an outcome study in the clinical sense. Nevertheless, I think it offers valuable insights into the treatment experiences of adults and parents. In clinical outcome studies, clinical viewpoints can be privileged. I hope that giving a voice to affected persons and parents may redress this imbalance to some degree.

In addition to conducting interviews, I undertook participant observation at a wide variety of public venues discussing intersexuality. To capture the clinical perspective, I attended (and still do) numerous medical meetings, grand rounds, conferences, and other professional medical meetings. Several clinicians facilitated my access to closed medical meetings. They also provided contacts to other clinicians and patients and informed me of new developments. I attended annual meetings of key medical associations (e.g., of the Lawson Wilkins Pediatric Endocrine Society, the Midwest Pediatric Endocrine Society, the Society for Pediatric Urology, the American Urological Association, and the American Academy of Pediatrics), smaller medical lectures and meetings (e.g., the John Duckett Urology Lectures), grand rounds at numerous teaching hospitals, and meetings of the North American Task Force on Intersexuality (NATFI), as well as some of the meetings of NATFI's Research Protocol Working Group.

To capture the perspective of affected adults and parents, I observed support group meetings, activist fundraisers, everyday operations at ISNA offices, patient-led grand rounds, intersex human rights awards, many of the meetings of the San Francisco Human Rights Commission, and other functions. I have also tried to keep abreast of the many Web sites and discussion groups for specific diagnoses, parent support groups, meetings, and newsletters.

Finally, I have been immersed in this issue since 1997, when I first began exploring the topic. Much of my time since then has been spent engaged in some aspect of intersexuality. So, in addition to what might be described as formal research, I have spent many hours in informal discussions with other clinicians, parents, and intersex adults whom I did not formally interview but whose perspectives are central to my overall picture of the debates under discussion here. More recently, I have begun giving grand rounds on the topic and presenting my work to diverse audiences, both clinical and nonclinical. My research sites are essentially unbounded, which adds both to the pleasure and the difficulty of ethnographic research.

Despite my attempt to capture the depth and breadth of experience, this study is limited in several important ways. Because I was interested in clinical decision making and parent-clinician interaction, I would have preferred to observe actual clinical encounters at the birth of an infant with an intersex diagnosis. Unfortunately, because of the sensitivity of this issue, and the nature of multisited research, such observations were fraught with difficulties. To compensate for this obvious limitation, I probed repeatedly for specific and detailed instances of clinical problem solving and parents' recollections of this period. Although I strove for racial and economic diversity among those I interviewed, my sample is overwhelmingly white, with only one African American adult. Because of my relatively small sample sizes, my ability to provide in-depth analyses of how understandings, decisions, and experiences are shaped by ethnoracial, class, and socioeconomic factors remains quite circumscribed. Finally, as noted earlier, I interviewed only one individual who identified as male.

Types and Frequency of Intersex Diagnoses

There are few reliable estimates of the frequency of intersexuality, in part because even the question of what diagnoses should count as intersex remains disputed and controversial. Estimating the frequency of intersexuality would require reconciling what somatic characteristics are strictly male or strictly female (a task made impossible because few, if any, physical characteristics are exclusive to one sex) and which deviations from this norm count as intersexuality. How small a penis or how large a clitoris count as sexually ambiguous? Does one count sex-chromosome anomalies as intersex in the absence of any apparent external sexual ambiguity? Even with these questions settled it would remain extraordinarily difficult to determine population-based statistics for intersex diagnoses. In the United States, there is no national census or method for keeping track of intersex conditions.

Despite these difficulties, a few studies have attempted to derive estimates of the frequency of intersex diagnoses.[11] In a meta-analysis using data culled from international studies, the authors conclude that the frequency of all causes of "non-dimorphic sexual development" may account for 1.7 percent of all live births and that roughly the same percentage of the population may undergo genital surgery (Blackless et al. 2000: 159). Elsewhere, one of the authors writes that those who undergo genital surgery may be between one in one thousand and one in two thousand live births (see Fausto-Sterling 2000). Both statistics have been widely repeated in the lay press, as well as in academic and scientific articles.

Some believe these statistics are overinclusive, including infants with diagnoses that many would not consider intersex. One critic, Leonard Sax, argues that the common definition of intersex as any "individual who deviates from the Platonic ideal of physical dimorphism at the chromosomal, genital, gonadal, or hormonal levels" (Blackless et al. 2000: 161) is too broad and includes, as its own authors acknowledge, individuals who present no symptoms or outward signs at birth (Sax 2002). He argues for a more "clinically focused" definition restricted to conditions in which the phenotype (outward, physical characteristics) is not classifiable as either male or female, or in which the chromosomal type (e.g., 46,XX) is inconsistent with phenotype. According to Sax, the problem with the estimate by Blackless et al. is that individuals with the five most common diagnoses (late-onset CAH, vaginal agenesis, Klinefelter and Turner syndromes, and other non-XX and non-XY karyotypes) are very rarely born with ambiguous genitalia and should not be considered intersex. Sax counters that only about two in every ten thousand infants are born with atypical genitals (Sax 2002).[12]

The most common intersex diagnosis is CAH, an inherited condition affecting the adrenal glands.[13] The most common form of CAH is 21-hydroxylase deficiency, which has varying degrees of severity. Classical CAH, the most severe form, is typically detected at or soon after birth, and it occurs in approximately one in fifteen thousand births. Nonclassical CAH (NCAH), the milder form, which may cause symptoms at any time from infancy through adulthood, is much more common than classical CAH and affects somewhere between one in one hundred to one in one thousand in the general population. The wide range is due to its higher frequency in some ethnic groups, such as those of Ashkenazi Jewish background.

A well-respected clinician and researcher argues that of the total estimate of 1.7 intersex births per 100 live births, 87 percent are due to the inclusion of individuals with late-onset or nonclassical CAH, who are not born with atypical genitalia, and also of infants with diagnoses such as Klinefelter and Turner syndromes, who are very rarely born with atypical genitalia. He argues that if one deducts these three diagnoses, the rate drops to one in one thousand live births, still based on very shaky estimates (Heino Meyer-Bahlburg, personal communication, April 12, 2002).

The second most common intersex-related diagnosis, AIS, occurs in individuals with a 46,XY karyotype. People with AIS have a functioning Y chromosome (and therefore no female-typical internal organs), but a mutation on the X chromosome renders the body's tissues (including genitals) completely or partially insensitive to the androgens the body produces. Depending on the degree of residual sensitivity to androgens, AIS will either be complete (CAIS) or partial (PAIS). In a fetus with an XY karyotype who does not have AIS, the genitals masculinize under the influence of androgens. In the case of CAIS, the external genitalia take a female form, whereas in PAIS, the appearance of the external genitalia ranges along a spectrum from male typical to female typical. Because AIS is a condition that affects androgen receptors, individuals have testes with the normal production of testosterone and a conversion to dihydrotestosterone, but the body cannot use the androgens the testes produce, and thus genital development is redirected. Another hormone produced by the fetal testes suppresses the development of female internal organs: thus individuals with AIS do not have fallopian tubes, a uterus, or an upper vagina. Data on the frequency of AIS in the United States are not currently available; however, the best available estimates suggest an AIS incidence of approximately 1 case per 20,400 infants with 46,XY karyotype.[14]

For many intersex diagnoses, prenatal diagnosis is rare either because it is not possible or not indicated. (If testing is possible it is often not done unless there is a family history of the condition.)[15] One route to prenatal diagnosis is fetal imaging. An ultrasound may occasionally reveal genital and gonadal atypicalities, but it cannot determine whether a fetus is likely to have an intersex diagnosis.[16] Amniocentesis, the sampling and analysis of amniotic fluid usually performed between the fifteenth and twentieth week of pregnancy, allows for a prenatal diagnosis of nearly all chromosomal atypicalities (but not all genetic ones). Because it carries a small risk of miscarriage (roughly one in two hundred), as well as the risk of nonfatal physical damage to the fetus, it is usually offered only to pregnant women with an elevated risk of chromosomal or genetic birth conditions.[17] Risk factors include maternal age and a previous child or pregnancy with a birth defect. The lateness of the test makes any decision on keeping the fetus fraught with difficulties and has ethical charges as women are put in the position of judging the quality of their fetuses and thus the standards for entry into the human community (Rapp 1999: 3).

At present, even if they are aware of an elevated risk of a child having such a condition, parents and clinicians have only limited and imperfect options for prenatal intervention. If CAH is known to run in a family, the mother may be treated with dexamethasone (DEX), a powerful steroid. It replaces the cortisol the fetus with CAH does not produce, thereby reducing the production of excess androgen by the fetal adrenal glands, with the hope of reducing genital masculinization in affected female fetuses. For this therapy to arrest virilization, it must be administered throughout the pregnancy, starting early in the first trimester.[18] At this stage, however, many women may not know that they are pregnant, much less the karyotype (or genotype) of their fetus. Moreover, the treatment itself may complicate diagnosis. Prenatal diagnosis of CAH due to 21-hydroxylase deficiency by amniocentesis is not possible if DEX has been administered because it suppresses the levels of the steroid that are measured for diagnosis. To use amniotic fluid for prenatal diagnosis, the woman must discontinue DEX for about a week prior to the amniocentesis. Clinicians have more recently turned to genetic tests to determine the diagnosis as a way to avoid this interruption.

This steroid also has a number of undesirable side effects for the mother, and the long-term safety of the treatment on the fetus remains uncertain. Because, statistically, only one in eight fetuses will be an affected female if both parents are known carriers, seven of eight pregnancies may be treated unnecessarily. As a result, treatment with DEX is hotly debated. Clinicians therefore recommend accurate genetic testing to minimize unnecessary treatment.

Yet because of the current limitations of prenatal testing, the birth of a child with an intersex condition typically comes as a complete surprise to first-time parents and their clinicians. This may change dramatically in the coming decade as genetic testing is refined and streamlined. With the possibility of cheap and easy early-detection home pregnancy tests and minimally invasive prenatal genetic testing as a routinized standard of prenatal care, some have conjectured that fetuses with intersex diagnoses will increasingly be aborted.

Overview

I have divided this book into three parts. Part 1 begins with a chapter discussing the history of understandings of and the treatment for intersexuality, examining the complex technological and social developments that shaped and transformed it. Chapter 2 traces the development of the traditional treatment paradigm starting in the 1950s, while chapter 3 introduces the contemporary controversies.

Part 2 marks the beginning of the ethnographic portions of the book and explores contemporary medical treatment practices and decision making. In chapter 4 I explore how clinicians and parents locate sex when the typical markers of sex are absent or ambiguous. This discussion necessarily examines the factors clinicians think important for the development of gender identity. Debates about intersexuality reveal normative ideas not only about gender but also about sexuality. In chapter 5, I examine how beliefs about supposedly normal masculine and feminine sexual desires and behaviors figure into gender assignment and surgical treatment decisions made by parents and clinicians for intersex infants. Numerous assumptions about sexuality and its relationship to gender are embedded in and drive treatment practices for intersexuality, including the belief that heterosexuality is the natural sexuality and thus the correct sexual outcome for these infants, and that penile-vaginal intercourse functions as the exclusive heterosexual sexual act. As a result, genital surgery seeks to produce genitals that not only match the sex of the infant but that also enable or preclude certain sexual acts and thus certain ways of being.

Intersexuality combines scientific and technical practice with lived experience. Despite temptations to privilege either clinical authority and technology or the experiences of adults with intersex diagnoses, I endeavor to show how these aspects are tightly and crucially intertwined. I am interested in how clinicians, as pragmatic thinkers, individualize social issues, how they both constitute and transform gender, and how their ideas about gender influence their treatment decisions. When I question these decisions, my aim is not to deny the relevance of somatic traits or of the material markers and signifiers of sex, nor to make light of what individuals and families experience or of clinicians' compassionate urge to use medical knowledge and skill to help them. Nor do I aim to portray medicine as a batch of bad intentions fueled by a desire to control people. Medicine is more benevolent than that, and its authority and power more diffuse. It is also certainly not solely responsible for the ways in which sex and gender are socially constructed. At the same time, after hearing of the lasting physical pain and psychological trauma of a number of those who have undergone genital surgery and other aspects of treatment, I find it impossible to absolve medicine of all responsibility, no matter how seemingly beneficent, well-intentioned, or sincere the clinicians' desire to help. Rather, my hope is to challenge what some understand as intersexuality in frameworks other than biomedical and to force some submerged questions to the surface: Why label some people as intersex? Whom or what does this serve and why?

I hope it is clear that I care greatly about those with intersex diagnoses and their families, about the difficult decisions parents face, and about the more difficult consequences of some of those decisions. These issues are explored in part 3, which represents my attempt to move away from clinical decision making to view the effects of medications, surgeries, and treatment decisions on the lived experience of those involved. Chapter 6 explores the experiences of parents of children with intersex diagnoses, their experiences of raising their children, and their relationships with their adult children. Chapter 7 explores the experiences of individuals with intersex diagnoses in childhood and adulthood. Finally, chapter 8 looks at the personal and social consequences of contemporary intersex management protocols by examining the efforts of activist groups and their influence on the current practice of medicine.

Following anthropological convention, all names used in the text are pseudonyms. Most of the individuals I interviewed preferred to remain anonymous; some agreed to talk to me only on condition of anonymity. Two affected individuals desired to have their real names used, viewing anonymity as a form of continued stigmatization. I respect this view and agree with it in many ways; however, it presented a problem when deciding how best to honor the wishes of all those I interviewed. Finally, those individuals who wished to have their names used said they would not object to pseudonyms. In giving each one, I do not mean to further stigmatize these individuals or to distance the reader from the life experiences of those who participated. In the end, I hope that the participants' words will help create a culture in which individuals can reveal their condition without fear of stigma or rejection.

PART 1

What Is Intersexuality?

History and Theory

ONE

Taxonomies of Intersexuality to the 1950s

Until the middle of the twentieth century, medical intervention in hermaphroditism (later known as intersexuality) remained relatively uncommon, in part because the technological capacity to intervene, especially at birth, was limited. Unable to change the hermaphroditic body, medical investigations focused instead on attempting to understand and classify it. These investigations both informed and were informed by legal and social understandings of sex, the biological features generally used to define femaleness or maleness, in the West.

Over the past several centuries, a binary model of sex as unequivocally male or female has remained an almost universal axiom, despite evidence from human and animal biology calling this distinction into question. Over time, the biological markers of male and female have shifted, yet not the notion of sex as binary and discrete. This belief in two distinct sexes is an example of a concept so patently obvious and "natural" to most people that pointing to its historicity, and thus to its inherent unnaturalness, proves confounding, even incomprehensible to many. Yet the history of understandings of and approaches to hermaphroditism repeatedly points to attempts to use the phenomenon to reinforce the binary model, rather than to question it, from what have been called the Hippocratic and Aristotelian attitudes of the early modern period to the long-standing gonadal diagnostic principles of the nineteenth-century German pathologist Theodor Klebs (Daston and Park 1995; Dreger 1998a).

During the first half of the twentieth century, medical approaches to intersexuality moved further away from observation and classification and more toward intervention. Advancements in surgical techniques, the discovery of "sex" hormones, new understandings of sex differentiation in embryology, and the ability to test for sex chromosomes—as well as the consolidation of medical and scientific authority—not only shaped how somatic sexual difference was understood but also how intersexuality could be treated by suggesting sites and modes of intervention. Such intervention was—and largely remains—aimed at enabling the individual to fit, more or less superficially, into the binary model of sex.

The labeling of certain bodies as intersex is intimately and inextricably tied to understandings of supposedly normal male and female bodily characteristics.[1] To call a body hermaphroditic, one must already have some idea of what normal male and female bodies look like, what traits they possess, and of the parameters of acceptable behavior for men and women. In this vein, medical practitioners have used knowledge, instruments, and technologies to read bodies, ascribe them a sex, and, when one is not evident, apply a sex in practice. In the case of intersexuality, sex is not merely conceived but enacted, thereby producing and reproducing the category sex and our understanding of what males and females are. Here I trace some of the earlier historical understandings of human reproduction, sexual difference, and hermaphroditism. While not meant to be exhaustive, this history serves as a background to later debates and shows that understandings of hermaphroditism, like much else, are subject to the constraints of time and place.

It might be tempting to dismiss earlier classificatory schemes for hermaphroditism (and for men and women) as quaint and inconsequential. Indeed, it is important to caution against drawing linear connections among conceptualizations of hermaphroditism in different historical epochs (Daston and Park 1995). Yet earlier strains of thought are echoed in current conceptualizations of intersexuality, including in tensions over what intersexuality is and how it should be treated.

A central point of concern and debate has been whether those classified as hermaphrodites are both male and female, neither, or some unique combination thereof. The range and intensity with which hermaphroditism has been an object of interest and speculation has fluctuated; however, over the past century, the cause, classification, and social status of hermaphroditism have proven of enduring medical and social interest. Heightened anxiety over hermaphroditism and the urgency of addressing it are linked to societal changes in gender roles and the corresponding associations between hermaphroditism and other moral and social concerns. Accompanying the intense fascination with and anxiety concerning hermaphroditism has been a need to classify gender-atypical bodies and normalize them—often relying on clinical medicine—to mitigate the threat they pose to the binary model of gender: in other words, the need to give them a single sex. Another issue is the enduring debate over what features should mark an individual's "true" sex, and who should make that determination. Since the late nineteenth century, medical discourses have exercised hegemony over political, legal, and even literary discourses, but this was not always the case. Finally, over several centuries of contemplating and addressing intersexuality in the West, theories for understanding and dealing with intersexuality have shifted considerably.

The Early Modern Period

Hermaphroditism has provoked curiosity for centuries because of its implications for the binary model of sex. Historians have demonstrated that understandings of sex differences before the modern period differed markedly from contemporary ones. Thomas Laqueur has argued that during premodern times, a one-sex model prevailed, in which understandings of sex focused more on similarity than on difference, and that "sex was a sociological, not an ontological category" (1990: 8).[2]

Other historians have sought to complicate the binary periodization of sex into modern and premodern by turning an eye to the heterogeneity of conceptualizations within early modern accounts of hermaphroditism (Daston and Park 1995). Instead of a one-sex model, Lorraine Daston and Katherine Park propose two competing views of hermaphroditism, which they characterize as Hippocratic and Aristotelian because of their roots in the respective authors' early works on human generation, arguing that these were widely debated in early modern times. The Hippocratic view held that hermaphrodites were an intermediate sex exactly between male and female. By contrast, Aristotelians viewed male and female as opposite poles, with no possible intermediate points. For them, hermaphroditism was a condition primarily of the genitals, and thus hermaphrodites did not have an indeterminate sex, but rather doubled or redundant genitalia. In the Aristotelian view, an individual's true sex was always either male or female and thus never truly ambiguous. Not only was the genesis of hermaphroditism subject to debate within medicine but, Daston and Park assert, medical understandings and discussions of hermaphroditism did not dominate legal and political discussions. It was only in sixteenth-century France that hermaphroditism became strongly associated with sexual ambiguity (1995).

Intersexuality has attracted much recent attention because it calls into question a series of intractable dichotomies—nature versus culture, and male versus female, to name two—but according to Daston and Park these dichotomies were not relevant in the early modern period. After 1550, however, a frenzy of interest in hermaphroditism and a newly urgent association with morally charged sexual behavior and sexuality—such as sodomy, transvestitism, and even pornography—explicitly sexualized those labeled hermaphrodites. This intense concern and sexual association led to a resurgence of Hippocratic explanations of generation and sex difference. Hermaphroditism became "emblematic of all kinds of sexual ambiguity and associated with all practices that appeared to blur or erase the lines between the sexes" (Daston and Park 1995: 424).

From the sixteenth to the eighteenth century legal understandings of hermaphroditism in France and England were not always in accord with medical understandings of the time (Epstein 1990). Legal definitions of biological sex (and thus of hermaphroditism) were rigid, assuming two sexes, male and female. In contrast, medical understandings of biological sex as revealed in medical texts were more varied, complex, and contradictory, recognizing a spectrum of sex rather than the mutually exclusive categories of male and female. Legal sex identity was important because of its implications for marriage, property ownership, and civic status. Once a person was assigned a sex, even if prevailing notion held that it was the dominant rather than the "true" sex, the person was treated as a member of that sex for all legal purposes.

A significant question, of course, was who determined a person's sex. In the sixteenth-century, jurists relied on outside testimony from medical experts to determine a person's sex, rather than allowing adults in such cases to make determinations about their own sex. In contrast to Michel Foucault (1980) and Laqueur, who both argue that so-called hermaphrodites could freely choose their sex in the early modern period, Daston and Park contend that the decision was not without conflicts, in part because hermaphroditism, newly associated with sexual ambiguity, sexual deceit, and fraud, was seen as challenging the model of binary sex (1995: 426).

In the eighteenth century views shifted, leading to an emphasis on biological difference, not similarity, between men and women (Schiebinger 1989). Medicine extended sex differences to every part of the body from bones to brains, and these differences were codified by language: male and female sex organs, which had previously shared the same names, were now distinguished by separate terms. According to Laqueur (1990), this shift marks the emergence of the two-sex model of the human body.

The new emphasis on sex differences was not prompted by scientific developments. Indeed, scientific work provided numerous examples of discoveries that could have strengthened the one-sex model: for example, the budding field of embryology claimed that reproductive organs begin from the same embryonic structure. But the political, economic, and cultural transformations of the eighteenth century created the context in which identifying radical differences between the sexes became culturally imperative. Incommensurable sex difference was created despite, not because of, new scientific discoveries (Laqueur 1990: 169).

The Nineteenth Century

Despite their differences of opinion on understandings of sex difference in the early modern period, historians agree that scientific and medical interest in hermaphroditism increased markedly in the nineteenth century. In *Hermaphrodites and the Medical Invention of Sex*, the historian of science Alice Dreger describes how if a person designated a hermaphrodite visited a doctor, it was usually to receive a diagnosis, to relieve discomfort or pain, or for sex determination or revision; in some instances, the patient sought to marry or avoid conscription into the army.[3] By the middle of the nineteenth century, there were two predominant classification systems for hermaphroditism, proposed by the French anatomist Isidore Geoffroy Saint-Hilaire and the British obstetrician James Young Simpson, with each system dominant in its author's respective country. By the mid-nineteenth century, the widespread expansion of hospitals enabled clinicians to see a greater number of individuals with rare conditions such as hermaphroditism, but except for a handful of the most notable clinicians, most saw very few.

In 1833, Saint-Hilaire presented his system for sex determination on the basis of the body's sexual organs and on whether or not they were concordant within an individual. He divided the sexual organs into six segments that he deemed the essential markers of sex. In this system of classification, one was truly male or female if all six segments were wholly male or female; any combination of male and female segments indicated hermaphroditism. Echoing earlier views, Saint-Hilaire argued that hermaphroditism was rooted both in excesses and arrests of otherwise normal sex development. Consequently he categorized individuals according to those segments in which they had excess and those in which they were lacking. Operating from an understanding of normal sex development, Saint-Hilaire believed, would enable an understanding of abnormal sex development. In turn, the study of abnormal variations could allow greater insight into normal sex development. As with many of today's investigations, his attempts to define the abnormal derived from the explicit desire to identify normal sex difference.

The long-standing British classificatory system emerged in the late 1830s when Simpson divided hermaphroditism into "spurious" and "true," categories that persisted through much of the century. Spurious hermaphrodites possessed genitals that were "approximate in appearance" to those of the opposite sex, whereas true hermaphrodites had a mixture of male and female organs. Characteristics of spurious hermaphroditism in the female could include a prolapsed uterus and an enlarged clitoris. In the male, spurious hermaphroditism included adhesion of the penis to the scrotum, hypospadias (an abnormal positioning of the meatus, the opening from which urine passes), and an extroversion of the urinary bladder (an abnormality in which the bladder protrudes outside the abdomen), because the latter were thought characteristic of the female sex. Examples of true hermaphroditism included the combination of an ovary and a testis, conflicting external and internal sexual organs, and a combination of male and female internal sexual organs.

Nevertheless, scientists and doctors did not agree on which traits conclusively determined sex, in part because few traits were sex exclusive. For example, ejaculate was an unreliable indicator because some men did not emit it, while some women did. Similarly, some women menstruated, but others did not. Cases of "doubtful" or even "mistaken" sex, where clinicians believed an individual was living as the "wrong" sex, elicited negotiation and disagreement among scientists and clinicians about which traits or body parts were masculine or feminine and thus about whether an individual should be considered male or female. Consequently, medical professionals relied on culturally derived and gendered notions about anatomy for sex determination. Clinicians, however, also made determinations about an individual's true sex with an eye toward the person's gendered behavior and sexual desire.

In the second half of the nineteenth century, doctors sought to build a broader consensus on true markers of sex. In 1876, the German pathologist Klebs was the first to suggest a taxonomy based exclusively on the analysis of gonadal tissue and on whether an individual had ovaries or testes. Klebs distinguished between pseudo- and true hermaphroditism, a classification system that remained until a 2006 call for a change in nomenclature. Those who had both ovarian and testicular tissue were called true hermaphrodites. Those who had a "doubling of the external genitalia with a single kind of sexual gland," that is, either ovarian or testicular tissue, were labeled pseudohermaphrodites (following Simpson's definition for spurious hermaphroditism) (Dreger 1998a: 145). Pseudohermaphrodites were individuals who were "really" male or female, as indicated by their gonads, but whose mixed external anatomy, especially their genitals, obscured their sex. The category of pseudohermaphroditism was further divided into masculine (testes present with female genitalia) and feminine (ovaries present with male genitalia).

Under Klebs's system, someone with a typically female body type and feminine external genitalia who also had testes would be classified as a male. Klebs codified an already widely accepted notion, namely, that gonads alone indicated an individual's true sex. If clinicians could find conclusive evidence of either ovaries or testes, the problem of sex determination was ostensibly solved.

Practically, gonadal sex determination proved difficult. Because surgical intervention on living patients was not common until the beginning of the twentieth century, the only way nineteenth-century doctors could assess the gonadal tissue of an individual was by palpation in living patients or by autopsy after death. In part because of the inability to assess gonadal tissue in living patients, many individuals were classified as hermaphroditic after death. Even autopsy did not always reveal an individual's gonadal type, as gonads could be ambiguous or malformed. Moreover, Klebs's model did not account for individuals found to have both ovarian and testicular tissue.

By the end of the nineteenth century, several important medical and scientific shifts began taking place. Clinicians could use microscopic analysis to positively identify the structure of the gonad, though more often for deceased than for living individuals. By the early 1900s, however, improved techniques of anesthesia and asepsis permitted exploratory surgery in living patients. As doctors began to perform more of these procedures, they increasingly and glaringly faced the limits of the classificatory system for hermaphroditism. They confirmed that some women had testes but phenotypically were completely feminine. Should these feminine-appearing individuals be classified as males because of their testes? Moreover, if an individual was diagnosed as a true hermaphrodite, was that person supposed to live neither as a male nor as a female? Socially, this stance seemed absurd and untenable.

Dreger (1998a) has argued that the appeal of a gonadal definition of true sex lay in its ability to preserve a clear distinction between males and females, not simply in theory but in practice. As social changes increasingly shifted and blurred the boundaries of masculine and feminine behavior, rooting gender in biology began to seem increasingly important. The period witnessed increasing attention to and concern with homosexuality, another phenomenon that disturbed conventional binary notions of sex and gender. Suffragists also began challenging the boundaries of gender-appropriate social roles and behaviors. Because the definitions of hermaphroditism necessarily touched on the supposedly true nature of male and female, managing hermaphroditic bodies would help maintain the social body (Matta 2005, Reis 2005).

The medical and technological developments of the late 1800s and early 1900s enabled the assessment and diagnosis of certain types of hermaphroditism in subsequent decades. Still some cases of hermaphroditism went undiagnosed and without intervention until adolescence, when secondary sex characteristics sometimes appeared that were incongruous with the individual's social sex. Although medical evaluation and intervention aimed to simplify the diagnosis and treatment of hermaphroditism, assigning a single true sex—a social necessity required to prevent inappropriate marriages and other social dilemmas—in fact became more complex in the early twentieth century, as the markers of and methods for determining this true sex became multiple, diverse, conflicting, and negotiable. Rather than simply displacing previous markers, each new discovery added to the complexity of sex determination.

The Early Twentieth Century

Developments in science and medicine during the first half of the twentieth century—such as advancements in surgical techniques, the medicalization of the birth process, new understandings of sex differentiation in the burgeoning field of embryology, the discovery of sex hormones, and the understanding of and ability to test for sex chromosomes—as well as the consolidation of medical and scientific authority not only shaped how somatic sexual difference, and thus hermaphroditism (or intersexuality), was understood but also how it could be treated by suggesting sites and modes of intervention. In the era before modern surgical techniques and hormone treatments, however, medicine's power and ability to intervene in hermaphroditism was largely restricted to observation, diagnosis, classification, and attempts at explanation and sex determination (Dreger 1998a). The study of hermaphroditic individuals proved, moreover, somewhat a matter of chance. Until the early twentieth century, the prevalence of traditional, woman-centered home birth meant that most newborns were not seen by a doctor. As male doctors gradually replaced midwives, and as women and their physicians believed that increased medicalization would make giving birth safer and more comfortable, childbirth shifted to the hospital (Leavitt 1986). This change gave physicians a larger role in the birth process. Still, some people classed as hermaphrodites never came to the attention of the medical profession because they had no visible anomalies in their external sex characteristics. Individuals without outward signs of their condition were assigned a sex at birth based on the configuration of the genitals. If contradictions arose when the individual reached puberty, patients could seek clinical help on a case-by-case basis. Although some individuals designated hermaphrodite later chose to visit a doctor about their condition, others never sought medical attention; some even independently chose the gender in which they lived (Dreger 1998a). Indeed, some individuals may not have perceived anything unusual about their development.

The turn of the century brought new possibilities of medical treatment, albeit crude ones, consisting mainly of rudimentary surgery to remove organs deemed incongruous with the true (i.e., gonadal) or social sex of the patient. Yet such possible treatments were accompanied by questions about the best course of action. A medical controversy erupted when case histories published in medical journals revealed that clinicians did not always assign sex according to the gonadal composition, and in some instances had not ascertained the individual's gonadal makeup before making judgments about that person's sex and "treatment options" (Mak 2005). Some clinicians, rather than ascertaining an individual's true sex, were increasingly preoccupied with erasing external, visible signs of sexual ambiguity, thus giving the patient one sex.

In 1902, for example, a Dutch gynecologist argued that when determining sex, the external genitalia and their suitability for copulation, as well as the "inner life" of the individual, were more important than gonadal composition (Mak 2005: 67). The physician was highly critical of the "inhumane" clinical practice of forcing people to live in a particular sex. Another case from 1904 sparked intense controversy when a professor of gynecology in New York performed genital surgery according to the patient's wishes, having paid scant attention to her "true" sex (Mak 2005: 67). The tacit exceptions which doctors made to the strict clinical diagnosis to accommodate social realities were, over time, explicitly incorporated into their protocols.

These case histories reveal a shift from the nineteenth-century model of gonadal sex to the mid-twentieth-century model of determining a "best sex" based on an assessment of one's likely gender identity (Mak 2005: 68).[4] The interest in this concept may have surfaced because of surgical advancements and the many cases in which assigning a sex to an individual based on his or her gonads would have proven socially impossible. During this period, psychiatric interest in cases of "sexual inversion," which at the time referred to both same-sex desire and having a gender consciousness at odds with one's sex of rearing, was also growing (Mak 2005: 85–86). The diagnostic techniques, explanations, and categorizations developed in the study of sexual inversion and sex-gender consciousness began to influence medical interests, observations, and the treatment for hermaphroditism, resulting in a growing interest in the latter's gender consciousness (Mak 2005: 86).

In the early 1910s, the British surgeon William Blair Bell, recognizing the possible incongruity between an individual's gonads and his or her phenotype, challenged the gonadal definition of true sex. Bell understood that an individual's gonads did not determine the development of that individual's physical and psychological traits: he thus pointed out that none of the biological markers of sex, whether gonads or genitals, acted as its sole determinants. To address this conundrum, Bell proposed a commonsensical model of sex determination: sex should be determined by the majority of the physical characteristics, with an emphasis on secondary sex characteristics (Dreger 1998a: 164). According to this model, if a woman came to see Bell and she looked "womanly," Bell declared her a woman, even if she had testes. Despite these innovations, Bell's model was, like those of his predecessors, aimed at eliminating physical ambiguity: as explained by his contemporaries Théodore Tuffier and André Lapointe, with whom he agreed, "The possession of a [single] sex is a necessity of our social order, for hermaphrodites as well as for normal subjects" (qtd. in Dreger 1998a: 161).

Increasing Awareness of Sexual Ambiguity

Hermaphroditism became an increasingly medical issue precisely at a period of heightened concern with and anxiety about the relationship among bodies, gender, and desire. As males and females came to be thought of as utterly distinct, hermaphrodites and homosexuals came to be understood as troubling these distinctions. Historically, heightened concerns about hermaphroditism and homosexuality coincided with and can be traced to nineteenth- and twentieth-century medical and evolutionary discourses on deviance (Katz 1995). The terms *homosexual* and *heterosexual* arose in the late nineteenth century—though with different meanings than their current usage—coinciding with an intense interest in hermaphroditism. In its earliest usage, *heterosexuality* actually referred to those who had sexual desire for both sexes, not the opposite sex only, and thus it was applied primarily to those who deviated from gender, erotic, and procreative norms, whereas homosexuals were those whose "mental state" was that of the opposite sex. Thus heterosexuality, what today we would call bisexuality, was associated with a condition called "psychical hermaphroditism" (Katz 1995: 20). Homosexuality and hermaphroditism overlapped in part because the hermaphrodite posed a threat to the moral order as someone who could practice nonprocreative sex. This, not the gender of the individual's partner, made for the predominant concern at the turn of the century (Matta 2005).

Although some sought social and psychological explanations for phenomena such as effeminacy in men, tomboyishness in women, and same-sex desire, others proposed biological explanations. Based on the observation that the human embryo in its early physical development does not have an apparent sex, sexual differentiation, they suggested, may be only partial. This view provided an early theory of bisexuality in which all bodies had aspects of both male and female and thus masculine and feminine traits.

In the late nineteenth century and the early twentieth, the life sciences were focused on the interconnected problems of heredity, development, evolution, and reproduction, which served as the basis for the emergence and realignment of the newly bounded disciplines of genetics, developmental embryology, and the reproductive sciences (Clarke 1998: 66–67). Researchers in a variety of disciplines sought to tackle the question of the inheritance of acquired characteristics. A series of discoveries between 1880 and 1910 suggested that chromosomes carried hereditary information and that sex determination was tied to a so-called sex chromosome. These findings ushered in the notion of a chromosomal basis for sex.

Endocrinology in particular stirred considerable interest in the early twentieth century because it seemed to promise unambiguous medical explanations for sexual ambiguity. By 1910, endocrinologists had discovered that ovaries and testes secrete chemicals and thus are glands, not simply producers of gametes (reproductive cells).[5] As scientists began to isolate and identify various hormones, they located the source of sexual differentiation not in the gonads themselves but in their secretions. Investigators believed the secretions of the testes and ovaries to be sex-specific hormones that functioned as the chemical messengers of masculinity and femininity (Oudshoorn 1994). These hormones, they believed, shaped not only the physical body but also the human psyche and behavior (Hausman 1995: 23).

Early endocrinologists believed they could identify bodies as female or male simply by the hormones present in them: females had estrogen and males testosterone. It was not until 1921 that researchers found the same hormones in both males and females (Oudshoorn 1994). Frank R. Lillie, a zoologist who conducted research on freemartin cows (sterile female calves whose external genitalia appear to be masculinized), first suggested a hormonal basis for hermaphroditism in 1916 (Lillie 1916, 1917). Lillie felt that an investigation of the embryology of the intersex freemartin calf would not only uncover the explanation for this phenomenon but might also shed light on the normal processes of sexual differentiation (Capel and Coveney 2004). He was the first to link the early stages of sexual development with the adult phenotype.

Freemartins occur when the affected female cow is one of a pair of twins, the other twin being male. Breeders were very familiar with the freemartin cow and knew that their infertility resulted from their atypical reproductive anatomy: female external genitalia and some degree of phenotypically male internal organs. Using the then prevalent gonadal criterion of true sex, many assumed that freemartins were insufficiently masculinized genetic males because of their more male typical internal organs. Lillie hypothesized that freemartins actually were genetic females that had been modified by their male twin during gestation; this masculinization resulted from male hormones traveling through placental blood from the male fetus into the female one.[6] Lillie's careful assessment of freemartin cases revealed that the morphological features of sexual development were not uniformly affected: freemartins' female traits were derived from their chromosomes, but their male traits came from the fetal environment (i.e., from the hormonal milieu).

This study affected a shift away from understanding the gonads as the sole locus of true sex. If the influence of hormones in utero could result in a substantial mix of sex attributes, then true sex must be regarded as more than simply the function of the gonads. The development of endocrinology also raised the possibility of using endocrine therapies to correct an individual's sexual ambiguity. Moreover, it undermined the concept of a true sex by demonstrating that the hormonal environment in utero could derail typical sexual development.

An alternative view was offered by Richard Goldschmidt, an unorthodox German geneticist and biologist who conducted research on the role of genetics in sex determination. While performing experiments on breeding gypsy moths, which are typically sexually dimorphic (possessing distinct primary sex characteristics), Goldschmidt found he could produce a sexual continuum in moths, which he called "intersex" (Dietrich 2003). In a 1917 article, he proposed a "time law of intersexuality," by which each individual contained the "factors" required to become both male and female. Intersexes were produced, he argued, according to the amount of time the individual spent developing as one sex before reaching a turning point at which they began to develop into the opposite sex (Dietrich 2003). Extending this theory to vertebrates, he proposed an intersexuality grounded in both genetic difference and hormonal action, what he called zygotic intersexuality and hormonal intersexuality. Intersexuality in invertebrates was grounded in genetic differences present in the zygote, whereas intersexuality in vertebrates included sexual differentiation later in life mediated by hormones and controlled by hormone-producing tissues (Dietrich 2003: 69). Not long thereafter clinicians adopted the term *intersex* and began extending it to humans.

Early Twentieth-Century Interventions

As biology revealed numerous conflicting and contradictory signifiers of sex, the clinicians of the 1920s to the 1940s expressed uncertainty and some anxiety about assigning sex to intersex individuals (Hausman 1995: 79). Treatment for intersexuality was thus highly contested. Even so, improved techniques of anesthesia and asepsis led to the increasing use of surgery to correct perceived genital anomalies. Hugh Hampton Young, one of the first to offer such surgical treatments in the United States, at times sought to reconcile an individual's biology with his or her identity as male or female.

Young, who became head of genitourinary surgery at Johns Hopkins in 1897, began publishing on hermaphroditism in 1921. As early as 1915, Young began developing surgical techniques for genitourinary diseases at Johns Hopkins, and in 1916 he established the Brady Urological Institute within the medical school, making Johns Hopkins a primary center for cases of so-called indeterminate sex. His classic 1926 urology textbook, *Young's Practice of Urology*, coauthored with David M. Davis, was based on his experience treating the urological conditions of 12,500 patients and represented an early effort to provide comprehensive coverage of the burgeoning field of urology (Young and Davis 1926).

Early in his career, Young's interest in surgical challenges led him to develop surgical techniques to aid hermaphroditic patients. At the time, surgical intervention in hermaphroditism was uncommon. As a result, his expertise garnered referrals from physicians across the country of "interesting" or "unusual" cases (Kenen 1998). By 1937, when he published *Genital Abnormalities, Hermaphroditism, and Related Adrenal Diseases*, Young had seen an impressive number of patients (Young 1937). This book, the first American treatise on the types of human hermaphroditism, provides detailed case histories for fifty-five patients. It also includes more than five hundred step-by-step, intricate illustrations of the highly specialized surgical techniques Young developed for the treatment of hermaphroditism, many of which became the basis for future surgical techniques.[7] In the mid-1930s, for example, Young pioneered genital reconstructive procedures, including clitoral and vaginal plastic surgery on children. During the same period, before clinicians understood the pathophysiology of congenital adrenal hyperplasia (CAH) or employed steroid replacement therapy, Young devised a technique to slow the progressive virilization of female patients by surgically "debulking" the enlarged adrenal glands (Meldrum, Mathews, and Gearhart 2001).

Young's work revealed not only unusual surgical dexterity but also the complexities inherent in determining whether a person was male or female (Kenen 1998). He would often determine a patient's sex by assessing the gonadal tissue. However, Young considered the presence of ovaries or testes alone insufficient for sex determination: he implicitly required evidence of their normal hormonal function, which he determined by assessing nonphysical attributes such as personality traits and sexual desire (Kenen 1998). Without perhaps intending to, Young deployed flexible definitions of sex. Nevertheless, his therapeutic goal was to fit his patient into one of two sexes.

The practices of these early medical pioneers demonstrate the variability in the diagnosis and treatment of hermaphroditism in the nineteenth century and the early twentieth. During this period, medical understandings of what should count as true sex shifted. Moreover, the determination of an individual's true sex was based on each individual observer's interpretation of a wealth of somatic data. Different clinicians using the same classificatory scheme might very well have arrived at an opposite determination of sex for the same individual.

One publication from this period was unique in its criticism of Klebs's gonadal model of classification. Alexander P. Cawadias, an English physician and the author of *Hermaphroditos: The Human Intersex*, viewed intersex as a constitutional disturbance involving multiple systems of the body. He argued against Klebs's category of the true hermaphrodite, which he claimed was "repeated parrot-like in all our textbooks," calling it the "non-existent third sex." He argued vigorously for seeing intersexuality as an exaggeration of a normal process: "Intersexuality is a normal phenomenon. There is no absolute male nor absolute female. Every male possesses latent female features, and vice versa. Thus the so-called 'normal' male and female represent the lowest degree of intersexuality, which is thus a physiological phenomenon" (Cawadias 1943: 3, 5). Although this characterization of all humans as intersex represented his attempt to complicate our understanding of sex, his views did not gain any traction. It would be another ten years before John Money would attempt to bring this disparate knowledge into a more nuanced and systematic understanding of gender development.

While theoretically gonadal sex was understood to indicate true sex well into the 1930s and even the 1940s, pragmatically, clinicians often turned to the sex of rearing when determining sex for adolescent or adult patients (Redick 2004; Mak 2005). Concerns centered not around whether intersex individuals should be assigned as male or female, but precisely how clinicians should arrive at the decision. By the early 1940s, the strict reliance on gonadal sex determination provoked anxiety: an intersex individual's direction of libido often conflicted with the gonadal sex (Redick 2004). Although clinicians were concerned with an individual's libido as early as the 1920s, it was regarded predominantly as an indicator of gonadal function. (The concern that sex-assignment decisions might result in individuals' feeling desire for members of their own assigned sex—might, in essence, create homosexuals—emerged only in the 1940s, coinciding with a more widespread societal concern.)

Later refinements in medical technology allowed for more systematic ways of understanding and categorizing bodies. In 1948 Murray Barr discovered a microscopic chromatin mass (now known as the Barr body), which is present in female but not in male mammalian cells, allowing for an investigation of the "genetic sex" of individuals with ambiguous external genitalia. Tissues from so-called true hermaphrodites and pseudohermaphrodites were examined and gave reliable results; but these results, while clinically useful, did not necessarily simplify sex determination. By the 1950s, scientists had developed reliable tests that used chromatins in skin samples to determine chromosomal sex (Moore, Graham, and Barr 1953; Moore and Barr 1955).[8]

By the late 1940s, physicians began to revisit the role of psychological factors in making decisions about surgery for intersex patients (Ingersoll and Finesinger 1947), and psychological sex began to compete with and even displace the gonads as the indicator of true sex (Redick 2004). Still, no unifying theory, principles, or method existed for the gender assignment for intersex persons; treatment practices could vary widely, with even the most seasoned practitioners expressing uncertainty about how to reconcile biology and psychosocial outlook. Advances in science and medicine complicated rather than clarified gender. The midcentury saw not only vigorous scientific debate over the extent to which hormonal or genetic processes determined sex attributes and the extent to which the psyche should be considered in treatment decisions; but clinical specialists, primarily urologists and endocrinologists, also found themselves at odds over how best to treat intersex patients. Urologists were inclined to resolve contradictions of sex such as ambiguous genitalia through corrective surgery. Endocrinologists, by contrast, tried to apply theories of hormone secretions to treat intersexuality with the administration of hormones (Redick 2004: 47). Endocrinologists' understandings of hormonal complexity undermined the notion of discrete binary sex, but this revelation did not undermine their belief "that two sexes existed, or that every person must belong to one of the two sexes" (Redick 2004: 47). Now that greater degrees of medical intervention were possible, the central issue became how best to intervene. These disagreements foreshadowed modern debates over medical treatment for intersex diagnoses as history takes us from ideas of true sex to those of one sex to, finally, those of best sex.

Despite several centuries of speculation and investigation, then, no coherent or unifying theory has emerged for understanding and dealing with intersexuality in the West, either socially or medically. The following chapter looks at the most ambitious attempt of the twentieth century to establish such a model, one that served as the basis for the treatment of intersexuality for the better part of fifty years and is still influential today.

TWO

Complicating Sex, Routinizing Intervention:

The Development of the Traditional Treatment Paradigm

By the middle of the twentieth century, the proliferation of biological markers of sex revealed through new scientific technologies and developments—markers that could conflict within an individual—only added to the complexity of sex determination and assignment. Clinicians had to weigh chromosomes, gonads and their function, and, by the late 1940s, psychological sex as well. In the late 1940s, the potential conflict between psychological and gonadal sex in intersexuality triggered heightened anxiety among practitioners, who expressed their deep discomfort with the potential for social disruption produced by individuals with contradictory genitals, gonads, and gender role behaviors (Redick 2004: 333–34). In the face of a highly variable and even ad hoc approach to sex assignment and determination, the time seemed ripe for standardization.

In the 1950s John Money and his colleagues thus introduced a systematic model of gender assignment and treatment for individuals with intersex conditions. Their protocols, which incorporated both biological and psychological variables, represented a radical departure from earlier work on hermaphroditism and intersexuality, which had been dominated by sciences such as genetics, embryology, clinical medicine, and surgery.

Earlier views about sex determination had often developed without the input from specialists in other fields and had overwhelmingly focused on the body, not the psyche. Money, by contrast, provided a nuanced analysis of the complexity of biological sex, as well as expanding the physical characteristics considered in gender assignment. In addition, he provided a method of gender assignment that put equal emphasis on the psyche, taking into account individuals' likely psychosexual orientation (their gender role and identity) and the physical development that could be expected at puberty in specific intersex diagnoses. His concept of gender-identity/role introduced psychological principles into the medical treatment of intersexuality and provided a link among the fields of psychology, endocrinology, and surgery in gender assignment and treatment. This protocol, which was rapidly and widely adopted, has remained the basis for much contemporary thinking about treatment interventions for intersexuality.

"A Very Vexed Question": The Development of Money's Model

Money, a native of New Zealand, immigrated to the United States in 1947 to undertake a PhD in psychology at Harvard University. Having begun with a broad interest in psychosexual theory, he narrowed his research to focus on hermaphroditism after learning of the case of a child who had been raised as a boy "despite having been born with, instead of a penis, an organ the size and form of a clitoris" (Money 1986: 6). The boy had feminized at puberty, but despite these changes, "psychologically he was a boy and could not entertain the idea of reassignment as a girl" (6). The case inspired research that eventually led to Money's 1952 doctoral dissertation, "Hermaphroditism: An Inquiry into the Nature of a Human Paradox."

Reviewing the relevant literature on the psychology of so-called hermaphrodites from the first half of the twentieth century, Money found that neither anatomy alone, nor genetics or hormones could indicate the status of psychosexual differentiation and identity (Money 1952). Money was critical of the treatment practices of the 1950s, which, he observed, were still based on the microscopic study of gonadal tissues. Of the gonadal theories of sex determination in hermaphroditism, he wrote, "In some quarters the microscope became a sort of tyrant controlling the destiny of ambiguously sexed patients," even when scientists knew that one could not predict pubertal and secondary sexual development from gonadal status (1952: 1:8). Many clinicians further took it for granted, he argued, that gonads would determine erotic inclination, and thus assigned sex with an eye toward ensuring heterosexuality. "Lurking in the background," he wrote, "there also seemed to be a moralistic horror at the possibility of errors of sex leading to marriages between persons of the same sex" (1:8). As a result, "the declaration of a hermaphroditic baby's sex at birth has been, and still is, a very vexed question" (1:9). In a later article he pinpointed the shortcomings of relying exclusively on the gonads for sex determination. First, he argued, this approach discounted the psychological disposition of hermaphroditic individuals. Second, it did not take into account physical developments in these individuals at puberty, which could take different trajectories depending on the underlying condition (Money, Hampson, and Hampson 1955b: 284–85).

Money was also well aware of the failings and inconsistencies of midcentury practices of sex determination for patients labeled hermaphrodites. New medical technologies and knowledge could help predict and control development, but some decisions, he said, were "made with the future taken on faith" (1952: 1:9). Individuals of ambiguous sex might be assigned a sex that contradicted their gonads but fit their social lives; conversely, some individuals might be assigned a sex consistent with their gonadal makeup in infancy which later developments of secondary sex characteristics would contradict. Only recently, he noted, had some clinicians correctly advocated giving primary weight to the psychological disposition of the patient in reevaluating gender assignment when it became obvious the original gender assignment was not working out.

For his dissertation, Money did a comparative analysis of 248 published and unpublished case histories (from 1895 to 1951) and patient files, as well as an in-depth assessment of ten living individuals classed as hermaphrodites. He was interested, first, in how people "adapt themselves to their sex of their rearing when their bodily form and physiology contradict it" (1952: abstract 3). Expressing an interest in what he called psychosexual orientation (gender-role behavior, including sexual orientation), he asked: "Are the physiological functions automatically in charge, or is the impact of rearing enduring?" His second set of questions concerned the mental health of such individuals: "Do they, with such manifest sexual problems to contend with, break down under the strain, as psychiatric theory may lead one to believe; or do they make an adequate adjustment to the demands of life?" (abstract 3).

Although his dissertation had direct relevance for the medical treatment of hermaphroditism, Money was more broadly interested in the opportunity such cases provided to test contemporary psychological theories in two respects. The first was the origin and the determinants of one's "libidinal inclination, sexual outlook, and sexual behavior." The second concerned the origins of "psychosexual role" (the social manifestation of sex differences or gender roles). On the first question, Money found that what was often referred to as sex drive or libido was hormone dependent, whereas an individual's sexual orientation or erotic inclination was not, instead bearing a strong relationship to "teaching and the lessons of experience" (Money 1952: abstract 5). Noting a low incidence of homosexuality in cases of hermaphroditism according to the sex of rearing, he broke from previous work in this area, concluding that theories suggesting biological origins of homosexuality were dubious and instead emphasized the importance of upbringing.

Money's second, and surprising, finding was that so-called hermaphroditic individuals generally fared well psychologically, with surprisingly low incidences of psychoses and neuroses despite their atypical anatomical configurations (Money 1952: abstract 6). This finding, he argued, suggested that the field of psychoanalysis had placed undue emphasis on psychosexual etiologies for mental illnesses and disorders. Moreover, most hermaphroditic individuals, though certainly not all, grew up to accept their sex of rearing, despite physiological contradictions. In general, those facing the greatest contradictions among their physical characteristics, psychosexual orientation, and sex of rearing fared the worst. These findings suggested that intersex individuals could be raised contrary to their "intended" sex without negative psychiatric sequelae.

These findings supported his emergent theory, expressed more fully in 1955, that the sex of rearing was a primary determinant of an individual's gender role and psychosexual orientation. Prior to Money's theory about the importance of rearing in the development of psychosexual orientation, scientists and clinicians typically credited hormones with a significant role in its development. This view made it more difficult to assign sex at an early age because hormonal fluctuations at puberty could potentially shift libidinal orientation and result in homosexuality. By showing that hermaphroditic persons conformed to the sex of rearing in both psychosexual role and orientation, Money effectively minimized this concern. Still, he thought it was desirable to mitigate the possibility of severe contradictions between somatic sex traits and the sex of rearing by employing hormonal and surgical interventions. This marked the beginning of a hypothesis that he was able to put into practice through a collaboration with the pediatric endocrinologist Lawson Wilkins.

The Protocols: A New Treatment Paradigm

At the time Money was conducting the research for his dissertation, few institutions in the United States were known for treating hermaphroditism. Johns Hopkins was one, owing to the pioneering work of the surgeon

Hugh Hampton Young in the early 1900s. Another physician, Wilkins, had founded the world's first pediatric endocrine clinic in 1935 at the Harriet Lane Home in Baltimore, a children's clinic associated with Hopkins. When Money met him, Wilkins had just moved to Hopkins and been credited with the important discovery that the newly synthesized hormone cortisol could remedy poor adrenocortical functioning and thus arrest both virilization and the potentially fatal salt loss in babies with CAH (Wilkins et al. 1951).

Although this development and the ability to test for an individual's sex chromosomes enhanced the ability to diagnose infants, predict how a newborn would develop at puberty, and arrest further virilization in girls with CAH, this information did not necessarily streamline gender-assignment decisions. In some cases additional information pointing to contradictory sex markers only complicated matters. A chromosome test, for example, made it easier to distinguish between CAH in girls, more severe cases of hypospadias in boys (when the urinary tract opening is not at the tip of the penis), and partial androgen insensitivity syndrome (PAIS)—all three of which could potentially look very similar at birth—but it did not dictate gender assignment.

Wilkins, seeing the merit of an interdisciplinary approach to intersexuality, assembled the first cross-specialty team to deal with infants with intersex conditions (Redick 2004). The team included Howard Jones, a young gynecologic surgeon who performed feminizing surgeries, and William Scott, a urologist who had followed Young and performed masculinizing surgeries. Recognizing the value of incorporating psychological expertise into this team, Wilkins also worked with two fellows in the department of psychiatry at Hopkins, John Hampson and Joan Hampson. In 1951, Wilkins invited Money to join him at Hopkins, where Money became the first pediatric psychoendocrinologist and founded the Psychohormonal Research Unit to study "all the different types of hermaphroditism in order to discover all the principles of psychosexual differentiation and development that they would illuminate" (Money 1986: 10). From the early 1950s through his retirement in 1960, Wilkins oversaw the treatment for intersexuality at Johns Hopkins (Redick 2004). As the director of the endocrine clinic and as the preeminent pediatric endocrinologist in the United States, he made treatment decisions that were surgically carried out by Scott and Jones. Money and the Hampsons provided psychological expertise and counseling and conducted longer-term outcome studies on the individuals treated at Hopkins.

Money and the Hampsons first introduced principles and protocols for the medical management of intersexuality in a series of articles published in the 1950s (Hampson 1955; Hampson, Hampson, and Money 1955; Money, Hampson, Hampson 1955a, 1955b, 1956, 1957; Money 1956). The early research and principles were largely derived from Money's doctoral dissertation.[1]

In these articles they set forth their groundbreaking hypothesis (further elaborated in Money and Ehrhardt 1972) emphasizing the role of rearing and socialization in shaping a person's "gender role," a term they first formulated and which they defined as "all those things that a person says or does to disclose himself or herself as having the status of a boy or man, girl or woman. . . . It includes, but is not restricted to, sexuality in the sense of eroticism" (Money, Hampson, and Hampson 1955b: 285). Although it is clear that *gender role* refers to both outward expressions and subjective, inner conceptions of masculinity and femininity, the term *gender identity*, which specifically refers to the sense of oneself as being male or female, emerged several years later and is credited to the psychoanalyst Robert J. Stoller (1964).[2] Eventually Money came to use the term *gender-identity/role* (G-I/R) to refer to what today we would distinguish as gender identity, gender behavior or role, and sexual orientation.

The first article by Money and the Hampsons opened with the observation that "the homespun wisdom of medically unsophisticated people confronted with a newborn hermaphrodite usually guides them to assign the baby the sex which it most resembles in external genital appearance." This method of sex determination, the authors argued, was an "extension of the age-old practice of inferring, on the basis of a single glance, that the reproductive system in its entirety is either masculine or feminine." The article went on to criticize clinicians who used the gonadal criterion, as well as those who had shifted to a chromosomal criterion with the advent of karyotyping (chromosome analysis), as the exclusive basis for determining true sex and gender assignment (Money, Hampson, and Hampson 1955b: 284–85).

Based on a study of sixty-five "ambiguously sexed people," the authors declared it inappropriate, even unwise, to rely solely on gonadal, hormonal, or chromosomal criteria for gender assignment.[3] Even in cases of severe physiological contradiction, they claimed, sex of rearing had prevailed over biology. Later assessments of larger numbers of patients (it is not clear to what extent the samples overlapped) seemed to confirm their theory. In another article published in the same issue of the *Bulletin of the Johns Hopkins Hospital*, they found that of the seventy-six hermaphroditic individuals followed, seventy-two had a gender role and orientation congruous with gender assignment and rearing, even though many showed incongruities between the multiple variables of sex and their sex of rearing (Money, Hampson, and Hampson 1955a). The sample had increased to ninety-four individuals by 1956, of whom five had an "ambiguous gender role" (Money, Hampson, and Hampson 1956: 43). In a study of 105 individuals with various intersex diagnoses, Money and his associates found that only a small number of subjects (five) had a "gender role and orientation [that] was ambiguous and deviant from the sex assignment and rearing" (Money, Hampson, and Hampson 1957: 333). Based on these results, they concluded that "the sex of assignment and rearing is consistently and conspicuously a more reliable prognosticator of a hermaphrodite's gender role and orientation than is the chromosomal sex, the gonadal sex, the hormonal sex, the accessory internal reproductive morphology, or the ambiguous morphology of the external genitalia" (Money, Hampson, and Hampson 1957: 333).[4]

Across these essays, Money and his colleagues argued that the gender role could not be attributed to any simple biological cause and instead proposed a nuanced and complex understanding of gender-role development that included seven variables, "which may operate independently of one another in hermaphroditism" (Money, Hampson, and Hampson 1955b: 299): chromosomal sex; gonadal sex; hormonal sex and secondary sex characteristics; external genital morphology; accessory internal genital morphology; assigned sex and sex of rearing; and gender role, including psychosexual orientation, as so-called male or female (Money, Hampson, and Hampson 1955b: 299; 1955a: 301–2). Hermaphroditic individuals might express sexual incongruities as a result of the various combinations and permutations of the first six variables. By appraising these incongruities with respect to gender role, one could ascertain the relevant importance of each of those six (Money, Hampson, and Hampson 1955a).

The researchers made a notable and novel proposition: they considered gender-role development a multistage process that relied on multiple attributes of biological sex and social variables but that could not be said to derive from these exclusively. Further, the components of gender development might be incongruous. The status and future development of each in the physical body had to be assessed to determine the best gender assignment for each child, but again these alone were not sufficient for gender assignment because gender-role development had no single determinant.

Gender role itself was not innate but learned, and the sex of rearing proved an important determinant. Gender role, Money said, is "built up cumulatively through experience encountered and transacted—through casual and unplanned learning, through explicit instruction and inculcation, and through spontaneous putting two and two together to make sometimes four and sometimes, erroneously, five" (Money, Hampson, and Hampson 1955b: 285). Moreover, the sex of rearing could be more important than biological variables for sex assignment because a gender role, once learned, was not easily modifiable: if an individual's gender role and biological sex proved incompatible, the best hopes for intervention were to alter the person's physical characteristics through surgery and hormones.

Money and his colleagues did not say, however, that the environment and rearing were the sole determinants of an individual's gender role. In the first article, they made clear that their findings "do not represent adherence to a theory of environmental and social determinism, for experiences encountered do not dictate experiences transacted in a simple, point-for-point correlation. Transactions are frequently highly unpredictable, individualistic and eccentric" (Money, Hampson, and Hampson 1955b: 285). They reiterated this point two years later, noting: "On the one hand it is evident that gender role and orientation is not determined in some automatic, innate, instinctive fashion by physical agents like chromosomes. On the other hand it is also evident that the sex of assignment and rearing does not automatically and mechanistically determine gender role and orientation" (Money, Hampson, and Hampson 1957: 333–34).

Money's research on hermaphroditism demonstrated that although it was critical to assess fully all the markers of physical sex—chromosomes, gonads, genitals, and hormonal functioning—none of these alone sufficed to predict or explain gender. Money's budding theory of gender development, which suggested that sex of rearing was critically important for gender acquisition and development, filled this gap and, when coupled with surgical and hormonal treatment, could ensure that the child avoided physical developments incongruous with the assigned gender.

Contrary to recent characterizations of Money's theory as exclusively social, he actually suggested a complex system of psychological and physiological interaction and development. Several biological factors were critical to this development: prenatal hormones, which provided the initial basis for gender development; circulating hormones in the postnatal period, which might need to be managed for a gender assignment to be successful; and finally, genital appearance. All of these factors would need to be evaluated and managed to ensure that intersex individuals would conform to their sex of rearing.

It is important to understand the complexity and innovation of Money's approach in the 1950s. It represented a departure from and in many ways an improvement on earlier work on hermaphroditism that had been overwhelmingly focused on usually one aspect of the body, and not the psyche, for gender assignment. He sought to help hermaphroditic patients, "who so often appear ludicrously dressed in the clothes of the wrong sex," by determining the gender to which they were, or would be, best adapted (Money 1952: abstract 3). To this end, he provided a nuanced analysis of biological sex, expanding the physical characteristics considered in gender assignment. In addition, he provided a method of gender assignment that put equal emphasis on the psyche, while also taking into account the physical development that could be expected at puberty given specific intersex diagnoses. He thereby introduced psychological principles into the medical treatment of intersexuality and provided a link among the fields of psychology, endocrinology, and surgery in gender assignment and treatment.

Given their belief in some flexibility and malleability in gender development and formation, Money and his colleagues proposed moving away from identifying an individual's supposedly true sex and toward a new model of gender assignment that would take into account multiple biological variables of sex and its future development at puberty to select the *optimal* gender for the individual.

CRITERIA FOR ASSESSMENT AND TREATMENT RECOMMENDATIONS

Based on their theory of gender acquisition, Money and the Hampsons proposed new recommendations for the treatment of infants born with intersex diagnoses (Money, Hampson, and Hampson 1955a, 1955b). They suggested a small window of flexibility and opportunity—until roughly eighteen months of age—during which gender assignment could be most successfully accomplished. They recommended, however, that for the best possible outcome, sex assignment be made within the first few weeks of life, before "establishment of gender role gets far advanced" (Money, Hampson, and Hampson 1955b: 288). This recommendation was based in part on the belief that successful gender assignment required complete certainty on the part of the child's parents as to whether the child was male or female. By age two and a half, Money believed, gender role would be so well established that changing gender after that time would prove difficult for both parents and children, who might risk "psychological disturbance." He based this recommendation on the study of eleven individuals who had changed gender; those older than eighteen months when the change took place had a much higher prevalence of psychopathology (1955b: 289).

Although Money cautioned against assigning the sex of a child with so-called indeterminate genitalia based on external genital appearance (1955b: 284), the net result of his own recommendations was often similar. Money and his colleagues recommended that in assigning sex to infants whose genitalia looked neither clearly male nor clearly female, primary consideration be given to the appearance and morphology of the external genitals and "how well they lend themselves to surgical reconstruction in conformity with assigned sex" (1955b: 290). This assessment was important, first, to remove parental doubts about the sex of the child. Second, it was also important so that this incongruity would not impede the development of a gender role consistent with the assigned sex. Finally it was important with respect to future sexual function. Based on this decision, surgery and hormone treatments could be prescribed. If, however, an infant's genitals could be surgically made to approximate either male or female genitalia, then the clinicians would weigh the presence and functioning of the gonads, because these would be indicative of pubertal development and fertility. For older children and adults, Money suggested that clinicians thoroughly assess the extent to which gender role had been established and maintain the sex of rearing (with appropriate surgical or hormonal interventions) unless an individual explicitly requested a change of sex.

Money realized that giving primary consideration to the genital configuration in gender assignment could result in an assignment and treatment that would render the child infertile. He downplayed this concern by arguing that many individuals with intersex diagnoses would already be infertile; the major exception would be girls (chromosomally speaking) with CAH, for whom he recommended female assignment. The prioritization of genital morphology also had a pragmatic component. According to the protocol, an infant with more feminine-appearing genitalia should always be assigned female. Genitals, he reasoned, could not be surgically enhanced to be more masculine (e.g., by lengthening a penis), but they could be made to appear more feminine: the exterior labia, if fused, could be opened and the vagina made to open exteriorly; the size of the clitoris (if deemed too large for the female phenotype and too small for the male) reduced. As Money put it: "If the external organs are so predominantly male or so predominantly female that no amount of surgical possibility will convert them to serviceably and erotically sensitive organs of the other sex, then the sex assignment should be dictated by the external genitals alone" (Money, Hampson, and Hampson 1955b: 288). Money and his colleagues believed that current techniques enabled surgeons to "make" females, but not males; even though the surgical removal of the penis (or clitoris) left no clitoral equivalent, they argued that erotic feeling and sexual climax were still possible (see, e.g., Money, Hampson, and Hampson 1955b: 288, 295).[5]

Babies with a 46,XX chromosomal type, most of whom would have CAH, should always be assigned female because subjects with CAH were known to be fertile, and Money felt that feminizing genital surgery was relatively successful. Cortisol treatments could suppress further virilization. In babies with 46,XY chromosomal type (aside from those with CAIS), clinicians should base gender assignment primarily on the size of the phallus.[6] A baby born without a penis should be "rehabilitated" into a female and the testes removed. The ethics of castration, Money wrote, should be "weighed against the ethics of preserving fertility by exposing a child to live as a boy without a penis, incapable of intromission." Again, the decision had to be made early in the child's life: after twenty-four months, gender role differentiation would already have taken place, and the only rehabilitative possibility was a strap-on penis (Money 1974: 217).

An infant born with what clinicians referred to as a micropenis presented a different concern: whether the phallus could grow.[7] A penis was deemed too small if it could never function as a "copulatory organ," no matter what the intervention. For Money, the assignment of males with micropenis to female was not unethical; their gonads, he said, were typically sterile, which to him rendered moot the concern about removing them and hence their fertility.[8] Moreover, leaving these children as boys would lead to an unsatisfactory life, "even to the point of sexual life being impossible." With a female gender assignment, he reasoned, the infant could be surgically and hormonally corrected "to function as a female in adult life. . . . She could expect to lead a normal girl's life socially, in romance, marriage, and sexual life with erotic feeling" (Money 1974: 217–20).[9]

According to Money's theory, once sex assignment was made, surgery should be done as soon as possible so that the genitals could be made to match the assigned sex (Money, Hampson, and Hampson 1955b: 291; Money 1974: 216). Although his dissertation showed that biological markers did not determine gender identity, Money favored surgery, feeling it would facilitate gender identification and rearing. The reason for this was twofold: first, Money felt that the child was more likely to develop a proper gender role with genitals matching those of the assigned sex; second, parents troubled by genital ambiguity might waver in their commitment to raising the child in the assigned gender. Money further sought to relieve children of any embarrassment they might suffer from having anomalous genitals. Stressing rapid gender assignment and surgical intervention in infancy, Money's work resulted in an increase in medical and surgical intervention for infants with intersex conditions.[10]

For female assignment, genital surgery could be performed at three to six months, unless there were indicators to suggest that the baby might be rejected by the parents, in which case surgery was done before the infant was discharged from the hospital (see, e.g., Wilkins 1966). More often, however, Money suggested that surgery be done in the neonatal period (generally the first six weeks of life): "Otherwise, the child will carry traumatic memories of having been castrated, and the parents will not be able to react appropriately to their baby as a daughter if she has a penis" (Money 1974: 219). Psychosexual development and orientation "begins not so much with the child as with his parents" (Hampson and Hampson 1961: 1417). Parents had to be firm in their conviction that their child was really a girl or a boy, and this "unequivocal definiteness" would help the child to develop the "proper" gender identity.

Along with surgery, successful treatment depended on the administration of hormones at puberty so that the body would develop the appropriate and anticipated secondary sex characteristics. If these interventions were administered, Money and his associates argued, the child would develop a gender role in accordance with the sex assignment (regardless of chromosomal, hormonal, or gonadal makeup). Surgical and hormonal interventions were to be accompanied by psychological care and support for the families and the child (see, e.g., Money, Hampson, and Hampson 1955b: 291–96). Money also advocated speaking frankly with the parents and the child about the diagnosis and the treatment to minimize psychological disturbance, observing that "truth is seldom as distressing as the mystery of the unknown" (Money, Hampson, and Hampson 1955b: 294). Although counseling and openness were typically recommended ("the more manifest the signs of hermaphroditism, the more necessary are psychologic guidance and follow-up" [Money, Hampson, and Hampson 1955b: 292]), it is not clear that these aspects of the protocol were implemented to the same extent as the other treatment recommendations at hospitals around the country.[11] Nevertheless, the advocacy of psychological support made for an important innovation in Money's protocol.

Because Money was aware of the stigmatization of persons with intersex conditions as "freaks" or "half-boy, half-girl," he suggested parents be immediately told either that their child was a boy or that it was a girl and that the child's sex organs were undifferentiated or "sexually unfinished" but would be "finished" through genital surgery (see, e.g., Money 1955b; Money and Ehrhardt 1972). Although intended to alleviate the parents' anxiety, statements such as these may have meant that doctors were not always fully explicit with parents about their child's anatomy or diagnosis.

Money also knew that parents often believed intersexuality to be the cause of homosexuality and suggested two ways to relieve this anxiety. First, parents who might "regard sex as an absolute" should be taught the development of the embryo, showing the "original hermaphroditism" of all human embryos (Lee et al. 1980a: 161). Second, parents should be shown photographs of other children with the same condition from infancy through maturity, including not only details of the genitalia and physique but also "ordinary portrait and group photographs in everyday social or recreational poses" (Money 1974: 219).

Because Money felt no child could be prevented from learning, sooner or later, about a birth defect of the genitalia, or fail to notice that his or her genitals were the focus of parental and clinical attention, he argued that children should be told the truth about their physiology, diagnosis, and treatment, and he developed suggestions for the age-appropriate disclosure of this information, keeping in mind that a full disclosure too early in life might jeopardize the gender assignment.

In prepuberty, he recommended that clinicians and parents begin to give the child age-appropriate knowledge of the condition and its proposed long-term treatment and prognosis, including information about reproductive potential, the onset of menses, and so on. "In childhood," he and his colleagues wrote, "prognosis becomes rehearsed in fantasy, imagery, and play, thus detraumatizing later transition to reality" (Lee et al. 1980a: 161). One could prepare the child psychologically for parenthood by adoption, for example, or a future surgery, or help set expectations for a future sex life. Counseling at puberty would build on the prepuberty discussions. Money and colleagues stressed candor and open-mindedness on matters of romance, dating, and sex life. "No sexual topics are taboo," he stressed, by which he meant masturbation, oral sex, homosexuality, and bisexuality (Lee et al. 1980a: 161).

Yet adherence to the chosen sex of rearing might require more circumspection in the disclosure of biological details to the child to avoid emotional turmoil and the confusion of the child's gender role. With regard to karyotyping, for example, it might be necessary to explain the child's anomalous chromosomal makeup to avoid an unplanned and alarming discovery in a high-school biology class, but Money advised caution. For example, girls with an XY karyotype could be told that their Y chromosome was "metaphorically equated with a broken-armed X. Even though their broken-armed X is technically a Y, it is not an effective masculinizer" (Lee et al. 1980a: 162). Ovaries or testes should be referred to inclusively as gonads. Somewhat confusingly, then, Money and his colleagues advocated both honesty and concealment—a fact that may have led many clinicians to assume that given the option, concealment might prevent more harm and engender less confusion.

The Implementation of the Protocols

In light of current controversies it is easy to forget Money's contributions, some of which were no doubt improvements, to the medical treatment of intersexuality at the time. He was the first to suggest a multistage model of gender development. He was also the first to present a more complicated view of intersexuality that recognized and argued for the consideration of the typical phenotype of various diagnoses and anticipated later physical developments. More important, Money's approach incorporated psychological principles and development into intersex management. Finally, he developed a method for gender assignment and set standards for treatment. Clinicians greeted the introduction of these innovations with interest and approval. Shortly after the initial publication of their theory and protocol, Money and the Hampsons received the Hofheimer Prize from the American Psychiatric Association for their "outstanding contribution to the advancement of psychiatry."

Following publication, the treatment protocols were quickly incorporated into medical practice and texts, and they achieved a remarkable dominance for the following forty years. Within three years of publication, Money's model had diffused in important ways. Johns Hopkins physicians authored two medical texts incorporating Money's treatment paradigm to redefine treatment for intersexuality on a broader scale. Although Money did not author either publication, his influence is unmistakable. Wilkins largely adopted these management suggestions in the second edition of the foundational text *The Diagnosis and Treatment of Endocrine Disorders in Childhood and Adolescence* (1957). They also appeared in an important surgical text of the time, *Hermaphroditism, Genital Anomalies, and Related Endocrine Disorders* (1958), written by the gynecologist Howard Jones and the urologist William Scott.

This was the first medical text in the United States devoted to the treatment of hermaphroditism since Hugh Hampton Young's 1917 book.

These texts provided practitioners with a precise method for making gender-assignment and treatment decisions, one backed by empirical data and a theoretical justification for the new standards. Indeed, in the preface to Jones and Scott's book, Wilkins identified recent innovations in a number of different fields, all of which were incorporated into the new treatment paradigm for intersex conditions: "This new work has required radical revisions in the methods of diagnosis, the principles governing selection of the sex of rearing and the endocrine and surgical treatment of patients with sex abnormalities. The individual researches have been published in many different journals of various disciplines, so that the subject is still relatively unfamiliar to the medical profession at large. The time has arrived when it is most important to bring together in a single volume the present knowledge of abnormalities of sexual development" (Jones and Scott 1958: vii). These publications further diffused Money's treatment standards to other specialists and institutions. They also reinforced Johns Hopkins as a primary center for the diagnosis and treatment of intersexuality. Consequently, those seeking training in the new field of pediatric endocrinology or in urological or gynecological surgery came to Hopkins. There they received detailed, hands-on training in implementing the protocols, which they would later implement at their new institutions. Over time, increasing numbers of specialists took this training to institutions around the country where they, in turn, trained others.

As the medical texts circulated, these, too, helped cement the dominance of Money's treatment paradigm. The protocol also became incorporated in other medical literature, in medical-school curricula, and in teaching hospitals across the country. Consequently, the number of clinics able to treat affected infants grew, and intervention became more routine, especially at birth. As other practitioners published their own case studies, surgical procedures, and outcomes, the protocols were disseminated through the professional medical journals.

Although Money had published his research findings in medical journals for almost twenty years, it was with the publication of his book *Man and Woman, Boy and Girl*, written with Anke Ehrhardt and published in 1972, that his work received wide professional and popular attention. The *New York Times* heralded it as "the most important volume in the social sciences to appear since the Kinsey reports" (Collier 1973). Money's research and theories, which asserted that gender identity and role were primarily the result of rearing, were eventually taken up by scholars and professionals in other fields including psychology, psychiatry, sociology, and gender studies, as well as by feminists arguing against a biological basis for gender differences (e.g., Millett 1970; Robertson 1977; Sargent 1977; Kessler and McKenna 1978; Kolodny 1979).[12]

Based on his clinical work and research, Money published voluminously, writing or contributing to about forty books and almost four hundred scientific papers that bolstered his theory and practice. Over the past five decades, his work has been cited in virtually every clinical article on the medical management of intersexuality. Indeed, while Money continued to publish scientific articles (many on topics other than intersexuality), other pediatric specialists (such as endocrinologists, urologists, and surgeons) incorporated his theory and management protocols into their own work on the medical treatment for intersexuality. Consequently, the amplification and promotion of Money's model far exceeded the boundaries of his own published work. Once this larger body of work, which extended across disciplines, specialties, and institutions, gained widespread acceptance, it became very powerful. By the 1970s, medical protocol almost exclusively reflected Money's paradigm; the model's tenacity results in part from its institutional embeddedness. Money's paradigm formed among clinicians "a consensus rarely encountered in science" (Kessler 1998: 136).

Money's protocol provided clinicians with ways to intervene to help individuals for whom previously they had been able to do very little. It did not, however, pass without later criticisms or adverse consequences. The following chapter explores some of the controversies arising from Money's paradigm and the interventions to which it gave rise, both within the scientific arena and in the public sphere.

THREE

From Socialization to Hardwire:

Challenges to the Traditional Treatment Paradigm

Money's protocol became the standard for treatment of intersexuality in part because it filled a vacuum: prior to his publications with the Hampsons, no systematic plan for the assessment and treatment of such cases existed, and practitioners disagreed widely over what constituted good treatment decisions. But, more important, Money provided practitioners with a detailed guide for carrying out the treatment, one backed by empirical evidence and a theoretical justification for overturning a single standard (whether gonadal or chromosomal) for gender assignment. His protocol was innovative in several other respects and thus superior to previous treatment methods, particularly in its suggestion of a nuanced model of gender development.

Money and his colleagues argued that gender identity and gender-role formation could not be attributed to any simple cause, but instead were complex processes involving both biological and social factors. Their framework for intersex treatment sought to improve the lives of infants born with intersex diagnoses and atypical genitals in the face of imperfect knowledge about gender-identity formation, itself a complex, multifaceted, and deeply subjective aspect of human experience. Rather than simply hypothesizing, they collected data by observing patients and classifying their findings to develop both new concepts and a practical plan for treatment. Their recommendations included provisions for the psychological support of individuals and parents. They also stressed the importance of thorough workups on infants born with ambiguous genitalia to identify the etiology, assess intervention possibilities, and determine the intervention most congruent with anticipated physical developments in puberty and adulthood. Major surgical intervention passed into acceptance because it seemed obvious to both clinicians and parents that a child would experience problems personally and socially with atypical genitals.

For four decades Money's protocol went virtually without overt challenge from within the medical profession; if clinicians disagreed with its recommendations and diverged from them in treating intersexuality, they did so discreetly. At the same time that Money's principles and treatment protocol retained widespread dominance, however, divergent ideas were developing, both in science and in academia, as well as in the broader society, that would force the rethinking of the traditional model of treatment for intersexuality. Changing cultural understandings of sex, gender, and sexuality (and their relationships), concomitant movements for the acceptance of non-normative sexualities, gendered ways of being, and bodies, the decreased authority of the medical profession, and the rise of principles of medical ethics were changing the context in which intersexuality was understood and treated. These disparate currents of thought would come together in the 1990s, in part because of a highly visible intersex activist movement and the widely publicized case of David Reimer, to pose a serious challenge to the traditional model of treatment—initially from outside the clinical community and later from within it (see chapter 8). The sometimes contradictory developments on how best to treat intersexuality pulled clinicians in different directions, calling into question any consensus that had existed. By the mid-1990s, the medical treatment of intersexuality was being seriously questioned; by 2000, the field was deep in crisis. In this context, the domination of Money's treatment paradigm for nearly four decades appears like an aberrant period of a largely uncontested viewpoint nestled between a century of debates prior to its formulation and the past twenty years of modern debates urging its revision and even its demise.

Early Challenges to Money: Milton Diamond

Much of the late twentieth century saw an enormous amount of scientific interest in purported male-female differences and in efforts to locate the origins of these observed differences in biology, especially in hormonal effects on the developing brain. Understanding the processes and mechanisms that shape normative sex differences has been critical to thinking about how these processes might shift in those with intersex conditions. Although research on brain differentiation has not been used specifically in any classificatory or treatment schema for intersexuality (Zucker 1996), this research has been at the center of recent debates about gender-identity development and about whether theories concerning brain differentiation should or even can be applied to gender-assignment decisions for infants with intersex conditions. The origins of this debate go back forty years, to the early work of Milton Diamond, one of the first and most persistent critics of Money's principles and theories.[1]

In 1965, with his PhD in anatomy newly in hand, Diamond published the first of several essays challenging Money's theory of gender development. Whereas Money believed that gender identity and behavior (including sexual behavior) were strongly influenced by rearing up to a certain age (even if they were partly based in biological factors), Diamond saw gender as determined primarily by biology, or what he called prenatal organization and potentiation, which postulated that hormones, not rearing, were its primary determinants. Hormonal exposure during prenatal development—and the critical period just after birth—encoded the brain as male or female, dictating whether an individual will think and behave in the world in typically masculine or feminine ways (1965: 168).

The debate between Diamond and Money, which persisted until Money's death in 2006, certainly has implications for the treatment of intersexuality; their disagreement, however, centered on human psychosexual development in general, which was tied to longer-standing debates over the relative roles of nature and nurture in determining behavioral differences between males and females.[2] In a 1965 article published in the well-respected *Quarterly Review of Biology*, Diamond challenged Money's idea that sexuality remained undetermined at birth, suggesting instead that humans were born with an "inherent sexuality" largely shaped by hormones (Diamond 1965: 148).[3] This paradigm, often referred to as organization-activation theory, was suggested in the late 1950s by several researchers including Robert Goy, a psychologist and one of Diamond's mentors who was instrumental in positing a role for hormones in observed behavioral differences between male and female animals.[4]

When Money published his gender socialization hypothesis in 1955, clinicians and scientists understood androgens (i.e., "male" hormones) as the hormones responsible for masculinizing genitalia and, later, for the pubertal development of secondary sex characteristics; they were not seen as fundamentally shaping the brain or psychosexual differentiation. In 1959, however, researchers found that female guinea pigs administered high levels of androgens in utero exhibited greater mounting behavior than untreated females (Phoenix and Goy 1959). (Researchers use the term *mounting behavior* to describe when an animal raises the anterior or front part of its body onto the posterior or back part of another animal. This is also known as the "top" position in copulatory behavior, which the researchers understood as male-typical behavior.) Androgens, the researchers posited, did more than masculinize genitalia; they directly organized the brain in early development. Pubertal androgens would further activate the already organized brain, resulting in the expression of masculine behaviors. In the absence of exposure to androgens, the brain and hence behavior would be feminine.

On the basis of this and subsequent research (much of which was carried out on rodents and involved looking for evidence of mounting or "rough-and-tumble play"), Diamond proposed that male and female human embryos begin with a predisposition for psychosexual development as either male or female; however, as development proceeds, this predisposition is limited by biological and cultural factors (during critical periods of pre- and postnatal development) on accepted behavior. He advanced this view in opposition to Money; however, his representation of Money's argument was something of a straw man, as he inaccurately ascribed a number of assertions to Money. Thus, Diamond argued, humans are not psychosexually neutral at birth, as he characterized Money as arguing (1965).[5] Diamond suggested that Money posited a psychosexually undifferentiated newborn, whose gender identity and role development stemmed *solely* from social and environmental factors. Money's theories shifted over time, but even in his early publications, he explicitly stated that both biological and social factors contributed to the formation of gender identity and role. As others have argued, in an effort to differentiate his own theory, Diamond characterized Money's work using the most extreme version of Money's at times vague or inconsistent ideas (Zucker 1996; Fausto-Sterling 2000: 284).

Diamond, it should be noted, did not take issue with Money's findings about the development of gender identity in persons with intersex conditions.[6] Rather, he rejected Money's theoretical conclusions that his findings provided insight into the psychosexual development in *non-intersex* persons. He argued that to support Money's "theory of psychosexual neutrality at birth," we would need evidence of "a normal individual appearing as an unequivocal male and being reared successfully as a female, or vice versa" (Diamond 1965: 158). Money, however, never proposed that humans were psychosexually neutral at birth. Moreover, this evidence, although perhaps of theoretical interest, would have little pertinence for the treatment of intersexuality, which Diamond did not dispute.

Diamond faulted Money with assuming that sex role is based on what he characterized as a "very elaborate, culturally fostered deception" and proposed a different explanation: cultural taboos and defense mechanisms acting in concert with a biological (i.e., hormonal) prenatal organization to produce sex roles at odds with morphological sex (Diamond 1965: 148). Although Diamond acknowledged a role for the environment and human flexibility, he held these factors to be of limited significance; the environment could only modify effects of an "inherent" and "previously organized soma" on which it was superimposed (1965: 160). Prenatal biological factors set limits on the extent to which environmental factors such as culture and learning determined gender identity. Gender identity, he argued, was virtually hardwired into the brain from the early stages of fetal development.

The crux of the disagreement between Diamond and Money concerns the question of whether pre- and postnatal hormones had a determinative role in psychosexual development. Diamond said they did; Money said they did not. At precisely the same time two opposing theories were thus put forth: one emphasized the importance of psychosocial factors on aspects of psychosexual differentiation, the other offered a biological explanation (Zucker 1996: 151).

When Money formulated his theory of gender development and his intersex management protocol, the effects of prenatal hormones on the developing brain were still unknown. However, when Goy and his colleagues published their research on organization theory, it had an important influence on Money, and he built on it, extending the findings to humans. In the debate sketched above, Diamond often omitted the fact that Money was the first to test empirically in humans the influence of prenatal hormones on behavior. In the 1960s, for example, Money and his colleague Anke Ehrhardt reported masculinized gender-role behavior (understood as a preference for toy cars and guns, and more frequent and vigorous outdoor play, fighting, and aggression) in girls with CAH who had been exposed to high levels of androgens in utero. They concluded that organization theory applied to humans to some extent (Ehrhardt and Money 1967; Ehrhardt, Epstein, and Money 1968; Ehrhardt, Evers, and Money. 1968). Yet they found that despite more masculine gender-role behavior, gender identity in these individuals was female, suggesting that gender identity and gender-role behavior were related, but in complicated ways. Moreover, they showed that gender identity could accommodate wide variations in gender-role behavior (Ehrhardt, Epstein, and Money 1968). In others words, females could exhibit tomboyish behavior but still identify as female.

After conducting additional studies on the effect of prenatal hormones on behavior in some intersex conditions, detailed in their 1972 publication *Man and Woman, Boy and Girl*, Money and Ehrhardt concluded that masculine and feminine behavior were partially influenced by the brain's exposure to prenatal androgens in utero. Hormones proved important in gender development, but they were not determinative. In the 1972 book, Money also provided the first application of his theory to an individual without an intersex condition: a young boy whose "penis was ablated flush to the abdominal wall" during circumcision and for whom he recommended reassignment as female (Money and Ehrhardt 1972: 118–23). Six years after the gender reassignment, Money reported a reasonably well-adjusted girl with "tomboyish traits." *Time* magazine reported the case shortly after Money presented it at a meeting of the American Association for the Advancement of Science, writing that "this dramatic case provides strong support for a major contention of women's liberationists: that conventional patterns of masculine and feminine behavior can be altered. It also casts doubt on the theory that major sexual differences, psychological as well as anatomical, are immutably set by the genes at conception" ("Biological Imperatives" 1973). The case of this individual, known as "John/Joan" in the medical literature (and later identified as David Reimer), would become a central focus in contemporary debates over gender assignment for persons with intersex diagnoses. Even though Reimer did not have an intersex diagnosis, his case had profound implications for the medical treatment and public understanding of intersexuality.

Although Diamond would later garner enormous publicity for his views, largely due to the John/Joan case, his criticisms of Money's theory—which he doggedly published in numerous articles over the following two decades —received little attention, as did his largely essentialist hypothesis about human "psychosexual predisposition." This may be in part because Diamond's theory, when first presented, was based on observations in rodents, and his arguments were not well grounded in empirical data on humans.[7] Moreover, unlike Money, Diamond had no clinical experience with intersex patients: he did not treat children and was not part of a team that treated them. (He would later conduct outcome studies on humans, but these were never as extensive as Money's, which were funded for the better part of four decades by the National Institutes of Health.) Perhaps in part because of these limitations, few others concurred with Diamond's criticisms. In fact, very few scholars challenged Money.[8]

The disagreements between Money and Diamond over psychosexual development persisted over three decades; they culminated in a rather public exchange in the late 1990s over the case of John/Joan. Diamond interpreted the case as a conclusive demonstration of the failure of Money's gender-socialization hypothesis, which he felt had tremendous implications for the treatment of intersexuality. The John/Joan case reignited long-standing nature-versus-nurture disputes both within and outside professional circles regarding gender-identity formation and psychosexual development.

Evidence of Failure? The John/Joan Case

Diamond's trump card in his criticism of Money was the case of an infant treated by Money who at puberty rejected the gender assigned to him in infancy (Diamond and Sigmundson 1997b). "John" was born in Canada, one of two identical twin boys. At roughly seven months, an accident during a routine circumcision operation destroyed his penis, leaving it flush to the abdominal wall (a condition referred to as "ablatio penis" in the medical literature). The parents, understandably distraught and desperate for a solution, sought advice from numerous clinicians; none could offer any solution. Prospects for the restoration of the penis were dismal because techniques of phallic reconstruction at the time were too crude to offer any real remedy. After suffering through a "long saga of finding no answer" (Money and Ehrhardt 1972: 118), the parents saw a television program featuring Money, and they decided to visit Johns Hopkins to consult with him about the possibility of raising John as a girl, which they eventually did (Colapinto 2000).[9]

The story of John/Joan's upbringing as a girl, transitioning to live as a male in his teens, and enduring psychiatric distress was heavily publicized in the United States and in Canada, both in the popular media and in psychiatric and pediatric medical journals (Money and Ehrhardt 1972: 118–23; Money 1975, 1998a, 1998b, 1998c; Diamond 1982; Colapinto 1997, 2000; Diamond and Sigmundson 1997b). When Diamond famously broke the news in 1997 that Joan was living as a male and triumphantly claimed that this case provided unequivocal support for his theory, it made the front page of the *New York Times*. The journalist John Colapinto, then a writer for *Rolling Stone*, obtained an exclusive interview with John/Joan, on which he based a widely read article (Colapinto 1997); he later expanded his account into a book. The case inflamed the old debate of nature versus nurture, directly reflected in the title of Colapinto's 2000 book, *As Nature Made Him*.[10] Through the John/Joan case, intersexuality, once a condition known only to physicians, researchers, and those affected, was showcased on television—on the *Oprah Winfrey Show*, *Dateline*, and *Primetime Live*—and in magazines and newspapers across the country.

Part of the reason that a specialized medical debate erupted into a cause célèbre was that issues arising in the John/Joan case dovetailed with other broad social changes, especially with the resurgence of biologically based theories of gender difference in the 1990s, which opened the way for debates about the treatment of intersexuality. The readiness of the popular media to seize on the issue—in the process oversimplifying and often misrepresenting the arguments—illustrates the popular hunger for a clear-cut biological explanation of what makes us male or female. It also demonstrates a reluctance to believe that both gender differences and the theories advanced to explain them are complicated, shaped by culture, history, politics, economics, family structure, and biological factors.

The enormous publicity the case received gave it and Diamond—who had been a minor figure in the field of intersexuality—a disproportionate weight in discussions of gender assignment and obscured some of the problems with Diamond's argument. Diamond based his rejection of Money's hypothesis (and, in turn, of the contemporary treatment paradigm for intersexuality) on one sensational and extraordinarily rare case. Other similar cases are less conclusive. A 1998 paper describes another case of sex reassignment for an infant whose penis was destroyed during an accident—the only other analogous case in North America at the time for which follow-up data were available (Bradley et al. 1998). This infant was assigned a girl at seven months, and at age twenty-six she was described as a bisexual woman with a female gender identity.[11] Thus in two cases of otherwise "normal" males with no penis assigned as female, one failed while the other seemingly succeeded.

That the failed case received enormous attention, while the second case received virtually none, points to the influence of journalism, the importance of a "speaking" subject (Reimer), and the desire to reaffirm male-female differences as "natural" more broadly in society. The underlying commitment to gender as natural is what makes the accusations of monstrous experimentation on John/Joan even make sense. Most of the popular debate over John/ Joan has focused on the idea of a true, underlying sex connected in a direct and predictable way with gender, and on the notion that intersex management protocols have been wrong chiefly because they were not based on the idea that biology determines gender.

This case also illustrates what the psychologist Kenneth Zucker has described as the exact opposite of the usual "file-drawer" problem in most scientific research—the tendency for unsuccessful experiments to remain unpublished. Cases of unsuccessful rearing are more likely to be published than successful ones because the unsuccessful ones are deemed newsworthy (Zucker 2002: 9). What makes unsuccessful cases newsworthy in the first place is that they reaffirm taken-for-granted ideas about gender differences as rooted in our biology.

Theories about gender difference and gender-identity formation, which have far-reaching social and political implications beyond the treatment for intersexuality, are frequently linked to these reported cases, of which there are surprisingly few.[12] A meta-analysis of all published cases of individuals with a traumatic loss of the penis in childhood and a subsequent gender reassignment to female yielded only seven, including John/Joan (Meyer-Bahlburg 2005a).[13] Of these seven individuals, five were living as females at the time of follow-up, whereas two had transitioned to live as males: John/Joan and a youth in Colombia who transitioned in his teens (Ochoa 1998). This hardly gives resounding support for Diamond's theory, but the other cases have remained absent from popular and even some medical debate and analysis.

The outcomes for males with catastrophic genital anomalies assigned female represent near ideal experiments, ones otherwise not ethically defensible. They generate intense interest and controversy precisely because the results are used to prove or disprove divergent theories and assumptions about the role of biology in the formation of gender identity, gender role, and even sexuality. These theories, often grounded in either predominantly biological or psychosocial explanations of gender-identity formation, get so much attention because many assume that these findings help explain so-called normal gender-identity development and that they might reveal something important about gender change in non-intersex individuals.

Yet with such rare and unique cases it remains highly questionable whether they should be seen as relevant to intersex case management or as supporting or refuting any theory of gender-identity development. John did not have an intersex condition. He was a biologically typical male until the accident, which made his case extremely unusual. He would never have been a candidate for sex reassignment had it not been for his lack of a penis and the dismal prospects for surgical reconstruction. The family first consulted with Money when the child was nineteen months old. The suggestion to raise him as a girl, though perhaps extreme, nevertheless seemed appropriate because nothing could be done medically for the baby, who, as a boy with no penis, would face the prospect of an extraordinarily difficult social and sex life. Money's theory of gender malleability suggested that reassignment was possible; his understanding of the effects of hormones on brain organization implied that they would not inevitably result in a male gender identity. He felt that the child, via surgery, could have a functioning vagina that would enable sexual intercourse and allow sexual pleasure and orgasm. In short, he could have a reasonably normal life.

Back in Canada, the parents consulted with more experts, again reaching no clear-cut answer. They then agreed to a female gender assignment, which took place when the child was at the upper limit of the period of gender flexibility (seventeen months according to Money, nineteen months according to Colapinto). At twenty-one months, surgeons at Johns Hopkins removed his testes and refigured his scrotum to resemble labia. Money and his colleagues recommended that surgery to create a vagina be performed when the child reached puberty. Money counseled the family on the future prognosis and treatment. The parents gave their child a female name ("Joan"), female clothing and toys, and they grew her hair long. They also sought to reinforce the assignment by encouraging supposedly female behavior and discouraging male behavior. It is easy to point to problems in some of the rationalizations given for a female assignment—for example, the presumed success of feminizing genital surgery—but the rationale behind raising John as a girl is more reasoned and complex than retellings of this case often imply. Although some have characterized Money's recommendation as a "monstrous experimentation," particularly with the hindsight that it failed, at the time he made it, Money's suggestion was based on the best knowledge available, in a situation in which, frankly, no good options existed.

Money monitored Joan through yearly visits to Johns Hopkins until the child was about nine years old; he reported a reasonably well-adjusted girl who was a "tomboy" (Money and Ehrhardt 1972: 118–23; Money 1974, 1975; Money and Tucker 1975). The apparent success of this remarkable case of a non-intersex child's sex reassignment prompted Money to claim in the original report of this case that "gender dimorphic patterns of rearing have an extraordinary influence on shaping a child's psychosexual differentiation and the ultimate outcome of a female or male gender identity" (Money and Ehrhardt 1972: 144–45). Here was the real-life example of a hypothetical scenario—the case of an otherwise typical child raised as the opposite sex—that Diamond had pointed to in his 1965 article. It is a strong statement regarding the influence of rearing in gender-identity formation, but not one in which he suggested psychosexual neutrality.

By 1980, however, Diamond had learned through a film that Joan was not doing well as a girl. In May 1979 the BBC aired a documentary, *The Fight to Be Male*, reporting that by age thirteen Joan's psychiatrists in Canada believed she would transition to live as a male. (The film never aired in the United States, but Diamond recapitulated the findings in a 1982 paper.) Evidencing an ardent commitment to this case, and the desire to expose it, Diamond sought out the psychiatrists treating John/Joan by taking out ads in the American Psychiatric Society journal. Diamond argued that this was the only known case to demonstrate the influence of rearing over biology in an otherwise "normal" male child (1965; 1982).

Several years later Joan's psychiatrist, Keith Sigmundson, contacted Diamond, revealing that Joan had received a prosthetic penis in 1980 and was living as a man. In 1997, Diamond and Sigmundson published an update to the John/Joan case, labeling it a failure and arguing it invalidated Money's gender-socialization theory (Diamond and Sigmundson 1997b). According to their account, Joan had begun to question her gender between the ages of nine and eleven, and by age fourteen she considered herself a male. Based on this finding, the authors wrote, "This update . . . completely reverses the conclusions and theory behind the original report" (1997b: 302).

Specifically, they challenged the notions that humans are psychosexually neutral at birth and that genital appearance is crucial to gender-identity development. "Normal humans," they argued, "are not psychosexually neutral at birth but are, in keeping with their mammalian heritage, predisposed and biased to interact with environmental, familial and social forces in either a male or female mode" (1997b: 303). Diamond suggested not only that this case had failed but also that it disproved Money's theory about human gender-identity development. With no further evidence, he and Sigmundson called for a reconsideration of Money's long-standing treatment paradigm for intersexuality and a moratorium on all genital surgeries that were medically unnecessary (1997b).[14] As one clinician assented: "Based on this case, the concepts that individuals are psychosexually neutral at birth and psychosexual development is intimately related to genital appearance were developed. This individual is now an adult who tells the rest of the story. . . . It should stimulate us to reexamine our current methods of deciding gender assignment" (Elder 1997: 1635).

A significant flaw in Diamond's and Sigmundson's criticisms of Money derives from the fact that Diamond's evaluations of masculine and feminine gender-role behavior embody rather than interrogate assumptions and stereotypes about gender difference. In the 1997 update, for example, Diamond and Sigmundson claim that Joan began rejecting her female gender assignment as early as at age six (Diamond and Sigmundson 1997b: 299). On this basis, they argue not only that the gender assignment failed but also that individuals are not "psychosexually neutral at birth." The evidence to support these claims includes statements from the parents to the effect that as a child Joan rejected "girls' toys, clothes and activities" (dolls and sewing), instead preferring boys' activities and games. She also preferred to play with her brother's toys and liked to "tinker with gadgets and tools . . . and take things apart to see what made them tick" (299). She was regarded as a tomboy and did not "shun rough and tumble sports or avoid fights" (299). Joan fought back forcefully when teased and as a teen tried to stand to urinate.[15] Sexuality and sex-typical behavior are thus subsumed under the category of gender, by which males are assumed to desire females exclusively. Perhaps it is no surprise that the authors conclude their update on John with a heteronormative ending that ostensibly provides the capping proof of the failure of the gender acquisition: "At age 25 years he married a woman several years his senior and adopted her children" (300). (Although the authors could not have known it when they published their account, the real ending of the story would prove a tragic one: Reimer took his own life in 2004, a mere thirty-eight years old.)

Each behavior that Diamond and Sigmundson read as unproblematically masculine or feminine is culturally derived and value laden, and the reasoning at times appears circular. They make sense of John's male gender identity by describing the degree to which his *behavior* was gender-stereotypical and attributing this behavior to biology, specifically prenatal hormone exposure. They note that "a male's predisposition to act as a boy and his actual behavior will be reinforced in daily interactions and on all sexual levels" (1997b: 303). Bernice Hausman points out the contradiction in this statement: "Diamond and Sigmundson assume that gender both precedes and follows from sex—the male's predisposition to 'act like a boy'—but it and the 'actual behavior' of the boy need reinforcement 'in daily interactions.' [They thus maintain] their claim about gender's innateness at the same time they acknowledge gender as the result of social forces" (2000: 122).

Moreover, both the subject (Reimer) and the researchers may have read his behaviors in light of the knowledge of his reassignment as a male and the desire to "demonstrate a consistent narrative history as a male subject"

(Hausman 2000: 121). Diamond further assumes that gender identity and gender behavior are unproblematically linked, so that females have "feminine behavior" (assumed to be self-evident) and vice versa. But this sidesteps an examination and discussion of the complicated ways in which the ties between identity and behavior are co-constituted, shaped by the contours of time and place and constantly rearticulated over a lifetime. Other research (not to mention lay observation) shows that females display a wide range of behaviors, some of which are commonly understood as masculine, that are not incongruous with (and may indeed bolster) their sense of themselves as females.

Throughout the period of intense media coverage concerning the John/Joan update, Money chose not to speak publicly about Reimer, but in a 1998 publication he proposed several reasons why his patient might not have grown up to accept his female gender assignment (1998b).[16] These included the post-traumatic effects of having his penis burned off; the late age at which he was assigned female (and the even later age at which his genitals were feminized); the effect of having an identical twin brother with whom to compare himself in childhood, and the knowledge that identical twins are always of the same sex; sexual play which might have highlighted his atypical genitals; the mental distress of parents, grandparents, or other relatives over his condition; the effect of not presenting lesbianism as a viable alternative to a sex reassignment back to male; and, in a jab at Diamond, the effect on the family of the investigative mission of "intrusive outsiders" and their explanation of their purpose (Money 1998a, 1998b, 1998c). In addition to these explanations, researchers have noted that the parents raised Reimer as male alongside his twin brother until seventeen months of age, creating a "gender-typed social niche," and that the parents may have had doubts about raising him as a female that would have contributed to an unstable female identity (e.g., Bradley et al. 1998; Meyer-Bahlburg 1999).

Unquestionably this case failed. It does not follow, however, that the case invalidates Money's other empirical research in intersex cases (or that of other researchers supporting his theory), nor does it follow that this case provides unequivocal support for Diamond's theory. In any other area of clinical medicine or science, researchers would reject as ludicrous the idea that a sample of one proved or disproved any theory. The willingness to go against this tenet of science may say more about the conceptualization of research on gender difference. Most researchers still use a causal model of gender-identity development that relies on a nature-nurture binarism. In this model, if nurture is demonstrated to be incorrect or inadequate, it is assumed that nature must be the cause. However, such a conclusion reaffirms the opposition between nature and nurture (by reducing Money's theory, for example, to nurture) more than it affirms or refutes claims about how gender identity develops. But even if its evidence or reasoning did not justify its prominence, Diamond's well-publicized attack on Money's theory and treatment recommendations called the entire intersex treatment protocol into question. What is of interest is less what Diamond had to say, though this is also important, than the amount of attention he got for saying it. His theory resonated at a historical moment when biological explanations for behavior were ascending once again.

Today, there is tremendous scientific and popular interest in locating biological causes for male/female difference. A common feature of such efforts is a desire to provide a complete theory of gender development that can account for all of the observed differences between men and women. Through biology, these projects aspire to solve once and for all the puzzle of sexual difference, and in so doing they often flatten the myriad complexities and variances of human behavior into a single, immutable physical fact of nature.

In assessing the John/Joan episode, it is important to note that Diamond did not single-handedly foster the media frenzy or the subsequent demonization of Money by a number of critics. Colapinto's 1997 *Rolling Stone* article amplified Diamond's viewpoint, as did his later book, which received enormous publicity and made the *New York Times* best-seller list. Colapinto, understandably unfamiliar with research on gender-identity development before writing his book, presents a sweeping vindication of Diamond. In my interview with Colapinto, he explained why he was drawn to the work of Diamond: "Money's reporting in his scientific papers was so anecdotal. It was dressed up and tricked up to look very scientific but it's amazingly unscientific. . . . I thought Diamond's stuff was way more solid . . . when he gave you a number, when he based his conclusions on 'I spoke to this many AIS kids and this many CAH kids and I saw this many outcomes went this way and that way' . . . it seemed to fit what you understood about humanity and it just seemed less sensationalistic. . . . I realize that Money was a glib and facile writer, but it seemed to me that he was spinning theories out of very thin evidence" (author's interview, New York, June 14, 2000).

Colapinto's comments regarding the quality of Money's research in comparison to that of Diamond's seem based more on impressions than on close readings of either's work: how else does one conclude that dozens of studies involving hundreds of individuals with intersex conditions, and decades of clinical experience, are anecdotal? More telling even is the journalist's tautological claim that Diamond's work appeared more rigorous and scientific because it fit Colapinto's preconceived ideas about humanity; namely, that men and women are born, not made. Unsurprisingly, the biological determinists emerge as the good guys in Colapinto's book, arguing that they would not have assigned Reimer a girl nor recommended feminizing genital surgery, observations made over thirty years after the fact and with the benefit of hindsight. (Colapinto—and Diamond as represented—barely touches on the enormous difficulties of raising the child as a boy with no penis, even though Canadian and American physicians told the family about the multiple drawbacks to an artificial penis.) The bad guy is Money, who stands in for those promoting a role for environmental and social factors in gender-identity formation: he was performing an opportunistic "experiment in psychosexual engineering" in an effort to further his career (Colapinto 2000: xviii). The science writer Natalie Angier, reviewing the book for the *New York Times*, noted that Money was represented as "almost too evil to be believed" (Angier 2000).[17]

Another Hardwire Theory: William Reiner

Around the time this case was playing itself out in the media, William Reiner, a pediatric urologist turned child psychiatrist, also began vigorously challenging Money's theory of gender-identity development and his treatment protocol. Like Diamond, Reiner has argued for a strong inborn bias for gender identity and sexual orientation. As Reiner tells it, he became interested in psychosexual development through seeing patients in his urology practice. In the mid-1980s, one of those was a fourteen-year-old girl who wanted to transition to living as a male. Reiner performed surgery to make the child's phallic structure more penislike, closed the labia to form a scrotum, and inserted artificial testes. The case sparked Reiner's interest in the development of gender identity, and he subsequently shifted careers to train as a child psychiatrist. In his published account of the case, he concluded that "this case gives evidence supporting the concept of prenatal organization effects of sex hormones on the neurons . . . leading to gender identity and sexual orientation" (Reiner 1996: 802). Unlike the John/Joan case, the article created few waves.

By 1999, however, two studies by Reiner and his colleagues on cloacal exstrophy and other pelvic defects were frequently cited in debates over gender assignment (Reiner 1999; Reiner, Gearhart, and Jeffs 1999). Cloacal exstrophy is a rare condition, considered the most severe birth defect in the exstrophy-epispadias complex (congenital abnormalities that affect the bladder, genitals, and pelvis). In boys and girls born with these conditions, the lower abdominal organs, such as the bladder and intestines, are frequently not sealed in the abdominal wall and are exposed. Boys may have short, possibly split penises, cryptorchordism (a condition in which the testes do not descend into the scrotum), and very severe epispadias (in which the urinary opening is located above the phallus). Although clinicians are adamant that these children do not have intersex conditions, many boys with cloacal exstrophy have been assigned as girls, because the reconstruction of the male genitalia is considered difficult or impossible. Because these conditions are frequently associated with other birth anomalies, many children born with this diagnosis until recently had a very short life span.

Reiner's published studies did not include his entire sample, but he frequently presented his findings at conferences and lectures on the full sample, which at that time comprised thirty-three males (twenty-six with cloacal exstrophy and seven with mixed pelvic defects, primarily aphallia) aged five to sixteen who were raised as females and treated at Johns Hopkins (Reiner 2000a, 2000b). Of these thirty-three patients, Reiner noted in a talk that nineteen had assigned themselves back as male (Reiner 2000b). Since then Reiner has published findings from his follow-up sample of males with cloacal and classic exstrophy and "severe phallic inadequacy" (Reiner 2004; Reiner and Gearhart 2004; Reiner and Kropp 2004). He has found that roughly half of these individuals live as male (either through an acceptance of their male assignment or through a rejection of the female assignment). Based on these findings, Reiner has argued for a strong biological basis for gender identity and sexual orientation: "In kids with a normal hormonal milieu, maleness is fairly solid. There's a wall of maleness in terms of male behavior, male activity, male interest, male orientation—it's fairly solid. It doesn't take a terribly large amount of androgen to produce that, but certainly more than what a female has" (Reiner, interview, November 1, 2000). His willingness to speak frequently and candidly about his research and his unwavering biological essentialism have made him a prominent figure in the debates.

Reiner elaborated on his "philosophy of gender identity" at the annual meeting of the Midwest Pediatric Endocrine Society (MWPES) held in 2000: "It's not variable. It's not a process . . . and therefore, it's certainly not a response to an environment" (Reiner 2000a). He related his theory of gender-identity formation to what he saw as the problem with gender assignment today: gender identity is "not alterable, no matter *what* we do. So, the error of sex assignment lies in assuming that gender identity is either socially or scientifically determined, rather than a pre-scientific principle that makes possible the very science of gender that we're interested in" (Reiner 2000a). Reiner claims to talk only about males with typical hormonal exposure; however, he presents his theory as a universal theory of gender-identity development, one based exclusively on hormones.

One way in which Reiner confirms a male gender identity is through an assessment of the frequency of "masculine" and "feminine" behaviors exhibited by the children he studies. In a 1996 article, the child's inherent male gender identity is inferred from behavior: after the surgery, the individual "enjoyed rough-and-tumble play . . . never played house, never played with dolls, and preferred male roles in imaginary play but would be willing to take gender-neutral roles" (Reiner 1996: 800). Of another case, he has observed: "He's one of the most macho kids I've seen. He's only been a male since the second week in September and he's a starter on the basketball team. That's how good he is in athletics. That's how *male* he is" (Reiner, interview, November 1, 2000). To Reiner, these behaviors apparently constitute unassailable proof of an underlying male biology, male temperament, male behavior, and male gender identity, which are all, he says, "built in." Others, however, might reasonably suggest that he is in fact assessing to what extent the child's behaviors are male stereotypical, which tells us nothing about what produced those behaviors.

In Reiner's paradigm, not only gender identity but also sexual desire depends on androgen exposure. When speaking about the individuals with cloacal exstrophy who reassigned themselves as male, he says: "All of those who were fifteen years of age or older . . . have a girlfriend, which is interesting. One actually is engaged to be married" (Reiner, interview, November 1, 2000). Operating within an exclusively heterosexual framework, Reiner works backward from the individuals' object choices to infer their gender—if they are attracted to girls, they must be boys—which is often how physicians of the nineteenth century confirmed gender. Desire for females is seen as conclusive evidence of a male gender identity.

Reiner relies exclusively on the explanatory power of biology, especially of androgen exposure, in the development of gender behavior, identity, and sexuality. Although only half of the individuals in Reiner's sample identified as male, he felt it would only be a matter of time before the rest switched to live as men. In some instances, he suggested that strong family and social constraints were inhibiting this choice (Reiner 2000a, 2000b). Other studies, however, have shown mixed results.[18]

Although Reiner ties gender identity straightforwardly to the amount of androgen exposure, one of the researchers I interviewed, Dr. Q, cautions against such a simple causal link: "We need to be very careful because people are making some broad leaps in the absence of data. There is not a nice linear relationship between degree of androgen exposure and gender identity. You couldn't say, 'This child has been exposed to a moderate level of androgen and therefore is going to have this gender identity.' We have not ruled out the role of rearing in gender identity." Dr. Q feels that the appeal of Reiner's theory resides in its simplicity: "We want a simple answer right now, and Bill Reiner gives people a simpler answer."

The Limitations of Hormonal Theories

The resurgence of an interest in the role of pre- and postnatal hormones in sexual differentiation has forced the rethinking of the traditional model of gender assignment. Perhaps more significantly, however, debates over gender assignment often emanate from sensationalistic and rare case histories involving a gender change in individuals without intersex conditions, such as John/Joan, and spotty evidence about gender-identity change in 46,XY individuals with conditions such as cloacal, bladder, or classic exstrophy who were treated according to the treatment protocol for intersexuality.

Debates over gender assignment hinge on theories about the mechanisms involved in the development of gender difference and on the application of these theories. Two fundamentally different explanations for how gender differences develop ascribe varying degrees of significance to hormones in the development of *behavioral* differences between males and females. While some, such as Money, downplay hormonal effects, arguing that socialization plays an important role in gender identity and role formation (i.e., that biology is not determinative), others, such as Diamond and Reiner, ground their theories of development exclusively in biology, arguing that there is a decisive biological criterion—the prenatal organization of the brain—that determines a range of observed gender differences. These theories are united by the claim that sex hormones play an important role in the production of sex differences, though researchers and theorists differ in their opinions about which behaviors are affected. Within the fields of psychology and behavioral endocrinology, some posit that androgens masculinize behavior such as children's play; others theorize that gender identity and sexual behavior are also masculinized, by which they mean that the individual will be sexually attracted to females.[19]

These are not simply academic arguments: treatment hinges on these theories. As a result, debates over gender assignment do not concern all intersex conditions, but primarily those exhibiting gender-atypical prenatal androgen exposure. Broadly grouped, this includes what physicians understand as hyper-androgenized genetic females (CAH) and hypo-androgenized genetic males (such as those with 5-alpha reductase deficiency, PAIS, and micropenis).[20] Not surprisingly, debates also center on 46,XY individuals with typical male prenatal androgenization but with severe genital malformation or a traumatic loss of the penis—conditions not considered intersex. Finally, they include syndromes for which gender assignment has been regarded as very difficult, such as so-called true hermaphroditism and various forms of gonadal dysgenesis.

Support for the brain-masculinization or brain-organization theory is based on several types of evidence. Because it would be unethical to test these theories on humans by injecting fetuses with hormones, much of the research supporting the thesis has been carried out on rodents. In these studies, researchers manipulate pre- and perinatal androgen levels to examine their effects on sex-related behaviors such as sexual behavior, exploratory behavior, aggression, play, and maze learning. Androgens are theorized to cause masculine behavior; their absence results in feminine behavior. This research is important for what it implies about human sexual differentiation and the link among hormones, the body, the brain, and behavior.

Beginning in the 1960s, researchers sought to examine analogous effects of hormones in human behavior using the same linear and causal model developed for studies in animals. Much of this research has been carried out on "human experiments in nature"—that is, on girls with classical CAH who have been exposed to very high levels of androgen prenatally. Girls with this form of CAH overproduce androgens (e.g., testosterone), and this excess often leads to the partial masculinization of the external genitalia. Researchers have attempted to assess the effects of excess androgens on the psychosexual differentiation of these children, testing the brain-masculinization theory in humans by examining behaviors, activities, and interests that are thought to be typical of males or females (e.g., boys playing with construction toys and girls playing with dolls).

Most of this research involves case-control studies of girls with CAH, who are compared to unaffected girls (in some cases their siblings). Although a large body of research on these girls has shown a relationship between androgen exposure and male-typical play, androgens have not been found to determine gender identity. Put simply, despite an exposure to high levels of androgens in utero and so-called male-typical play preferences, girls with CAH very often identity as female. In scientific terms, observations show a correlation between hormones and behavior but have not proven that hormones cause the behavior. Androgens cannot be said to determine gender identity in any simple way. As Dr. T (a clinician and researcher I spoke with informally) wryly observed, "If Reiner was so right, all 46,XX infants [with classical CAH] would declare themselves male by age two, irrespective of gender assignment."

Other conceptual problems exist with these theories and with the studies to which they give rise. One problem is that the studies often model the biological and the social as distinct, rather than as interactive. The social has been further construed as a possible confounding factor for which the researchers must control. (When researchers control for a variable, the aim is essentially to eliminate its influence.) The effect of purporting to control for social effects is to strengthen any effects posited as biological. Moreover, researchers have frequently constructed a simplified notion of the social that neglects the effects of race, class, culture, and history. This makes things deceptively simple. With the focus on biology, much less attention is paid to the family and the world into which the child enters, how the parents react, whether they or the child need and receive counseling, the association of gender-atypical behavior with rejection, peer reactions to the child, other medical problems, family conflict, education, wealth, family functioning, early difficult and averse experiences, the effects of chronic illness or disability on the child and family, neglect, abuse, or attachment.

Even the biological factors in some of this work can come across as oversimplified. Researchers and clinicians all too often presume a linear theory of cause and effect in which the human brain is treated largely as a black box in which a prenatal hormonal input affects a later behavioral output (Doell and Longino 1988: 59). This model suggests that the effects of prenatal hormone exposure can either be quantitatively assessed as contributing a specific amount to the observed behavior or that it is wholly determinative of that behavior. Researchers and clinicians in this field have tended to assume that the neural and biological body is established early and remains remarkably similar over time, both across subjects and within individuals. However, recent research has shown a level of brain plasticity and the ability of the nervous system to adapt to changing circumstances, such that the adult brain, once thought to be a fixed collection of matter, is now viewed as malleable. Moreover, an exclusive focus on the perceived effects of hormone exposure fails to acknowledge or explore the ways in which nerve cells and genes can also affect behavior (Kandel, Schwartz, and Jessell 2000). It is overly simplistic to view hormones or genes as acting on their own to produce certain characteristics because their expression in the body depends on the actions of other genes, chemicals in the cell, and other factors. Although no geneticist suggests that genes act independent of their context, the relative importance of different influences is always in dispute.

Beyond Theory: Creating New Guidelines

Despite these limitations, having published their update of Reimer's case in the *Archives of Pediatrics and Adolescent Medicine*, Diamond and Sigmundson followed up with revised guidelines for assigning gender for newborns with intersex conditions. Much of what they suggested could hardly be objectionable, and with hindsight it seems almost surprising that their recommendations would need to be given at all: they emphasized that intersex conditions were not shameful, as well as the need to involve patients in decision making; to consider more than the size of the phallus in gender assignment; and to make the disclosure that genital surgery might impair sexual function and sensation. At the time, these were novel and important suggestions. The authors also provided revised ideas about how to assign gender, citing the effect of prenatal androgens on the sexual differentiation of the brain. Breaking from traditional practice, Diamond and Sigmundson suggested male rearing for XY infants with a micropenis, less severe androgen insensitivity, Klinefelter's, and 5-alpha and 17-beta reductase deficiencies, as well as for genetic females with CAH and "extensively fused labia and a penile clitoris" (1997: 1047). Because of prenatal androgen exposure, the rationale went, these infants would grow up with a male identity, male behavior, and male-typical sexuality (i.e., attraction to females).

In the same journal issue, Reiner proposed that "46,XY neonates with . . . testes and a somewhat small phallus" should be raised as males and suggested it might be advisable to raise "otherwise normal males" with a micropenis or an absent penis as male, also citing the masculinization of the brain (Reiner 1997: 1044). Elsewhere, he expressed concern about the chromosomal status of the infant: "If there is a Y chromosome, you have to be very worried about raising the child as female" (qtd. in Melton 2001). The chromosomal criteria would apply to a much broader range of syndromes and would theoretically include those with CAIS who identify exclusively as female.[21] His point was clear: Reiner, who once believed 46,XY infants could be raised as girls, now unequivocally said that they should be raised as males. Like Diamond and Sigmundson, Reiner also suggested that very masculinized females with CAH should be raised as males.

These suggestions proved controversial and engendered debate in part because they went against forty years of medical knowledge and practice. They were also based on little empirical evidence. Diamond relied on the John/Joan case to make his argument, as well as on earlier studies involving rodents and humans. Reiner, who in 1997 had not yet completed his research on cloacal exstrophy, based his recommendations on the same body of research as Diamond and an earlier case study he published. Moreover, the data that do exist on the relationship between androgens and gender identity do not support the guidelines; this discrepancy led one researcher to remark: "If biological factors fully determined the sexual differentiation of the brain and of gender identity as a functional outcome, female-assigned 46,XY persons . . . should have a male gender identity. . . . In fact, there should not be any difference in gender identity outcome between female-assigned and male-assigned patients of [*sic*] the same syndrome" (Meyer-Bahlburg 2005a: 424). Yet for many syndromes infants have been able to develop a gender identity as either male or female, suggesting that no single or specific biological factor overrides all others in gender identity formation and determination (Meyer-Bahlburg 2005c: 372).[22]

Although gender assignment has always proven more complex in practice than in theory, these recommendations complicated it in new ways. Even if one were to accept the brain-masculinization theory, there is no reliable way to assess an infant's degree of prenatal hormone exposure. Because it is impossible to know the level of androgen exposure in utero, researchers have used proxy measurements such as hormone levels sampled shortly after birth or the degree of genital virilization, which is thought to correspond to the degree of brain virilization. However, neither of these proximate determinants reliably assesses androgen exposure prior to birth, in part because the hormones primarily responsible for genital and brain virilization are thought to differ—prenatal masculinization of the genitalia is believed to be primarily influenced by testosterone, dihydrotestosterone, and anti-Müllerian hormone, whereas brain virilization is thought to be caused by testosterone and estradiol (Meyer-Bahlburg 2000, 2003). Moreover, researchers have found either a moderate relationship or no relationship between the degree of genital masculinization and supposedly masculinized play behavior in girls (Dittmann et al. 1990; Berenbaum and Hines 1992; Leveroni and Berenbaum 1998; Berenbaum, Duck, and Bryk 2000; Hall et al. 2004). Others have found no correlations between the degree of virilization and gender identity (Berenbaum and Bailey 2003; Meyer-Bahlburg et al. 2004), with one concluding that "the relationship is not strong enough to serve as a reliable basis for gender assignment" (Meyer-Bahlburg et al. 2001). In short, gender identity and role are not always consistent with either prenatal hormone exposure or chromosomal type. Nevertheless the belief in a biological basis for gender identity and gender-typical behavior is deeply entrenched.

Reiner and Diamond offered a theory conveniently reducing the complexity of the relationships between biology and identity, and a way to make complex human processes and bodies coherent and manageable. As the debate continued, clinicians began to cite these considerations in their decision making. One pediatric endocrinologist argued that the primary consideration for male assignment was "the degree to which prenatal exposure to androgens influences the eventual gender role in humans," adding that "in other vertebrates . . . postnatal sexual behavior is also masculinized" (Van Wyk and Calikoglu 1999: 538). Thus it is not only the role of androgens in identity formation that is of concern but also that of their influence on sexual behavior and object choice.

Others arguing for a male assignment for infants with a micropenis, aphallia, or any of the exstrophies, however, do not mention imprinting, which they argue "is by no means established." Rather, they argue that gender assignment "should be based on the underlying diagnosis, even if sex of rearing does not coincide with the size and functionality of the phallus" (Phornphutkul, Fausto-Sterling, and Gruppuso 2000: 136) and cite research showing that these individuals may fare well despite having a small penis (e.g., Reilly and Woodhouse 1989; Bin-Abbas et al. 1999). Echoing this sentiment, some have suggested that although the "potential for penis growth in relation to sexual function . . . must be considered," the capability for sexual response must be considered also, and "relative penis size plays no obligatory role" in sexual response (Lee 2001: 119). Some clinicians, and now the recent "Consensus Statement on Management of Intersex Disorders" (Lee et al. 2006), have also raised the issue of male fertility.

The principal challenges to Money's protocol by Diamond and Reiner have targeted his belief that gender identity and sexual behavior are not fixed at birth but can be influenced by rearing. Posing, as it does, a strong threat to the binary model of sex, Money's model has been challenged by researchers and clinicians searching for definitive biological markers of sex deeper and deeper in the body, not only in the genitals and gonads but also in gonadal secretions and the molecular assembly of chromosomes. The theories of gender-identity formation offered by researchers such as Diamond and Reiner are based on the notion that prenatal androgen levels contribute to a masculine sexual differentiation of the brain and to subsequent gender difference. Some posit that androgens masculinize behavior; others theorize that gender identity and sexual orientation are also masculinized, by which they mean that the individual will be sexually attracted to females. Ian Aaronson, the chair of the now defunct North American Task Force on Intersexuality (NATFI), a group that had hoped to carry out follow-up studies on the outcome of treatment for people with intersex conditions, suggested that in the interim, management protocols should assume that for more severely virilized infants, the brain is comparably masculinized (see Guttman 2000).

Gender assignment for infants born with intersex syndromes is one of the most contentious areas in the field of intersexuality (the other being early genital surgery). Although a degree of consensus exists for some diagnoses—for example, CAIS and CAH with so-called moderate virilization, in which gender assignment as female is uncontested—gender assignment for several other diagnoses, syndromes, or symptoms—5-alpha reductase, PAIS, and aphallia, to name a few—has remained highly contested. Most professional groups have no official policies regarding gender assignment specifically or treatment more generally.[23] This, then, constitutes the context in which clinicians and parents in recent years have faced the decision of whether and how to treat a baby born with an intersex diagnosis and what to tell the child about the condition as he or she grows. The following two chapters investigate this experience from the perspective of those involved in making decisions about treating infants born with intersex diagnoses.

PART 2

Making Decisions for and about a Baby with an Intersex Diagnosis

FOUR

Boy or Girl?

Bodies of Mixed Evidence and Gender Assignment

Sara and Jim Finney are an easygoing and thoughtful middle-class couple who live on the outskirts of a major city on the Eastern seaboard. In their early thirties, they already had one daughter when Sara gave birth to their second child. Although the doctors said they had a girl, the Finneys immediately questioned the gender announcement; their newborn's genitalia looked too masculine for a girl. The parents describe their experiences directly after the birth of their daughter:

Sara: After she was born, he [the delivering clinician] said, "It could be mixed gonadal. . . . She could be a hermaphrodite. She could be. . . ." [to Jim] What's the boy one?

Jim: Androgen insensitivity?

Sara: Right. "We don't know. We have to do a karyotype."

Jim: Meanwhile, we called up the family and said we had a girl, because some of the doctors were telling us, regardless of what we saw, it was a girl, so we told everyone that we had a girl.

Sara: Yeah, we called my ten-year-old, my mother, and our workplaces and said, "We had another girl." Later that day, [the pediatric endocrinologist] came in and said, "Don't name this baby."

At this point in the retelling the couple disagrees on what sex they initially told people the baby was. After some back and forth, they remember that on the day their baby was born, they told everyone they had a girl, but that on the next day, they told people they had a boy. Jim explains that they wanted to have a boy and that "visually, it looked like she was closer to a boy." They describe what happened next:

Sara: A nurse came in and said, "What did you name your baby?" That was the first time I cried, because [the pediatric endocrinologist] had come in and said, "Don't name this baby." I mean, the first question everybody asks is, "What did you have?" We had both! That's what we had. We had "what." I would never refer to her as "it." She was always "he." Even the nurses would say, "Oh, she looks like a boy down there."

Jim: So we said, "That's it. We're not going to go through this 'it' crap. As far as we can tell, it's a boy," and we named her. [The pediatric endocrinologist] was very upset.

Sara: Yeah, he got mad at me! She was born Wednesday, and we named her Thursday. My thinking was, she didn't have an identity, and I didn't know if she was going to live or die, but she wasn't going to die without an identity. She was referred to as "the baby," and I resented that right away. By the time she went to [the tertiary care hospital], she was Sam James because Jim said, "We have to fill out a birth certificate." We said, "She looks like a boy, we're going to raise her as a boy. If worse comes to worst, all we have to do is change the birth certificate." [The pediatric endocrinologist] found out we named her . . . him. He saw we had blue blankets and blue this, and . . . she needed an identity. She was not an "it." And he came up to me and said, "I told you not to name the baby! How *dare* you name the baby!" And I started crying and said, "How dare you tell me *not* to name this baby! This baby is not an 'it'! If I have to change the birth certificate, I will. This is Sam." (original emphasis)

After telling everyone they had a girl and then a day later that they had a boy, clinicians told the Finneys they had to wait for the test results and a diagnosis to determine a gender assignment. Several days later they learned that Sam's chromosomal makeup was mosaic (45,XO/46,XY) and that she had a testicle on one side and a what is called a streak gonad (a poorly functioning, immature gonad) on the other. Her diagnosis was mixed gonadal dysgenesis. Sara describes the distressing consultation with the pediatric urologist and pediatric endocrinologist about their newborn:

> They came in with their books, and they mapped out the pros and cons. We had been so insistent that we wanted a boy. They said that they could have made her a full-functioning male, but there was a risk of cancer [in her gonads]. Ninety percent, they told us. I cried and cried and *cried!* Jim cried. It was very, very emotional. Jim listened to everything and explained it to me later. I kept hearing, "Sitting on the fence," which, I *hate* that term! She was male and female, literally sitting on the fence. We had a decision to make. And also "God's plan B." That's what he said! Because I said, "We could have a boy; you can make me a fully sexually functioning boy?" And he said, "Yes, it will take many surgeries, but we can give him a penis, and he will be a boy, but with a 90 percent chance of testicular cancer, or he doesn't have testicles." Or we could have a girl, "God's plan B. We could give her a vagina, we could give her all the stuff that comes with one and, with the help of estrogen, she can grow breasts and she can get a period. She's one surgery and done. And there's *no* chance for cancer because we'll take the gonads out." So here they are throwing, "If it's a boy, many surgeries, cancer, but for a girl, one surgery, estrogen, growth hormone." Before they said "cancer," Jim was still thinking boy. But I'm thinking, "I am going to have a short, feminine, small-sized boy." Kids are cruel. If he takes a shower in high school and they see a small penis, that's it. They told us, "Think about it over the weekend." I just cried. They said she's going to be short, and I said, "What do you mean short? Dwarf?" That's when I thought, "Oh my God, what the hell did I give birth to?" I do not admit that easily. That was probably the worst day of my life, that day in the consultation room. . . . Finally, they said, "Raise her as a girl." And I was like, "What does *that* mean?" (original emphasis)

Feeling distraught and overwhelmed, the Finneys went home to make sense of what they had heard and to decide which gender to raise their newborn. As it would be for most parents, making this decision proved terribly difficult for them. For a week Sara and Jim had the baby boy they had wanted so much, but eventually they decided to raise Samantha as a girl. As Jim describes it, they felt they really had no choice on the basis of the information given: "The way I looked at it was, 'If we've got to, we'll reverse ourselves. It's pragmatism if we take emotions out. These are the facts, let's get on with it.' I remember [the doctors] said they had talked by phone over the weekend a couple times—they were both concerned that we would make the wrong decision, they felt the right decision was girl."

The couple's decision to raise Samantha as a girl meant the loss of a son. It was an acutely painful choice for them, as Jim describes: "When we named her Samantha, it felt like our son had died. Almost like we'd had twins. It was the only way for us to deal with it; Sam died, and Samantha was born. We told people, 'We mourn the loss of our son, but we welcome the birth of our daughter.' All these decisions are in the heat of battle. When you're trying to make decisions, you need to settle this to get on with her life."

Most families facing such a difficult decision regarding their newborn undoubtedly share these feelings. The stress of trying to make the "right" decision is overwhelming, eliciting utmost sympathy for families who must make these choices and deep respect for their resiliency and honesty about their concerns and fears for their child. On another level, their experience is striking in its demonstration of the degree to which clinicians control the process of gender assignment by the way they frame the terms of the decision. Although Sara and Jim were given the option of raising their newborn as either a boy or a girl, they were told that raising Samantha as the boy they so wanted would have high costs, including multiple genital surgeries, a very high risk of testicular cancer, and infertility. (Interestingly, in this case the clinicians did not mention chromosomal status or exposure to androgens in utero as a justification for male assignment.) Raising her as a girl, the clinicians said, would involve only one genital surgery (vaginoplasty), no risk of cancer, and a normal sex life. Given this information, as Jim says, there was no real choice to make.

Sexing the Brain: Nature or Nurture?

Although researchers are wont to express their theories and findings as seamless and unambiguous, the evidence on which certain treatment decisions or guidelines for intersexuality (e.g., Milton Diamond's or William Reiner's) are based is often shaky at best. In these debates, nuance and the exploration of an issue's various sides often loses sway to polarized positions where one's flag is firmly planted in the nature or nurture camp, despite the fact that much research on gender-identity formation reveals that its development is complicated, largely still unknown, and deeply debated, even for those who specialize in this field.

Despite these limitations, clinicians have increasingly begun to cite concerns about brain virilization and chromosomal status in the gender assignment for newborns with intersex diagnoses. This shift in part reflects a broader contemporary resurgence of enthusiasm for the idea that we are born, not made. Hormonal theories also appeal for another reason: because clinicians are faced with such enormously complicated and difficult decisions about gender assignment, they are grasping for something definitive to guide them, to help them feel they are making the right decisions. These theories provide a simple rationale.

Although clinicians have expertise—about the endocrine system or surgical techniques, for example—many have received only basic training in determining gender assignment. This is especially true outside major treatment centers, of which there are about a dozen in the United States. Comprehensive training has been rare in part because of specialization—although surgeons frequently participate in gender-assignment decisions, gender identity development is not seen as an important part of their training—and because the traditional paradigm made gender-assignment decisions relatively straightforward, at least in theory. Thus many clinicians handling these cases are specialists with extensive knowledge about the body's parts and processes, but they are not specialists in the development of gender identity. (This is made more problematic because there are few psychologists or psychiatrists specializing in gender identity more generally and intersex diagnoses specifically. Moreover, unless they work at a major center, they are unlikely to be part of the treatment team.) As a result, a good number of clinicians handling these cases know only the most basic outlines of contemporary theories and research on gender-identity formation. As one pediatric urologist said: "I was trained how to make a male or a female; no one trained me on how to make gender-assignment decisions," leading him to suggest that *he* needed a support group for how to deal with these cases. A colleague added: "We are swimming in a sea of conflicting and contradictory information. I don't know what to do. *We* don't know what to do. There's been a sea change in terms of how we think about gender assignment, but what to do about that is anybody's guess." Dr. A, a pediatric endocrinologist, summed up the state of gender assignment: "When you look at it right now, it is a guessing game." Dr. G, a fellow pediatric endocrinologist, noted: "There's just one diagnosis we're sure about, which is CAIS. Nobody knows enough about the other diagnoses to say what's going on." This uncertainty was reflected in the title of a recent article in a medical journal: "Possible Determinants of Sexual Identity: How to Make the Least Bad Choice in Children with Ambiguous Genitalia" (Mouriquand 2004).

With newly controvertible theories, evidence, and recommendations, clinicians must rely on their training and clinical experience, as well as on intangible and deeply subjective factors such as their own ideas about what makes a man or a woman. Although it is certainly true that clinicians making gender-assignment decisions weigh biological and physical traits, they also base their decisions on cultural ideas about the physical features and behaviors understood to be the natural signs of one's sex. Gender, then, is not so much an innate feature as something variously imagined and enacted. The idea that certain features cluster together to indicate a sex rests on sorting bodies according to a particular interpretation of natural properties that begins with the presumption of gender and works backward toward the body (Warnke 2001: 130).

Clinicians construe gender assignment as an enormously complicated medico-clinical decision. There is no question that it is. Detailed attention to bodily parts and processes helps clinicians assign gender, for example, by taking into account how the body will develop at puberty and whether the child will be fertile as an adult. On another level, however, gender assignment constitutes a process of interpretation and enactment based on what somatic traits and configurations clinicians believe are necessary (or even allowed) to be male or female. The clinician's role in gender assignment thus involves a peculiar balance of discovering sex and determining gender (Kessler 1998).

Although complicated ideas about biological sex enable physicians to rectify the breach that intersexuality poses, these also serve as the basis for invoking and reconstituting gender difference. The conditions that make the category "intersex" possible directly influence the medical practices resulting from this categorization. Medicine and medical disciplines are of primary importance in protecting the lifeworld distinction between men and women in part through its continual reconstruction (Hirschauer 1998: 13). Sex, then, is something that is fashioned, negotiated, practiced, and enacted. The apparent simplicity of the question "Is this a girl or a boy?" conceals the fact that to ask the question, one must already know what boys and girls are (Kessler and McKenna 1978). This not to deny biological differences between humans, but to ask how a difference already made in everyday life (i.e., gender difference) is both drawn on and reproduced in medicine (Hirschauer 1998: 13).

In the case of intersexuality the usual role of the medical practitioner—to observe, diagnose, and treat—is stretched beyond its normal boundaries because it involves the very personhood of the patient. Intersex bodies force clinicians and parents to make sense of seemingly unnatural gender combinations—for example, an infant with XY chromosomes who has testes and female-typical genitalia. What are the primary characteristics to which clinicians look for identification of an individual's sex? How do clinicians resolve the uncertainty and complexity of sex?

In this chapter and the next (on genital surgery) I examine the complexities inherent in the medical process of defining and determining sex which rub against seemingly objectified tools, techniques, and markers; the ways in which social concerns about gender are translated into medical concerns; and how beliefs about gender difference and its origin organize clinical practice.[1]

Genitals as Signifiers

No sooner than a baby is born its sex is announced by the attending clinician, based on an inspection and understanding of the external genitalia as either male or female. The process of sex identification at birth is one in which genitals are granted the power of synecdochic representation. Genitals, and the sex designation to which they give rise, create gender expectations for almost every aspect of an individual's life. Not only are they usually the sole factor of sex determination but they are also assumed to correspond with fully and uniformly differentiated internal sex organs and are further charged with the task of signifying and predicting gender (whether identity, role, or behavior) and even sexuality. Put another way, if a baby is labeled "female" at birth, it is assumed that that person will grow up to understand herself as a woman, to dress and act like a woman, and to desire and have sex with men. Because this is the usual course of events, it is assumed natural. At birth genitals are thus viewed as symbolically and literally revealing the truth of gender.

At no time are the connections between genitals and gender more evident than when the genitalia of an infant either do not signal or else missignal sex. In these instances, atypical or, in clinical terms, ambiguous genitals are seen not as the representation of sex, but as the signal of a misrepresentation of sex (Kessler 1990). Without legible genitals, and thus without an evident or stable sex, an infant with "ambiguous" genitals flutters not simply between sexes but between genders and sexualities: such infants are neither readily male nor female, neither masculine nor feminine, and consequently neither readily homosexual nor heterosexual. So-called ambiguity is posited as the ground of sexual and gendered difference: a prediscursive, precultural dimension of bodiliness rather than an effect of a social system that requires a binary and incommensurate set of two sexes (Morris 1995).

Bodies with atypical or conflicting biological markers of gender are troublesome because they disturb the social body; they also disrupt the process of determining an infant's place in the world. Gender-atypical genitals (and bodies) create anxiety about the borders of properly gendered subjects and a desire to reaffirm these borders. In a culture that requires clear gender division—a culture in which, to paraphrase Michel Foucault, we truly need a true sex—gender-atypical bodies threaten an entire system of laws, rights, responsibilities, and privileges built on notions of discrete and binary gender.

As a result, clinicians often rush to stabilize the sex of infants with intersex diagnoses. The urgency of this undertaking, to which parents no doubt contribute, all too often overrides the joy of the birth, as an infant may be whisked away for medical tests before the parents have had any chance to bond with their baby. Parents may be discouraged from naming their baby before a gender assignment is made. To avoid using gendered pronouns, clinical caregivers may refer to the newborn as "the baby." Because the announcement of sex is usually considered a prerequisite to naming a child, which is in turn a prerequisite to filing a legal notice of the birth, there is a sense in which biology determines—or confuses—the newborn's entire social and legal identity. Physically alive but denied a sex and a name, the infant has no social existence. Personhood depends on gender assignment.[2] Monica Cole, whose daughter has CAH, describes living with this uncertainty after the birth of her baby: "The doctor said we would need an ultrasound to determine our baby's internal sex organs, and a genetic test, which could take a week. Well, how could we not know the gender of our baby for a week? I had a hard time not being able to say 'he' or 'she,' and 'baby' was so distant. The hospital had only blue-striped or pink-striped baby hats, and a nurse asked which we would like to use. I picked a blue hat and decided to use a male pronoun. The nurses followed our lead of what pronoun to use, but they also placed both an 'I'm a boy' and 'I'm a girl' cards on the baby tub."

The birth of a baby with an intersex diagnosis is thus considered a social emergency in which medical experts are called on to intervene. The entire process could be understood as what the anthropologist Victor Turner has called a "social drama" with four stages: breach, crisis, redressive action, and reintegration (Turner 1974). The breach or schism in the social order caused by the birth of a baby with atypical genitals (and thus no obvious gender assignment) produces a crisis that must be addressed because it threatens social norms. The redressive action is the culturally defined process through which gender is assigned. Although not all parties may agree about the correct gender assignment for a particular infant, all agree that the resolution of indeterminate sex is necessary, and thus some accept a particular decision as final simply to bring about closure. Reintegration eliminates the original breach that precipitated the crisis. Treatment decisions remove biological or phenotypic atypicality, recreating a particular gendered world.

As this chapter and the next will reveal, clinicians and parents typically share the same goal, though their opinions on how to attain it may be diametrically opposed: to use the best medical technologies available to adapt the infant to life within the binary gender model; living as much as possible as a "normal" male or female.

Birth: *When Genitals Vex*

The birth of a child with atypical genitals sets in motion a team of specialists that might include a pediatric endocrinologist, a pediatric urologist or surgeon, a gynecologist, a geneticist, and, more rarely, a psychologist, psychiatrist, or social worker.[3] Some institutions, usually teaching hospitals, may have a team to deal with intersex births, whereas at others the assembly is typically ad hoc. Often the first specialist to see the family will have primary authority in organizing care and making treatment decisions. At a small, rural, or primary-care hospital, the family may be referred to a tertiary-care medical center, where clinicians order tests to assess the syndrome that produced the genital ambiguity and to interpret the multiple somatic markers of sex. (In some cases, physicians have already assigned gender before referral.) These tests may include karyotyping and chromosomal analysis; serum electrolyte, hormone, and steroid evaluation; ultrasound; laparoscopy; renal imaging; and a genitogram, a form of x-ray examination that uses a dye to reveal the structure of the internal genitals. Obtaining the results from these tests can take anywhere from several days to weeks, during which time the newborn has no gender assignment.

Given the current method of gender assignment, technology has its benefits. Today's technology enables a much quicker assessment of somatic traits than was possible even ten years ago. Prior to ultrasound and other technologies, for example, the clinical assessment of the internal organs could take much longer, might involve surgery, and could produce inconclusive findings. Ultrasound provides a more accurate result than the older method of palpation to assess the gonads, and it enables doctors to avoid invasive exploratory surgical procedures. The proliferation of highly technical tests, however, also reinforces the power of biological and clinical authority in the determination of sex. Clinicians, as specialized interpreters of the body, transform a problem of social gender into a biological anomaly, and having done so, they hold out the promise of both revealing an answer—the newborn's "true" sex—and providing the solution through medical intervention. Parents are thus instructed and led to view intersexuality (or gender atypicality) as a medical issue that can be resolved through medical treatment.

Clinicians may mask the extent of the biological anomaly through the determined use of language and narrative. For example, while clinicians and researchers routinely use the term *intersex* in the medical literature and among themselves, most of those I interviewed told me they refrained from using this term with parents because, as Dr. B explains, "it carries the connotation of circus freak." To avoid this connotation, they tend to use the term *ambiguous genitalia* and present the situation as a solvable problem: "There's some ambiguity, but, by doing these tests, we'll be able to figure out whether your child is really a boy or a girl" (Dr. I). Such phrasing appeals to the notion of a true sex, which must be uncovered through the specialized knowledge and expertise of the clinician.

Whereas some parents complain of the pressure to agree to tests and a gender assignment for a newborn more quickly than feels comfortable for them, clinicians I spoke to described intense pressure from parents to act fast to determine the baby's sex. Dr. D, a pediatric endocrinologist, observes: "There's nothing more challenging or stressful than an infant with ambiguous genitalia. The stress is not that in some instances the infant is acutely ill. The stress is, 'What's the sex of the baby?' The parents will *not* tolerate no sex assignment. I feel like I need to give them a sex assignment within a week" (original emphasis).

Parents understandably want a rapid gender assignment to allay their own anxieties, but they are also often unsure and nervous about what to tell others during the period prior to gender assignment (see chapter 6). The pressure to assign sex quickly drives some clinicians to make a preliminary assignment before they have all of the test results, in effect taking a chance that they may have to change the gender assignment later. Dr. A, a pediatric urologist, tells of such an encounter: "They [the parents] were really pushing me with 'What do you think it will be?' I opened my mouth and said, 'I'm 95 percent confident that this is a girl, but there's a 5 percent chance I'm wrong.' Then the karyotype came back 46,XY just before the family was going home. Luckily, I was able to come back and say, 'The 5 percent came through.'"

Imperfect Exercises: Assigning Gender

After conducting a battery of tests on the infant, how do clinicians and parents interpret and apply the results to identify or else select the baby's sex? Such tests stand as a testament to the belief in biological markers as reliable indicators of sex, identity, and a person's being. The focus on "objective" data obscures the fact that doctors and parents rely on folk rules about gender when interpreting biological information to make treatment decisions for intersex infants—rules resulting in treatment decisions and medical practices that both buttress and reconfirm binary gender. Despite the relative clarity of the prevailing theories (e.g., those of John Money, Milton Diamond, and William Reiner), clinical choices are never straightforward. These theories collide with the practical concerns of case-by-case management. Gender assignment at birth is informed by heterogeneous factors including the diagnosis, the type and degree of physiological ambiguity, the options for genital surgery, the attending clinicians' training and diagnostic sophistication, the hospital's medical policies, and the parents' desires. But perhaps the most important factor influencing clinicians' decisions is their own view of gender theories and ideologies: that is, their own beliefs about what makes us male or female.

Contemporary cultural beliefs about gender assume an agreement among a series of somatic characteristics (chromosomes, gonads, genitals, and secondary sex characteristics) and more phenomenological processes, such as gender identity, gender role, and sexuality. In this model, gender identity and behavior derive unproblematically and naturally from biological sex. Intersex infants violate these gender rules, forcing clinicians and parents to make sense of seemingly unnatural gender combinations. To be female is to be able to engage in the activities and practices of the feminine gender, but it also means having the appropriate body to fit cultural ideas of the female sex (Warnke 2001). The complexities of gender assignment hinge on the question of what—chromosomes, hormones, genitals, and rearing—makes us male or female. In trying to pinpoint the signs of gender in the body, clinicians can contradict one another and even themselves. These conflicts are important because they drive decisions about gender assignment and, ultimately, shape the lives of individuals with intersex diagnoses. They also call into question the naturalness of sex and gender distinctions. Gender is necessary for the interpretation and reconstruction of bodies as sexed (Warnke 2001).

MEASURING FOR PHALLIC INADEQUACY

Traditionally, penis size has driven gender assignment, much more so than genital morphology has driven decisions for genetically female infants, who overwhelmingly have been assigned female irrespective of the degree of genital masculinization.[4] If this concern reflects a privileging of masculinity, it also reflects the greater stigma attached to bodies that fail to measure up to social criteria defining a virile male and the limited possibilities for "correcting" such an abnormality. Because penile-vaginal intercourse is deemed the paradigmatic sexual behavior—it is what most people mean by the word *sex* and what Carole S. Vance calls "the meat and potatoes in the sexual menu, with other forms, both heterosexual and homosexual, arranged as appetizers, vegetables, and desserts" (1991: 878)—a male who is unable to perform it is considered to express an inadequate or failed masculinity. In the treatment of infants with intersex condititions, such potential inability has been deemed sufficient cause to question not simply the child's masculinity but his very maleness (Kessler 1990). In cases where males with a small phallus have been assigned and reared as female, the decision has been based more on infant penis size than on fertility, exposure to androgens (though some might argue such exposure must have been minimal because the phallus is small), or even chromosomes.

Why the focus on phallus size? The decision to raise an infant as male has centered on the potential for the phallus to "function satisfactorily for sexual activity" (Newman, Randolph, and Parson 1992: 646). But how is function conceptualized? Clinicians state in no uncertain terms that the phallus must be large enough for intromission or intercourse. As my interviews make clear, the only penetration that concerns clinicians is vaginal penetration. Often left out of this formulation is the ability of the phallus to become erect and to have sensation, or the individual's ability to produce and emit sperm and to reproduce. Clinicians have often overlooked the possibility that people may be pleased by erotic contact other than intercourse. One surgeon I interviewed summarized the conventional view:

Dr. K: If they have good sexual function, then we've done our job.
KK: What do you mean by good sexual function?
Dr. K: I am talking about intercourse.

Several clinicians subscribed to the general view expressed succinctly by Dr. B: "Our discipline [pediatric endocrinology] is male-dominated, and most men think the most important thing about yourself is the size of your phallus. If there'd been more women endocrinologists for a longer time, then maybe it would have dawned on them that men with small penises can still be happy. It's so important, it can take priority over fertility." However, in this respect clinicians also frequently respond to the concerns of parents. According to one clinician I interviewed, "A common scenario is for a child to be brought into our clinics in mid-childhood, where the concern is that the phallus is too small, and is there something wrong, or can we do anything about it?" (Dr. H).

At the heart of this concern lies the question of whether a boy with a small penis will be able to have penile-vaginal intercourse as an adult. In 1985, John Money and his colleagues published a follow-up study of nine men with a micropenis between the ages of twenty-two to thirty-one (Money, Lehne, and Pierre-Jerome 1985). It was the first article of its kind. In evaluating the "erotosexual" life of these patients, time and again the researchers used coitus as the defining sexual behavior: "His penis in adulthood, though thin and short . . . was long enough to enter the introitus in heterosexual intercourse." "Another . . . had a penis too short for coital penetration." "His penis was too thin and short in adulthood for effective coitus." Money writes that the last two patients became "homosexual in the submissive erotosexual role" (40). He continues: "The stigma of growing up with a micropenis . . . increases the chances that ordinary heterosexual G-I/R [gender-identity/role] differentiation and role will be dislocated" (40). Without an adequate penis with which to identify, gender identity and role are supposedly disrupted, making it likely that the individual will, in Money's view, assume a female role. The concern is a dual one: that the penis enable penile-vaginal intercourse, but also that a "too small" penis may result in gender-atypical desires. With an inadequate penis, the fear is not simply that these individuals will not identify as men but that they will not act like men.

Although statistical means for infant penis size have existed since 1942,[5] it was not until 1980 that numerical criteria for micropenis were established. By then several clinicians, including Money, defined a micropenis as a penis with a stretched length two and a half standard deviations below the mean for the infant's age or stage of development (Lee et al. 1980a), or smaller than 2.5 centimeters. Other authors have used a threshold of two centimeters (Donahoe 1991). Defining the micropenis is only part of the issue; the more critical question is how small is too small for male gender assignment. Money and his collaborators set this threshold at 1.9 centimeters (Lee et al. 1980a). In contrast, Patricia Donahoe recommends reassigning to female any "genetic" male with a phallus of 1.5 centimeters or less. Others argue for yet different standards for micropenis, in some cases simply as an "abnormally small penis" or one that measures "significantly less than normal." According to the available definitions, a micropenis could be 2.5 standard deviations below average, or less than 2.5, 2, 1.9, or 1.5 centimeters.[6]

Although the introduction of size criteria could be viewed as an important safeguard against arbitrary or subjective assessments of penile size and consequent gender assignment, tools of measurement to create or determine body norms can also constitute tools of normalization.[7] Norms for penis size, though intended as a diagnostic aid, have had the effect of creating size ideals based on a rigid gender ideal. They also fail to take account of the facts that neonate phallus size is not always predictive of adult phallus size and that phallus size varies a great deal among adult men. Most important, although different studies establish different criteria, they offer no rationale for the distinctions, no meaningful explanation as to why 0.4 centimeters might be a significant variation in the penis size of an infant. Dr. N, a pediatric urologist, suggested the rationale for the 2.5-centimeter criterion: "It was intuitive, not scientific. If you look at the [1942] nomogram and take the 2.5 centimeter [measurement] out to maturity, it is insufficient for intromission, and that likely was the rationale for that number." Using the available guidelines, an infant with a 1.8 centimeter penis might or might not "require" a female gender assignment, depending on the reference source. Conceivably, then, treatment differs based on the criteria used, which varies among institutions and individual clinicians.

With so much riding on these measurements, clinicians have to confront the awkward fact that the practice of measurement itself is imprecise. How do you gauge the length of a small, fleshy, spongy organ to an accuracy of 0.2 centimeters? Several authors have tried to address the problem of consistency and reliability in measurement by arguing that the penis should be fully stretched when measuring, suggesting that flaccid penile length was too variable (unreliable and irreproducible) and that there is a high correlation between infants' fully stretched penile length and the erect length of the penis. (To measure by this technique one engorges the penis by pinching it at the base, measures along the dorsal side, and includes the glans but not the foreskin.) Clinicians disagree about the ease of obtaining reliable measurements of an infant phallus. The pediatric urologist Dr. I felt confident: "It's an exact measurement. If you've done it often enough you know how to do it. Very straightforward." Another surgeon agreed, "Measurement is really standard, and it's an exact measurement. I don't think that you can stretch it too much or not enough. If you've done it often enough, you know how to do it." Their comments, however, suggest both a learning curve and a self-referentiality to the process: does experience beget greater accuracy or simply the validation through reiteration of one's own technique?

Most clinicians I interviewed were much more equivocal. Dr. D, for example, says:

> There's definitely error in measuring. It's not terribly precise. You take your index finger, press down to the baby's pubic bone and then stretch the penis up the length of your index finger, and then mark the end of the index finger with your thumb. You get your ruler, and that's the stretched length. The problem in a baby with ambiguous genitalia is they often have chordee [bent penis], and so it's tethered down by skin and you can't stretch it fully, so you have to guess. There are other problems. The penis is small, the measurements are close together, it's a crude ruler, you get different measurements for the same individual. So you do the best you can. You measure, but I wouldn't place any bets on it.

THE DIMINISHING IMPORTANCE OF PENIS SIZE

In medical writings and in my discussions with clinicians, the biological functions of the penis are often conflated with its social ones. As noted earlier, a primary social concern is that the penis be capable of vaginal penetration. Another concern is that the penis enable a proper urinating position. One medical article states: "Generally speaking penile length needs to be sufficient for the boy to stand up to urinate, to allow penetration during sexual intercourse, and, on the social standpoint, to avoid embarrassment when viewed by others" (Lee et al. 1980a: 156). The authors effectively define the first two requirements as biological needs. Dr. H, a pediatric endocrinologist, describes how this concern was presented in his training: "The classic way a senior physician would relay the decision making was to ask, 'Is this boy going to be able to stand up to pee?' In one brief sentence that was designed to acknowledge that 'If you can't stand up to pee, you're going to view yourself as less male, people are going to view *you* as less male. Society, whether it be boys in the locker room or whatever . . . it's just going to be an endless series of heartaches for that guy and, therefore, wouldn't it be better to take the other path?' " (original emphasis).

Not all clinicians are convinced that a micropenis necessitates a life without sex or life as a female. In the unfortunately titled article "Mutually Gratifying Heterosexual Relationship with Micropenis of Husband," the authors evaluate the sex lives of three individuals with micropenis, two of whom were genetic females with CAH (Van Seters and Slob 1988). (Both individuals with CAH were raised as boys.)[8] As proof of the successful treatment of the first individual with CAH, whose ovaries had been removed, the authors note, "He was married at the age of 28 and 3 years later became a father by donor insemination of his wife" (102). The second patient with CAH also "succeeded in pursuing a heterosexual relationship," was married, and became a father through donor insemination (102). The authors used these two examples to highlight to the third individual, a small-phallused but fertile man, and others like him, that "a satisfying and rewarding sexual relationship is possible for men with micropenis" (103). The mention of new reproductive technologies that enabled these individuals to become fathers is refreshing and highlights how little weight fertility, as opposed to sexual performance, has been given in male gender-assignment decisions.[9] Once again, however, the proof of success is the individual's ability to function within the heterosexual paradigm by marrying a woman and raising children.

Additionally citing the case of a patient with a micropenis who experienced pleasure with a female sexual partner through a variety of sexual activities other than vaginal penetration, the authors conclude, "Intravaginal intercourse is not a necessary prerequisite for a happy relationship" (103). Although this observation may seem obvious to many laypeople, it had scarcely been expressed in any clinical writings on intersexuality. The authors did not, however, cite this view as justifying the reconsideration of female gender assignment for males with microphallus. Instead, they stressed that the paper's conclusions concerned the management of the condition in adults only. While advising against genital surgery for adult males with micropenis on several counts, they maintained that the management of micropenis in children was less controversial.

Another study of men with micropenis (Reilly and Woodhouse 1989) sustained the clinical bias prioritizing heterosexual intercourse while dissenting from the mainstream view suggesting female gender assignment. The authors interviewed twenty males between the ages of ten and forty-three and reported that all were heterosexual and had erections and orgasms. Yet it is unclear how they assessed sexual orientation, as eight of the participants were still quite young (ten to thirteen years), and the researchers asked no questions about sexual attraction, fantasy, desire, or behavior. Moreover, questions about sexual activity were limited to penile-vaginal intercourse and masturbation. In a nod to Money's assertions that a penis of sufficient size was necessary to the development of a male gender identity and role, the authors noted with surprise that all of the adult males, despite penis size, "felt male." "The small penis has not," they wrote, "deterred them from a male sexual role" (571). On this basis they concluded that a small penis should not necessitate a female gender assignment.

A policy statement on the evaluation of infants born with ambiguous genitalia issued by the American Academy of Pediatrics in 2000 echoes Money's treatment paradigm when it states that the "size of the phallus is of paramount importance when one is considering the male sex of rearing" (American Academy of Pediatrics 2000). The consensus statement issued in 2006, however, suggests male rearing in all patients with micropenis, but taking into account equal satisfaction with assigned gender in those raised male or female and the potential for fertility in patients reared male (Lee et al. 2006).

It would be an oversimplification to assume that clinicians' more expansive understandings of masculine sexuality alone are responsible for the recent reevaluation of female gender assignment for infants with a micropenis. More important is the recent scientific and popular resurgence of biologically based theories of gender difference that, generally speaking, emphasize the importance of chromosomes, genes, and hormones in making us men and women. Although researchers acknowledge that gender identity cannot be predicted from karyotype, genotype, or prenatal androgen exposure alone, there is an overriding sense among both the researchers and the clinicians I interviewed that they contribute to gender-identity formation in important ways. At the same time, these researchers and clinicians reject the notions that rearing significantly determines gender identity and that genital appearance per se plays anything more than a small role in gender-identity formation. As a result of these two shifts, clinicians increasingly consider chromosomes and the degree of prenatal androgen exposure in their gender assignment decisions. Because science has yet to describe specifically how genetic and hormonal factors might shape gender identity, clinicians are left to surmise their supposed effects in any one case.

THE POWER OF CHROMOSOMES

Among many specialties, karyotyping to reveal the newborn's chromosomal type carries considerable symbolic weight in the process of gender assignment, despite medical knowledge about the importance of other biological markers of sex and no direct evidence of how chromosomal (or genetic) configuration contributes to gender-identity formation. Expressing what I think of as chromosomal essentialism, many clinicians I interviewed view chromosomes as indicative of the essence of the person and construe one's identity as inborn, natural, and unalterable, thus establishing gender categories based on a seemingly essential truth about the body (compare this to the genetic essentialism in Franklin 1993: 34). As a result, they are increasingly hesitant to gender-assign an infant contrary to its "genetic sex," even though the symbolic meaning of the chromosomes in this case is quite independent of their "scientific" meaning (Nelkin and Lindee 1995). In the case of a baby with an intersex diagnosis, chromosomal essentialism rests on the belief that men and women are made distinct by and through the chromosomes, and that their essence is defined in terms of their biological traits (Grosz 1994).

In his earliest work, Money argued against the use of chromosomes as the sole determinant of gender assignment. Although he suggested that "genetic males" who were "undervirilized" should often be raised as female, he recommended that infants with a 46,XX karyotype (primarily those with CAH) always be assigned female. Even though an infant's chromosomes were important, Money argued that chromosomal type should not dictate gender assignment. Indeed, beginning in the 1950s, he was a vocal critic of those who deemed chromosomes the essential marker of sex.

Marking a major shift away from Money's view, which rested on a fluid understanding of both sex and gender development, both the pediatric endocrinologists and the urologists I interviewed expressed grave reservations about making a gender assignment contrary to an infant's genetic sex. As Dr. H puts it, "Going against the karyotype is something you're fairly hesitant to do; it's the last straw." Dr. D concurs: "We once thought you could assign a genetic male female—that's what I was taught—but you'd be hard pressed to find someone who would do that now. John/Joan showed us that we just can't override things in that way." The newsletter of the department of pediatric urology at one institution advises: "In almost all cases [of ambiguous genitalia], genetic testing can quickly determine the child's gender." It continues, quoting the chief surgeon, who adds, "In most of these children, it is clear which is their intended sex," suggesting that chromosomes are the primary determinant of gender (Baskin 2003: 3).

The fact that so many clinicians I interviewed regard chromosomes as an individual's true sex, despite an awareness that chromosomes cannot guarantee a particular gonadal, hormonal, or phenotypical outcome, is surprising, because many of these same clinicians believe that prenatal androgen levels largely determine gender identity. In other words, their penchant to weigh chromosomes more heavily in gender-assignment decisions operates independently of their beliefs about how gender identity develops. That chromosomal type should be so important for 46,XX infants should not come as a surprise because they have always been assigned female. The heightened concern about chromosomes reflects an increased concern about assigning 46,XY infants with a small penis female such that biological constructs like the karyotype foster a chromosomal essentialism in the way they are interpreted and understood. This essentialism obscures the fluid nature of the supposed boundaries between male and female chromosomal types and the genetic variation within all groups. Indeed, these very same clinicians readily accept that infants with CAIS are assigned female; but this, they argue, is based on scientific evidence showing that they identify as female. However, the scientific evidence to support the gender assignment of non-CAIS 46,XY infants as male is highly variable (irrespective of diagnosis); this trend is tied to what the chromosomes *mean* more than what they *do*. Significantly, this shift in thinking about gender assignments for these infants took place in the absence of new scientific data on the issues.

A prominent pediatric urologist told me that in more than twenty-five years of practice he has never assigned an infant with a microphallus as a female because "chromosomes are too important." A surgeon also says of cases of microphallus: "I've always presumed they were male, and ought to be male" (Dr. K). This view suggests that Money's model was not always followed to the letter, highlighting the fissure between what clinicians are taught and what happens in their clinical practice.[10] Dr. K also said that he had gender-assigned as female infants born with no phallus, adding, "I've seen a lot of trouble no matter what you do, but I believe to undertake, in this day and age, gender reassignment other than chromosomal sex, is a huge step. That's something that I take more seriously today than I did when I started. I used to think you could make anybody female. I don't believe that anymore."

Whereas once these clinicians' views were in the minority, they are now much more common. Every clinician I interviewed said that he or she would consider a male assignment for a 46,XY infant, and quite a few insisted on it, citing the importance of chromosomes or else the effects of androgen imprinting on the developing brain. Demonstrating another departure from the traditional treatment protocol, many clinicians said they would seriously consider a male assignment for males born with no penis, either through lack of growth (agenesis) or through accidental or purposeful destruction.[11] Dr. J, a pediatric urologist, says, "These boys should be raised as their genetic sex," adding: "I can picture all of the painful, emotional scenarios down the road because he has no phallus, but it's not any easier the other way. At least this way you're working hard to accept the truth, whereas the other way you're constantly doing things to *hide* the truth that the sex is a genetic XY. They're boys. It's really as if my penis fell off. I would be a very frustrated male, but I will not die. But God forbid that somebody tried to give me a girl's name and a vagina just because my penis fell off" (original emphasis).

The simplicity of the chromosomes as a possible marker of future gender identity for 46,XY infants belies the complexity of gender-identity development and goes against research findings. As one researcher observes: "The majority of 46,XY intersex patients seem to develop a gender identity commensurate with the assigned gender and do not change their gender later. . . . there is . . . no unambiguous evidence for or against female sex assignment of 46,XY patients, even in the prenatally most androgenized conditions" (Meyer-Bahlburg 1999: 3457).

Echoing the importance of chromosomes for gender assignment, many diagnostic trees for intersex conditions begin with karyotype: it is one of the first tests ordered for a newborn with atypical genitals. However, because the test results often can take several days at minimum, results from other tests and the family's medical history are often available before the karyotype is known. Consequently, clinicians often know a great deal about the infant's physical body before they receive the karyotype. Thus they often have a hunch about gender assignment, but they tend not to announce it before assessing it relative to karyotype information, even in relatively straightforward cases. Gloria Jackson, the mother of a girl with CAH, says: "[The urologist] pulled us aside and said, 'Your baby is all female on the inside, but we have to wait for the chromosomal tests.' Up until the chromosome test results, they called her 'your baby,' even when they knew all of her internal organs were female. They did not call her a female until she came back XX." Although the chromosomal tests results were important for clinicians to label this infant female, the urologist's ambivalence about the infant's sex upset and confused the parents. To them, their newborn was a girl whose sex was confirmed by the initial announcement of girl at birth and by the presence of ovaries revealed by ultrasound.

Many of the clinicians I interviewed who were adamant about not assigning gender contrary to genetic sex nevertheless did not believe in a genetic basis to gender identity. One exception was Dr. F, a pediatric endocrinologist, who comments: "There is clearly determination of gender identity that we haven't got a clue about," adding, "I'm wondering, is there a whole bunch of gender genes?" In other words, clinicians' reasons for assigning gender in accordance with chromosomal status have little relationship to its believed impact on gender identity and everything to do with the idea that chromosomes make us male or female.

CHROMOSOMES VERSUS HORMONES

In addition to karyotype, another important consideration for many clinicians in selecting a gender assignment is the infant's prenatal exposure to androgens. Most clinicians who expressed concern about assigning a gender contrary to chromosomal type believe that androgen exposure in utero influences or determines a range of behaviors and attributes, including gender identity. As Dr. D explains it: "Exposure to androgens in utero flips a switch so that the developing brain is more virilized, more of a male pathway in how you think of yourself, your behaviors, and who you're attracted to." Another says: "What the science tells us is that the prenatal hormonal environment and exposure is critical in whether we see ourselves as a male or female" (Dr. R).

The trend toward male assignment for infants with a 46,XY karyotype also reflects concern about the infant's exposure to androgen levels in utero, which are believed to create differences in the brain that cause behavioral differences. One pediatric endocrinologist argued that the primary consideration for male assignment is "the degree to which prenatal exposure to androgens influences gender identity and role in humans." Expressing a similar concern about brain virilization, Dr. N, a pediatric urologist, says: "When we watched the CAH patients start to grow up, we realized the androgen imprinting made them quite aggressive. Something had happened with very high androgen levels in utero in these genetic females. If that was the case with them, might not that be the case with the boys who had testes and phalluses and all the imprinting in utero?" A recent survey of members of the Lawson Wilkins Pediatric Endocrine Society and the Society for Pediatric Urology noted a similar trend. That study, which surveyed over three hundred pediatric endocrinologists and one hundred pediatric urologists, found that a large number of both groups viewed prenatal exposure to androgens as the major determinant of gender identity (44 percent for pediatric endocrinologists versus 60 percent for pediatric urologists) (Sandberg et al. 2004).[12]

A pediatric urologist I interviewed said he changed his mind about female assignment for 46,XY infants after he attended a symposium on ambiguous genitalia at the annual meeting of the American Urological Association: "I was new to this community and had a kid with severe micropenis. We got the tests back and the kid had normal testosterone levels and XY karyotype, testes. Well, at the meeting, they said we don't make little girls out of all these cases anymore because of the imprinting. So we didn't."

Dr. J, the surgeon quoted above as a strong advocate of assigning microphallus cases as male on the basis of karyotype, expressed deep ambivalence about the gender assignment of a five-year old born with testes and a fully formed scrotum, who was gender-assigned female because "there was a flat surface where the penis should have been." The surgeon who initially counseled the mother was adamant that the child could not be raised as a boy without a penis, so the mother agreed to a female assignment. The surgeon removed the testicles and repositioned the urethra to enable the child to urinate sitting down. After a fallout with the initial surgeon over what she felt were poor surgical results, the mother consulted with Dr. J about performing vaginoplasty. He recalled, "The first thing I said was, 'Children who developed as a boy in utero, their brain's been exposed to testosterone and many grow up and want to remain a boy.' "

Yet some clinicians are more cautious about making conclusive statements about the relative role of androgens in shaping behavior and identity. Dr. H says: "I don't think androgen exposure of the brain of a 46,XX female is necessarily going to make it more likely that she'll be lesbian or that she'll be a tomboy . . . but I do think there are impacts of androgens on the brain." One clinician illustrates the role of hormones in sex determination: "Even with a testis in place, there are a number of instances in which the production or action of testosterone isn't completely normal, in terms of a little boy looking like a little boy. And the same is true of girls, or individuals who have a female chromosomal sex, a female gonadal sex, and yet, if exposed to high levels of androgens, are virilized and follow along a spectrum from very minor changes that would clearly be female in anyone's eyes, all the way to the other end of the spectrum where they can look very much like a male newborn" (Dr. A). This clinician is typical in his ambivalence about where to locate sex, particularly when so-called chromosomal and gonadal sex do not correspond to genital sex. But he also raises another issue, the tension between a belief in chromosomes as the true sex and the understanding that hormonal effects are required to bring the true sex into being. While karyotype reflects the baby's intended sex, clinicians are also attentive to whether the newborn has had enough androgen exposure for a male assignment and potentially too much for a female assignment. In these instances, concerns about the degree of hormone exposure can displace initial concerns about assigning according to karyotype.

Unease about the degree of androgen exposure has left some clinicians trying to estimate the extent of exposure to help determine gender assignment. Dr. R, a pediatric psychiatrist, explains the rationale: "It's pretty obvious that children with a normal hormonal environment in utero have a male identity, male behavior, male sexual orientation. It does not waiver much. We don't know how much androgen it takes to produce that so the question becomes, 'How much is enough?' We don't know." Another clinician says that if test results show low androgen levels or if the phallus is "very small," ostensibly also an indication of reduced androgen exposure of the brain in utero, he is more inclined to consider a female assignment "because the brain imprinting will also be less." Dr. F also said he considers the degree of androgen exposure in gender assignment decisions for 46,XY infants with atypical genitals: "Does male hormone make you male? I don't know for sure. It may be the quantity that's important, but it could also be when you get a spike. I try to consider the quantity in my decisions, but it's hard to estimate. Penis size gives you an idea of how much the kid has been exposed to." These statements assume a direct, causal, and linear relationship between hormone exposure in utero and one's sense of oneself as male. To a surprising degree, myriad aspects of personhood are attributed to largely unexamined biological effects, which are the lens through which we interpret and understand humans and their differences. Gender is configured as an on/off switch triggered exclusively by hormones—its complexity lost to biology's complexity (Kessler 1998).

One researcher who examines the relationship between androgens and behavior and gender identity, however, criticizes the quantity theory: "We need to be very careful, because people are making some broad leaps in the absence of data. There is not a nice linear relationship between degree of androgen exposure and gender identity. You couldn't say, 'This child has been exposed to a moderate level of androgen and therefore is going to have this gender identity.' We have not ruled out the role of rearing in gender identity" (Dr. Q). This researcher may not have ruled out the role of rearing, but many clinicians I interviewed have. This divergence between research and clinical understanding and practice raises a critical concern about the extent to which clinicians making gender-assignment decisions know and understand research on gender-identity development.

With some diagnoses, the conflict between chromosomal type and hormone exposure has created uncertainties in what were once regarded straightforward decisions. For decades there was very little argument about whether 46,XX infants with CAH should be raised as girls, in part because fertility is generally preserved in such cases. A recent consensus statement advised against male assignment for these infants (LWPES/ESPE CAH Working Group 2002). Some clinicians, however, have questioned whether the degree of genital virilization should influence gender assignment for these individuals. It is not the masculine genitalia per se that are of concern, but what this bodily virilization supposedly reveals about levels of androgen exposure and a possible masculinization of the brain.

Milton Diamond and William Reiner have both suggested that although most girls with CAH should be assigned female, more virilized girls (levels IV and V on the Prader scale of evaluation) are "a completely different scenario" (Reiner 1997: 1045).[13] Reiner argues that careful consideration must be given to gender assignment in these individuals, "remembering that the brain is the most important sex organ" (1045). Diamond expresses similar concerns regarding the importance and effects of brain masculinization, and Diamond and Keith Sigmundson recommend a male gender assignment in cases of "XX persons with CAH with extensively fused labia and a penile clitoris" (1997a: 1047).

Even though clinicians express concern about the effects of androgens, these effects do not generally warrant a male assignment, particularly if clinicians have removed the ovaries, thus rendering the individual infertile. This view was confirmed in a recent survey of almost two hundred pediatric urology fellows of the American Academy of Pediatrics, in which they were hypothetically asked about the gender assignment of a very virilized (Prader V) newborn with CAH and 46,XX chromosomes. All but two favored female gender assignment; 94 percent cited the potential for fertility as the most important factor (Diamond 2004).

Several clinicians I interviewed, however, were willing to consider a male assignment, thus weighing hormonal effects over chromosomes. One pediatric urologist, for example, says: "The problem with CAH girls is that we don't know what happens in utero with the imprinting, so that child's gender may be more appropriately male, even though chromosomes may be female. But if we assign male we're going against our principles to preserve fertility" (Dr. O). Of course, fertility would be preserved if they left the ovaries intact. Clinicians sometimes treat fertility as though it is more straightforward than other criteria for assignment, but this example highlights the way in which fertility is itself gendered. In theory, a fertile XX man with CAH could become a biological father with his male or female partner via egg donation, which would upset the meaning of fertility and reproduction in a heterosexual context.

The clinician cited above feels torn in regard to competing factors for gender assignment; namely, chromosomal status, androgen imprinting, and the potential for fertility. Another suggested genetic females with a "well-developed phallus" could be assigned as boys. Both suggested a male assignment might be preferable if the family "really desired a boy," or when the neonate's sex was announced as male at birth. Indeed, the same clinician can read the effects of androgen imprinting on the brain quite differently depending on the infant's chromosomal type. That is, for some clinicians, androgen exposure is used to justify a male gender assignment where androgens are constructed as that which makes one male; whereas for genetic females, androgen exposure and its ties to masculinity are minimized to support a female assignment. In both instances, clinicians express a view that men's and women's uniqueness is only a matter of our biological characteristics, our hormones, and our chromosomes.

PRAGMATISM: FERTILITY, SURGERY, AND PARENTAL WISHES

Despite the increasing belief in the importance of chromosomes and hormones in gender-assignment decisions, sometimes the overriding factors in clinicians' decisions about gender assignment are based not on biology but on practical concerns: predictions of a child's quality of life living as one gender or the other given the individual's diagnosis; the feasibility of the reconstructive surgery for the chosen gender assignment; and familial concerns. These concerns are still particularly prevalent in cases of males born with small or morphologically abnormal penises. Although clinicians increasingly cite chromosomes and the degree of androgen exposure as important for gender assignment, the smaller the penis, it seems, the greater its importance in gender assignment.

Although female gender assignment on the basis of small penis size appears to be less common, all the clinicians I interviewed endorsed the importance of assessing penis size. Dr. B, a pediatric endocrinologist, says: "Those measurements mattered quite a bit because they determined which way the kid was going to go, male or female. I remember standing over a bed not too long ago, measuring a phallus length in an undervirilized boy, and we made a decision about assigning gender based on the phallus size. In that case, it looked small, but it was a little over the number that we needed to allow ourselves to assign male. If it would have been smaller, we would have assigned female." As noted earlier, elaborate criteria exist for measuring an infant phallus and predicting whether it will grow sufficiently to enable "normal" sexual function (that is, penile-vaginal intercourse). Dr. F comments on this practice: "It always struck me as a little pragmatic, as far as what you do with ambiguous genitalia. It's easier to create female genitalia, so that's the way you would push an individual. And that does seem a little bit cavalier. But that's the way things have gone."

Others calling for a male assignment for 46,XY infants with a micropenis argue that gender assignment need not be linked to the "size and functionality of the phallus" (Phornphutkul, Fausto-Sterling, and Gruppuso. 2000: 136) and cite research showing that these individuals may fare well despite having a small penis (e.g., Reilly and Woodhouse 1989; Bin-Abbas et al. 1999). As one pediatric endocrinologist noted, although the "potential for penis growth in relation to sexual function . . . must be considered," the capability for sexual response must be considered also, and "relative penis size plays no obligatory role" in sexual response (Lee 2001: 119). Others see the John/Joan case as instructional, blaming the medical profession's "compulsive need to surgically alter male genitalia" through circumcision and surgery for hypospadias and microphalluses and the general "disregard we have for the genitalia of male infants in our culture. . . . The obvious lesson . . . is to leave the genitalia of infants alone. Only bad things happen if you do not" (Van Howe and Cold 1997: 1062).

In the case of penile agenesis or aphallia, surgeons have not had much to offer. One surgeon I interviewed said phallic construction is "disreputable and ineffective." Dr. M, a well-known pediatric urologist, says: "The dogma is it's easier to make a female than a male, because reconstruction of the phallus is very, very difficult. Only two people in this country can do it, and even what they can do is very limited. I would never do a total penile reconstruction. The success rate is not good, and the results that I've witnessed were abysmal."

One surgeon, Dr. P, disagrees: "With the modern knowledge in genital reconstructive surgery, I mean, heaven's sakes, I can make a penis! We can take a cutaneous skin flap off the arm and make a fine penis. It can't be done till they're a little older, but it can be done." However, other surgeons did not rank him as among the few most capable practitioners of penile reconstructive surgery.

Money's protocol has not always been followed with regard to gender assignment on the basis of penis size. Clinicians gave numerous examples of how they accommodated other concerns and thus bent the rules of the criteria for phallus size and thus female gender assignment. Dr. B, for example, says: "I have definitely been swayed by other things, only some of which have to do with the actual penis. Sometimes it's curved, and I take that into account, but other times, it is a feeling about the family's ability to cope with certain decisions. The rule is there, but I haven't always followed it." Dr. L, a pediatric urologist, explains more explicitly how family preference for a boy has influenced his decisions: "I remember Hispanic families who obviously valued maleness and did not want their infants reassigned as female. Not assigning them girls made me very nervous and *really* worried that these children would become adults who were males in orientation, but would feel very inadequate because their phallic tissue would be quite small" (original emphasis).

Although clinicians read certain somatic markers as unequivocally masculine or feminine—these physicians are not questioning that the child has a penis—competing social and familial concerns can sway a physician toward a different recommendation, demonstrating flexibility in determining not only who is male but also in who should remain male and why. When these factors influence a decision, physicians generally feel unconcerned about not following the traditional treatment paradigm because they are using clinical judgment to integrate knowledge and accommodate individual circumstances and concerns. The treatment paradigm and the theory on which it rests are thus not sufficient. What the doctors feel they are providing is a hard-won dimension of expert knowledge known as judgment. Although decisions about gender assignment may depend on somatic facts, they also fundamentally depend on social context and clinician judgment.

Other clinicians, however, continue to use phallus size as the primary determinant for gender assignment. Dr. I says, "I follow the two-standard-deviation-size rule. If there's limited potential for reconstruction, then your primary driver is not the 46,XY or testes but that the genitalia's insufficient, and you can't do much about that. You're not God." When pressed on what he meant by "insufficient genitalia," he replied, "A very poorly virilized male that's not likely to have normal appearing genitalia and not likely to have traditional intercourse."

As is true with all cases of gender assignment, clinicians are motivated by a strong desire to help parents and children. This desire, however, leads them to different treatment suggestions about what is best for the child. The same infant could be assigned male or female depending on the treating physician's ideas about the importance of phallus size or of prenatal androgens for gender-identity development and ideas about what counts as "sex." Dr. L said he had been involved in "horrendous, knockdown disagreements" about whether a phallus had sufficient size for a male assignment. The idea that a newborn baby's phallus size should be the sole determinant of maleness constructs a highly reductive and narrow vision of the importance of the organ for subjectivity and masculinity. Moreover, it assumes that intercourse with a surgically constructed vagina would be better than with a small penis, if either were possible at all. The assumption is that it would be better to be a woman without a vagina and no reproductive capacity than a man with a small penis. Dr. E, a female pediatric endocrinologist, offers a response: "It doesn't make any sense. When I discuss this with nonclinical men, they are furious about sex reassignment because they think it's castration. If you're lucky enough to be born a man, you do anything to hold onto that. Society's view is that we've been far too intrusive and we should leave well enough alone."

Research on the effects of androgens on behavioral differences between male and females has shaped clinical practice in complicated ways. One effect is that many clinicians have moved away from the wholesale idea that sex-specific genitalia are necessary for proper gender-identity development, especially for individuals with a 46,XY karyotype (excluding those with CAIS). Given some evidence that males with micropenis do well when raised male and develop a male gender identity, these infants are now more often assigned male. This change in approach reflects a growing belief in a genetic and hormonal, not social, basis for gender difference and gender-identity development.

What, then, is required for male assignment? Many of the clinicians I interviewed argued that genetic males with a small phallus should be raised as male, though their reasons vary. Dr. J explains that if the infant has normal testosterone levels, testes, and an XY karyotype, he now assigns them as males. Dr. H suggests that "a lot of adult men in the world" have been raised male with "a phallic structure that is below normal in size. Yet they have a female partner and are able to have enjoyable sexual relations and a family." Others cite fertility as important: "If it's important for females, why wouldn't it be important for males? We kind of assumed they could not have intercourse, but now there's also all kinds of reproductive technologies and they could have children, even if they can't the traditional way" (Dr. P).

Dr. K, a pediatric urologist, explains why he would assign infants with a small phallus as male, but not an infant with no penis: "Ten years ago, I would have said gender assignment ought to be female. Now I say, if at all possible, you assign the chromosomal gender. But, ablatio or penile agenesis, it's better to have these patients with a female gender assignment." When asked why, he explained that a female assignment was "better than having them go through years of discrimination and disability with a nonreconstructible phallus," adding: "Phallic reconstruction still is so rudimentary. So, so barbaric. I was at [a hospital] where we did sexual reassignments, changed men to females and vice versa, seeing how mutilating it was to make a floppy, flaccid skin tube and put in a prosthesis for somebody; it seemed horrendous." Interestingly, his colleague cites *improvements* in the surgical construction of the penis as a reason not to assign an aphallic male infant as a female: "If a kid that lost their penis in a circumcision accident today, they're not going to get gender-converted because reconstructive surgery has grown in quantum leaps in the last number of years and, by God, we can make that child a fine penis" (Dr. P).

Dr. K, who opposes female assignment for males with microphallus, says he still suggests female assignment for infants with no penis despite poor outcomes: "We've had three cases over the time I've been here, where I've done gender reassignment. . . . And they've all done badly. Horribly." When I asked what he meant by this, he appeared truly pained, saying that the surgical results were "just horrendous" and that the kids have had a very difficult time adjusting socially. He adds that with males with micropenis, "I've always presumed they were male, and ought to be male, and with a lot of androgen stimulation we're going to have some end organ growth with microphallus, versus none for the aphallic." In one instance, maleness is located in the chromosomes and androgen imprinting, in another in the presence of a penis.

Most clinicians I interviewed remain deeply conflicted and uncertain about how they would assign these infants given the poor options. Dr. N regrets assigning newborns with aphallia and cloacal exstrophy (described in chapter 3) as females, adding it was in these cases "where I think many surgeons, reflecting back, realize that we made wrong decisions" and that "these decisions were not made in a vacuum, the options were pretty miserable." He mentions that later in his practice "with the cloacals, I chose to keep them as males. It was easier doing clitorectomies, and they had small funny phalluses, that were sort of sewn together with no urethra or anything, but I felt they were male." Deflated and resigned to the fact that there was no easy answer, he adds: "Many of them had such devastating anomalies that they didn't live very long."

In a survey of 185 members of the section on urology of the American Academy of Pediatrics, clinicians were asked how they would assign a 46,XY cloacal exstrophy patient. Although clinicians do not count cloacal exstrophy as an intersex condition, many boys with this diagnosis have been treated according to Money's protocol and assigned as girls. (This is also one of the groups that Reiner has evaluated in his research.) Seventy percent of clinicians said they would assign male and 30 percent female (Diamond 2004). The factor cited as the most important one for a male assignment was brain imprinting by androgens, and the most important one for a female assignment was the likelihood of surgical success. The respondents who recommended a female gender assignment tended to be older and have more experience in pediatric urology. These survey outcomes dovetail with my finding that physicians are increasingly citing an imprinting of the brain as an important consideration in the gender assignment of 46,XY infants, motivating their shift away from female assignment for these infants.

NO RIGHT ANSWER: THE MOST COMPLEX CASES

Whereas with some diagnoses, such as CAIS, gender assignment may be straightforward despite the family's understandable anxiety and confusion, it becomes infinitely more complicated with other diagnoses. In what clinicians have long referred to as true hermaphroditism, a very rare condition, both ovarian and testicular tissues are present: a testis on one side and an ovary on the other, or a histological mixture of ovary and testis—an ovotestis. The internal and external genitalia vary over a range, from typically male to typically female; most individuals have a uterus, but other ductal structures are more variable. Many, although not all, people with this diagnosis are infertile. Dr. A describes a clash with a urologist over how to assign an infant with this condition:

> I had a true hermaphrodite with an ovotestis and a testicle, a uterus, and a blind-pouch vagina that did not open exteriorly. This is one of the ones I fought the surgeon over. The kid had a normal testis with normal testosterone and dihydrotestosterone levels. My feeling was that the kid could identify as male, because of the hormonal influence of the testicle. The surgeon went, "There's a uterus; this kid could have kids as a girl, and it's less likely that that testicle will be spermatogenic." He recommended the testis be taken out and the kid surgically feminized and converted to a little girl. The surgeon recommended female assignment and reconstructive surgery. I recommended male assignment because of the functional testis and levels of testosterone. The family took a couple of days but decided on a male assignment.

This example captures the complex set of concerns that each somatic trait raises for the clinician, hence the competing criteria for gender assignment, as well as the deep materialism on which physicians rely to make gender assignment decisions. One practitioner is overwhelmingly concerned with androgens and their effect on the future gender identity of the child; the other with the preservation of fertility, which he assumes would be more likely as a female and more important for a female gender role. (His conceptualization of fertility further restricts the mode of reproduction to penile-vaginal intercourse.) These concerns are not easily reconciled. Neither clinician can be sure that the child will be fertile, irrespective of gender assignment. Moreover, the capacity to reproduce has little bearing on whether the child will identify as male or female. Although the clinician assumes that androgens will produce a male gender identity, this is uncertain as well. In the face of such imperfect knowledge, clinicians are relying on the logic of how different bodily parts and functions (genes, ovaries, external genitals, normal intercourse) and ways of being "normally" go together, and what different cultural logics can be used to determine the next best outcome. Several clinicians have told me that some of their colleagues have stopped practicing in this field because the decisions are too difficult. This difficulty may also limit those entering the field.

In complete or partial AIS, the infant is born with 46,XY chromosomes and testes, which produce androgens, but the body tissues respond to these androgens only to a limited degree or not at all. As a result, the genital configuration of these individuals varies enormously. The external genitals are ranked on a scale from 1 ("fully virilized") to 7 ("fully feminized"). Infants with CAIS are often not diagnosed at birth because their genitalia appear typically feminine. Often this condition is discerned in early childhood, frequently because the testes have descended. Although these individuals have an XY chromosomal configuration, no clinician I spoke with argued that those with CAIS should be raised as male. This is because it is widely recognized that individuals with CAIS grow up to identify as female (Mazur 2005).

Interestingly, the diagnosis that presents the least difficulty for gender assignment from a clinical perspective can be one of the most difficult for parents to accommodate. Tina Haywood gave birth to an infant with CAIS who at two weeks of age developed abdominal hernias. During surgery to repair the hernia, the surgeon noticed that the gonadal tissue looked unusual and took a small sample to test it. Tina recalls that the doctor called her and her husband with the pathology report and said, "The first thing I want to impress upon you is that your child is a girl." In a panic, she replied, "What? I *know* my child is a girl" (original emphasis). The doctor then told them their baby's gonads had testicular tissue.

Once they learned her karyotype was 46,XY, the Haywoods had to wait while the doctors tested their daughter's responsiveness to androgens. Tina describes her concerns while waiting for the test results: "Even though she was genetically male, we knew she couldn't live as a male. She was insensitive, partially, to the hormones that *make* you a male. Ideally her sex would be concordant with her genetic sex and it's not, so that's a bad thing. The next best thing would be to be squarely in one camp or the other. . . . I didn't have a lot of technical questions at first. I had life-quality questions." Pamela Healy, another mother of a child with CAIS, was reassured by doctors that her child was a female despite her XY chromosomes: "Over and over the doctors kept saying, 'She's still your daughter, she's still your little girl . . .' It was so hard to understand that she could be a female and have male chromosomes. It took me a long time to come to grips with that, because it goes against everything you are taught about biology."

Whereas CAIS individuals identify as female, the gender identity of individuals with PAIS is variable. Reiner has noted of PAIS cases, "The possibilities of identifying as male are unknown. Some of these children may demonstrate more male-typical behaviors and may be capable of increased masculinization of the brain if treated with testosterone in the first few months of life" (1997: 1045). He seems to suggest that a male rearing may be advisable. Diamond and Sigmundson suggest that individuals with AIS grades 1 to 3 be reared as male and grades 4 to 7 as female (1997a). This recommendation, of course, raises the question of where exactly the clinician should draw the line between grades 3 and 4. Moreover, it is far from clear that the effect of hormones on behavior or even identity can be accurately assessed on the basis of small visual differences in the external genitalia, differences which themselves are subjective. (This scale was not intended as a proxy gauge to infer the degree of brain masculinization.) Today clinicians may test how responsive an infant with PAIS is to testosterone, and thus whether the phallus (if considered small for a male assignment) is likely to grow. Testosterone response may also help predict the extent to which the child will virilize at puberty. Clinicians also tend to infer the degree of brain masculinization from these tests as well.

The Friedmans' daughter was born with PAIS, a fact they discovered when her testes descended at nine months. When a chromosomal test revealed that she had a 46,XY karyotype, clinicians set about determining how sensitive she was to androgens, which would determine her pubertal development. For three months, the family and clinicians waited to see if her phallus would grow in response to topically applied testosterone. During those three months, the Friedmans assumed that they would raise her as a boy because of her chromosomal type. When it was established that the androgens were having little effect, they realized a decision needed to be made about gender assignment. Shirley Friedman recalls: "Gender assignment was just thrown at us. That was very, very shocking. . . . We ended up saying, 'If it was your daughter, if it was your child, what would you do?' And both doctors said, 'We would recommend going as female.' They felt that the chance of a successful sex life was far greater as female than male. Although she had a penis, it would be very small, and she'd pee like a girl. . . . We went with what they said because we had absolutely nothing else to fall back on." The parents, however, were worried about the effects of their daughter's "male" chromosomes and androgen exposure in utero: "It's one thing to say 'Raise her as a female,' but is she going to act like a boy, is she going to look like a boy? We were worried about that. They said, 'We can't tell you, because there's been no research.' They said they strongly felt that children were a product of their environment, so if we raised her as a female, there was a good chance she would act as a female, but they couldn't give us any guarantees. For years I worried about that, looking and thinking, 'Wow, she looks so female.' I was always watching that because we had no reassurances." There is still not a great deal of research about the gender identity of individuals with PAIS. Some have been assigned as males and some as females. A recent meta-analysis of studies reporting on gender change in individuals with PAIS (as well as with CAIS and with micropenis) found that just under 10 percent do not identify with their sex of rearing, irrespective of whether they were raised as male or female. These findings led the psychologist Tom Mazur to conclude "the best predictor of adult gender identity in CAIS, PAIS, and micropenis is initial gender assignment" (Mazur 2005). A recent study by Reiner, however, seems to suggest a male gender assignment for PAIS individuals (Reiner 2005).[14]

Clinicians also agree that gender assignment for patients with 5-alpha reductase deficiency is very difficult. This condition affects 46,XY individuals who are unable to convert testosterone to dihydrotestosterone, the hormone that masculinizes the external genitalia in utero. These infants may be misdiagnosed as having PAIS because of the way the genitalia form, which can vary from typically male to predominantly female. The infant has testes, as well as other male-typical internal reproductive structures. If there is no intervention, the production of greater levels of testosterone at puberty will cause the individual to develop secondary sex characteristics typical for a male: the phallus will grow, the shoulders will broaden and the hips narrow, muscularity and body hair may increase, facial hair will develop, and the voice will deepen. Dr. A explains: "You can take two siblings with 5-alpha reductase deficiency, and because their enzyme levels are going to be about the same, their hormone levels are going to be about the same—they're raised in the same household, and one will go male, and one will go female. [A prominent pediatric endocrinologist] looked at that and found that they could not predict who would go which way by any available piece of data."

Mixed gonadal dysgenesis (MGD) is another condition for which gender assignment proves very difficult. The majority of those with MGD have a 45,XO/45,XY mosaic karyotype. Those with MGD have asymmetrical gonads: usually one gonad remains at the unidentified stage of development and the other forms a testis, though many other combinations are possible. These infants can respond to androgens but because of the malformed testes, they are often incapable of producing androgens. (This diagnosis contrasts with AIS, in which individuals possess functioning testes but cannot respond to the androgens their testes produce.) The internal genitalia are female-typical, and the appearance of the external genitalia is variable; however, the majority appears more female. Dr. N notes: "Mixed gonadal dysgenesis are far and away the most difficult patients. The true hermaphrodites could often be difficult, but most of those were detected quite late, and they had literally been fixed in a gender, and what you were really doing was adapting their anatomy to their gender identification. Very rarely would any switch. But the MGD patients and the 5-alpha reductase deficiency patients are difficult." Note as well the reference to individuals with true hermaphroditism already fixed in a gender, suggesting an important role for rearing in the development of gender identity.

Understanding individuals as males and females is tied to biology more than to any other factor. Gender assignment proves difficult because clinicians need to assess the infant's body for signs that it may do better with one gender assignment than with another. But it is impossible to infer any infant's future gender identity from his or her biology (this is true for individuals without intersex conditions as well). In the face of biological traits that clearly "read" male or female, but that do not necessarily result in individuals who feel male or female, physicians hazard a best guess—one that they realize has enormous implications for parents and those treated.

Parents and Gender Assignment

Clinicians and parents alike refer to the period after the birth of an infant for whom gender assignment is unclear as a "nightmare." Not only does a child with "no sex" occupy a legal and social limbo, but surprise, fear, and confusion often rupture the parents' anticipated joy at the birth of their child. One medical article begins dramatically: "Next to perinatal death, genital ambiguity is likely the most devastating condition to face any parent of a newborn" (Low and Hutson 2003: 406). The birth of an infant with an intersex diagnosis, however, need not be devastating—the attitude reflected in this statement no doubt contributes to its devastating effects—but there are several factors that make these births difficult. Parents have often never heard of an instance when sex could not be determined; they have expected a routine birth, often know nothing about intersexuality in general, and are shocked to learn that their newborn's sex is not immediately clear. As Tamara Dawson says: "We wanted to be surprised, and we were very, very surprised." I describe the experience of new parents in the hospital more in chapter 6; here, I explore the role of parents in the decision about gender assignment.

INFORMATION DROUGHT AND INFORMATION OVERLOAD

Parents faced with a decision about gender assignment for their baby frequently recount great difficulties in gaining access to information to help them understand what is happening and in some cases to assist in making gender-assignment decisions. Perhaps in part because these births are relatively rare, hospital staff may be unable to answer questions at a level appropriate to the parents' emotional state and level of medical understanding. Parents may further receive conflicting information from nurses, pediatricians, and other specialists.

The heightened emotion and lack of clear communication can affect the relationship between parents and physicians, and this may inadvertently influence the gender-assignment decision. Frightened parents may feel that they have been confronted with a choice they are not equipped to make or that their child's future has been taken out of their control by high-handed physicians. Clinicians, conversely, may feel that they must make a decision on behalf of parents who cannot grasp the medical complexities of the diagnosis, or else that the parents' emotional reactions will force the clinicians to acquiesce in a choice with which they are not entirely comfortable.

The information delivered to parents about their child's condition at the time of birth may be highly clinical, piecemeal, and delivered with inadequate consideration of the parents' state of mind. Sara Finney recounts her thoughtless treatment by the hospital staff at the birth of her daughter: "A doctor came over and said something. I mean, first of all, she got me less than five minutes after I had given birth. I was exhausted, and she comes over and starts spewing something out of her mouth. I don't know who she was. I never saw her again. Then a nurse took the blanket off Samantha, looked at her genitals, and said, 'What you have here is not right. This is not normal.' " Justine Good recalls a similar experience at the birth of her daughter:

> They brought her back in all bundled up, and a whole bunch of people came in. The doctors, the nurses, and everybody and their mother. The pediatrician, the gynecologist, they all came in, and I hadn't really met the pediatrician, so he introduced himself and said, "There's a problem." . . . I really don't remember what he said to me. I totally zoned out. I got all hot. I just stared at him. All I remember is something about salt, something between the legs, and "Do you want to see what I'm talking about?" He undid her diaper and he said she didn't have any vaginal opening. He said she needs to get to a specialist right away. I remember, after they left, we [she and her husband] just kind of held each other and cried.

The nature of hospital routine, with its procession of nurses changing at each shift and different specialists and medical residents and often students making rounds, also contributes to the feelings of confusion, as Dr. L acknowledges: "We make rounds every day, and if the parents are at the bedside, they say, 'What do you think?' to whoever comes by. It can get very confusing for families. 'Well, this person said this,' and you say, 'Who said that?' 'I don't know, it was one of the other doctors.' " This experience is confirmed by Tamara Dawson, whose daughter has CAH:

> We were listening to what these experts were saying to us, but each one was coming in at different times saying different things. We had confidence in the endocrinologist because she came over and said, "This is what she has. Don't worry, we're going to get it under control." The problem we had was the genetic counselor who came over from [a nearby teaching hospital] who scared the living daylights out of us. . . . "It could have what the endocrinologist said, but it could be a double-X with a Y. . . ." He's going through all this stuff, and we're like, "What are you talking about?" He was making us feel just awful and totally inundated.

Moreover, the information provided by hospital staff may be inaccurate or out of date. Although this is less likely to happen at a large center or teaching hospital, it is not unheard of, especially as the family comes into contact with so many individuals during their time in the hospital, not all of whom may have the most up-to-date knowledge. Monica Cole recalls: "Understanding the medical reason for her [daughter's] condition was important to us. The nurses gave us information from a nursing manual about ambiguous genitalia and adrenal conditions, but the information was at least five years old, so I asked a friend to look up CAH on the Internet. By the time we met our endocrinologist, we knew 75 percent of the information he explained to us."

The discourse of gender assignment is necessarily medico-scientific. Clinicians often, but not always, attempt to tailor the depth and complexity of the information they present to what they assume to be the parents' scientific literacy—and also on the parents' perceived ability to cope emotionally with the information. At the initial consultation, most doctors begin by explaining the typical development of a fetus to the parents, partly as a way of explaining how fetal development went "awry," but also to normalize what may be upsetting and even shocking information about the baby's body. Next, some clinicians attempt to convey all significant information: an explanation of every possible diagnosis, the possible physical permutations that accompany those diagnoses, treatment options, and prognoses for the future. This information is frequently expressed in complex language about chromosomes, hormones, gonads, genitals, and gender. Tina Haywood recalls: "The doctor told us he was making a preliminary diagnosis of complete androgen insensitivity syndrome. And that's when my husband [who has a science background] said, 'Is she XY, or XX?' And, I'm like, 'What are they talking about?' "

Others keep the information to a minimum, stressing that they do not yet know the diagnosis (or gender assignment) and will not hazard a guess about it. Moreover, they may assume that parents are not competent to process complex information because of their lack of medical knowledge or their distress: "The parents are absolutely floored by this. They've never heard of it before. They're just shocked, and they want to know the sex of the baby. It would make our jobs a bit easier if people had already heard about it, because you have to *totally* educate them, and you have to start with development of the fetus" (Dr. D; original emphasis). Another clinician adds that "parents' anxiety is no different than it was thirty years ago. They have the same issues: 'What do I have here? Do I have a freak? Do I have a kid who's going to have some weird sexual preferences?' The things that go through their mind are unbelievable. It's *very* confusing to them" (Dr. N; original emphasis). Part of the clinicians' role, one for which they may be woefully unprepared, is to help parents adjust to this situation; the extent to which they are successful at doing this can vary a great deal from one clinician to the next.

Understandably, parents are thinking of the future and about whether their child will be able to attain what Dr. B calls "Ozzie and Harriet–type happiness." In some cases, parents deeply anxious about having a baby with gender-atypical physical traits are "talked down" by the clinicians. Tina Haywood, whose daughter has CAIS, describes how her doctor tried to reassure her: " 'Normal, normal, normal.' That's what the doctor kept impressing upon us. She was 'normal, normal, normal.' We'd come home and go, 'Normal, normal, normal.' "

The amount and type of information clinicians share with parents also depends on their assessment of the parents' attitudes. While some parents I spoke with were overwhelmed by the information they received, others were eager for more information. Tina Haywood turned to her scientist husband to answer her early questions. "The medical people," she observed, "just were not focused on our needs." Parents with lower scientific literacy or for whom English is not a first language may be at a disadvantage in getting much-needed information and explanations. They are more likely to ask fewer questions, and clinicians may be less likely to volunteer information to them.

Parents I interviewed feel a tension between wanting to know more and being overwhelmed by too much information. They are eager for a sex announcement, but in the case of CAH, for example, they may also be trying to understand a condition that can be life threatening. For these parents, the issue of gender assignment, if it is an issue at all, is only one of many concerns. Gloria Jackson says that some clinicians who had examined her daughter had given her and her husband the impression that their daughter was abnormal, even grotesque. She goes on to describe how their experience with a pediatric endocrinologist reassured them:

> From the moment we met him [the endocrinologist], he was very calming. He took really good care of me, particularly. The first thing he asked us was, "How are you doing?" Some of the doctors wouldn't even take her diapers down. She had a wet diaper and he changed it, doing a lot of "hmms" and flipping her back and forth gently. He said, "We won't

know for sure what's going on, but it looks like your daughter may have this thing called CAH. I can't tell you for certain, and there's a few different versions of CAH, and I can't tell you which, but I will tell you that, provided she gets the medical attention she needs, your daughter is going to be fine."

At the birth of their children, parents typically wonder if they did something to "make their child this way" and feel afraid about the child's future, which they fear will be bleak. In the face of these concerns, the pediatric endocrinologist's emphasis on the well-being of the parents and his sensitivity to their experience was very reassuring. As I discuss in chapter 6, these early hospital experiences profoundly shape how parents come to understand not only the child's condition but also the child him- or herself. One of the most profoundly alienating experiences for parents is the way in which their children are objectified. What most parents I interviewed longed for, and were comforted by, were interactions with medical professionals who emphasized the joy of the birth, who allayed some of the parents' fears by stressing that this was one aspect of the child and nothing to be overly worried about, and who provided referrals to other families and psychosocial support. Unfortunately, the positive experience described above, with the clinician taking the time to ask about the parents' well-being, was unique among the parents I interviewed.

WHO REALLY DECIDES GENDER?

Ideally, the gender-assignment decision is made once all of the tests have been completed and the results are in hand. Who makes the decision, however, is a complex and sometimes contested issue. Some clinicians believe they should decide the gender for the parents, relying on their clinical knowledge and judgment. For these clinicians, gender assignment entails a sober weighing of test results and bodily characteristics to arrive at an assignment. More common now is the view that the clinician's role is to offer a set of considerations or recommendations and then to let the parents decide. The story of the Finneys at the beginning of this chapter highlights how parental choice may represent more an ideal than a reality. Others let the parents decide after presenting what might be called the "biological facts." Indeed, parents are involved in gender-assignment decisions more than ever before, in part because of shifts in medical practice that emphasize shared decision making, but another effect is that clinicians and parents share responsibility if the child comes to reject the initial gender assignment.

Gender assignment is complicated by the fact that multiple specialists are involved, each of whom may have a different opinion on how the decision should be made. For example, clinicians on the same team may disagree about the role of the parents in decision making. Moreover, they may disagree about gender assignment for a particular infant. In these instances, professional rivalries and power struggles can influence the outcome. Some surgeons and pediatric endocrinologists believe surgeons should not decide gender assignment. One pediatric urologist I interviewed says: "Dr. [pediatric endocrinologist] deals with the family and chooses the sex of rearing. After he has counseled them, it is up to me to modify the child [surgically], to help the family, so the child's genitalia resemble the sex of rearing. I don't make choices with gender; I'm a surgeon. I'm not qualified to do that" (Dr. P). Dr. N, a pediatric urologist similarly situated at a different institution, comments: "I work for parents. What parents decide, if I agree with that, is what we do," which suggests that parents' may be involved in decision making only insofar as they propose what clinicians feel is right, effectively limiting their participation to agreeing with the specialists.

While some clinicians feel it is ideal to involve parents in gender-assignment decisions, most I spoke to perceive that parents' ability to be involved varies a great deal. One clinician elaborates on what he views as the constraints of fully involving parents: "It would be ideal to present the family a balanced argument, and then let them decide. But there are all of these other issues to deal with, like their ability to understand the information, to cope with the risks that are involved, and with the consequences of their decisions" (Dr. I). This clinician suggests that not all parents are equally prepared to make decisions about gender assignment, but also that he is not prepared to let some parents make these decisions. Moreover, parents' understandable difficulty in grasping biology, medical terminology, and fetal development, for example, can limit their level of involvement. As one clinician notes, "Intersexuality is complicated even if you have medical training. This is where we get into the arguments, 'Are you being honest? Are we using euphemisms?' We talk about chromosomes, and some families don't have a clue of what you're talking about. You try and describe it in ways that they can understand, but to explain in any deep fashion is impossible. This is where informed consent breaks down sometimes" (Dr. L).

In these instances clinicians become the de facto decision makers because they feel parents cannot make a truly informed decision. Two clinicians explore other reasons why gender-assignment consultations differ from family to family:

> Some families will develop a trust in you and say, "I understand what you're saying, but you decide." I'm happy to live with that. Other families absolutely don't want you to decide. They want to know every detail so that they can decide. In the end, though, they almost always ask your recommendation. More often than not, they'll accept your recommendation. (Dr. L)

> Education makes a difference, but so does anxiety. You sense what they want and need. They're not always words; they're gestures or "I've heard enough." Or "Tell me what to do," or "I need to know more before I can make the decision." You read that while you're talking to a family, and you might change your approach 180 degrees. You might start off saying, "I'm going to give them all the information they want and help them make the decision about this," and you come out telling them exactly what they need to do. (Dr. I)

Underlying all of these interactions are the power inequities between physicians and parents. Dr. A bluntly says, "Most families follow your advice. It's hard to argue with the white coat and the authority of physicians in our society." He adds, however, that he does not necessarily agree with physician authority: "We're better off enlisting the family as our allies in making the decisions, as opposed to our telling them what's going on. . . . It's hard for people with very high IQs, let alone somebody who doesn't have the biologic background to understand. You're trying to bring it down to them. But never underestimate a mother's power to define the situation!"

The parents' role is also variable. Some parents want to be involved in decision making; others prefer to let clinicians decide. Parental desire to be involved in gender-assignment decisions can depend on their levels of education, their preference for a particular gender, and their feelings about clinicians' authority, among other factors. Clinicians mentioned working with parents whose cultures valued male infants over female ones and who thus were highly resistant to a female assignment for their newborn. Some parents object to what they understand as a medically unnecessary intervention in their child's anatomy for religious or spiritual reasons: as one woman writes, "Early invasive treatments literally change the destiny, the identity, the 'me' of your child."

HOW PARENTS DECIDE

For some parents, choosing their child's gender is an incomprehensible task, or one they do not feel qualified to undertake. Monica Cole, whose daughter has CAH, observes: "We were scared that we would have to make the godlike decision of choosing a gender. . . . I was deeply concerned we would choose the wrong gender, and that at some point our child would tell us they felt more like the other gender. I didn't want to leave her genderless very long, yet it was such a monumental decision to make so quickly."

For most parents I interviewed, gender ambiguity felt extremely distressing, and they look to biological markers, particularly karyotype, that they hope will resolve the issue conclusively. To some parents of girls with CAH and virilized genitalia, chromosomal results are crucial to understanding and accepting their newborns as females. Gender assignment is finalized by knowledge of the XX chromosomes. Pamela Heath says: "I got a call from the genetic counselor, who said, 'I begged the doctor to let me call you with the results. She's all XX.' I started crying and said, 'Thank you for calling!' Every day waiting for those results was just horrible! She's not dying from this, but it was so important to me that she really be a girl and when I found out she was, I was so, so relieved."

In an interesting twist, some parents of girls with CAH feel that perhaps they should choose a male assignment, but are reluctant to do so because the child's genitalia seem insufficiently masculine. Monica Cole explains: "My husband especially had a hard time thinking that we may have a baby boy that would not have complete sexual organ anatomy. It made him really sad, and he used the term 'eunuch.' He thought it would be easier to choose female because there are restrictive ideas in our society about masculinity and maybe more leniency about female traits." Dorothy Powers gives the following story: "I was worried when they mentioned a male assignment for her. She definitely looked male, but not that male. I thought it would be easier if she was a girl, at least because they could do corrective surgery and make the area look more female, easier than male."

Some parents appear to regard test results, especially karyotype, as a confirmation of their innate conviction about their child's sex. Gloria Jackson relied on her own instincts: "For my husband, the hardest thing for him to stomach was potentially having to raise a child as a different gender than their actual sexual makeup. For him, that was really difficult. And for me, it was a concern, but I just knew that she was a girl and that wasn't going to be an issue." Justine Good recalls, "Every test that came back, we'd get a further indication that yes, she's a female, and we kept going, 'Yes, yes, yes! Thank God.' The last one was whether she was XX, or XY, because I remember that endocrinologist going, 'What she's got, there's no way we can make a penis out of it, and no matter what, she'll have to be raised as a girl.' I remember thinking, 'Whoa!' Because if she's a boy, having to be raised as a girl, that just doesn't work! . . . So the last test came back XX, and we were very happy!"

THE FINNEYS

Returning to the Finneys, the situation was perhaps not as straightforward as the clinicians made it seem. The risk of testicular cancer was one of the most important factors in the parents' decision to raise their baby as a girl. This risk is inherent in any undescended testicle (not only in cases like Samantha's). There is no question that there is increased risk of germ-cell tumors for children with this diagnosis (MGD) and that in this instance gonadectomy is likely indicated. However, it is not the gender assignment as male or female that puts Samantha at risk of cancer, but leaving her testicles in her body too long. As an adult male, Samantha would not have testicles, but as an adult female, neither would she have functioning ovaries. Whether assigned male or female, Samantha would be infertile.[15]

The other reasons the clinicians gave the Finneys for a female assignment are also open to debate. The clinicians implied that whereas phalloplasty would require many surgeries, only one relatively simple surgery would be needed to create a vagina. In truth, both types of surgery prove difficult. As the next chapter shows, multiple surgeries for vaginoplasty are common because of complications like scarring and stenosis (where the vaginal opening closes). The surgeon also suggested that whereas Samantha could not have a normal sex life as a boy, she could as a girl. Yet mounting evidence shows that women who have had genital surgeries have problems with sexual function and sensation, and in some cases are inorgasmic. Jim summed up their thinking about surgery and gender assignment like this: "Once you make those decisions—she got the surgery, she's a girl—end of story. Plus, the surgery we were going to have later, had we elected the male option, would have been not just one surgery, but several. That was another negative to going that way."

Jim's "end of story" comment suggests that he anticipates no further adjustment issues for Samantha, that surgery might be a panacea. However, he also notes that surgery alone cannot guarantee that Samantha will grow up to have a female gender identity: "The big determining factor is not the surgery, but how she develops intellectually, mentally, emotionally. It doesn't make any difference . . . your physical, with today's technology and, and surgeries, you can do whatever you need to do. But it's how she sees herself, how she behaves, what *she* sees herself as." Jim understands that for all their efforts, Samantha herself will determine her identity—social, gendered, and sexual. What he does not understand fully are the complexities and shortcomings of genital surgery. The next chapter explores these issues.

FIVE

Fixing Sex:

Surgery and the Production of Normative Sexuality

Once clinicians and parents have made a gender assignment for an infant, the next steps are often to plan and carry out genital surgery to "normalize" atypical genitals by making them congruent with the child's assigned gender. This might include reducing what is perceived as a too-large clitoris or lengthening a "short" vagina. John Money recommended these surgeries take place early in life, the earlier the better, and for nearly four decades it appears that few physicians questioned their benefit.

Beginning in the 1990s, some adults on whom genital surgery had been performed began to challenge publicly the necessity and success of early genital surgery. They variously charged that surgery damaged their sexual sensitivity, sometimes making sexual contact painful and orgasm impossible; that their genitals did not look normal; that the surgery (as well as other aspects of their medical care) made them deeply ashamed of their bodies, harming their ability to enter into and maintain relationships; that the surgery was medically unnecessary; and that the choice to undergo surgery was not theirs (e.g., Chase 1994, 1997, 1998; Holmes 1994; Coventry 1998; Dreger 1999). Some even called the surgeries mutilation and the doctors who performed them butchers. Their criticisms, frequently iterated in both the clinical literature and in the popular media, have led to a crisis over early genital surgery both within and beyond professional medical circles. (There is no dispute over surgeries necessary for the child to live, such as those that enable the child to eliminate urine.)

In the wake of such criticisms and a mounting but still small body of clinical evidence of poor outcomes, clinical opinion is deeply divided over whether surgery intended to improve the psychosocial adjustment of the child fulfills its promise. Some have called for a moratorium on surgery until the patient can consent, arguing that these surgeries are not medically necessary, that there is insufficient proof that surgery yields better results when performed in infancy (rather than later in life, when the individual can participate in the decision), and that mounting anecdotal and clinical evidence suggests results are too often poor. Others counter that outcome studies evaluate outdated techniques (the same techniques from which adults have suffered), that complications are due to surgeons' inexperience, and that there is no proof that *not* doing surgery eases the child's psychosocial adjustment.

Despite ten years of debate over early genital surgery, surprisingly little has changed either in terms of the arguments put forth or the practices themselves. Many of the clinicians I interviewed saw early genital surgery as necessary, most surgical articles take it for granted that surgery will be performed (and hence simply describe how to do it), opinions at professional meetings appear to agree that surgery still makes for good care (although surgeons preface these statements with medico-scientific rationales that I explore later), and discussions with parents and visits to Web sites for support groups suggest that a fair number of parents continue to choose genital surgery for their children. Surgeons and other specialists certainly urge greater caution in surgical decision making, but few appear willing to question outright the wisdom of early genital surgery. This ambivalence is reflected in the section covering genital surgery in the "Consensus Statement on Management of Intersex Disorders" (Lee et al. 2006). The authors acknowledge that there are minimal systematic surgical outcome data about genital surgery (whether on cosmetic appearance or sexual function), that orgasmic function may be harmed by surgery, and that there is little support for the belief widely held among physicians and others that surgery performed in the first year of life relieves parental distress about atypical genitals (Lee et al. 2006). These findings no doubt led the authors to narrow the range of acceptable surgery, saying that unless medically indicated, surgery should "only be considered in girls with severe virilization." This constitutes an important shift, but the authors fail to provide a rationale for why, given the problems they noted, it is still advisable or acceptable to consider surgery for girls with greater genital virilization. Although it might be tempting to get caught up in debates over what degree of genital masculinization might warrant surgery, it is a secondary issue: if there is little to no evidence to support the benefits of early surgery, a child's genital configuration is irrelevant.

Perhaps the most obvious change in thinking about genital surgery and surgical practices since these debates began is that surgeons are performing far less feminizing surgery on males with a small phallus. Clinicians' increasing willingness to embrace ideas about a strong biological role in gender-identity development have made them less willing to consider female assignment in these cases, minimizing the specter of feminizing surgery for these infants. Clinicians are thus moving away from the wholesale idea that sex-specific genitalia are necessary for "proper" gender-identity development.

Whereas this shift might appear to disrupt the profound belief in the inseparability of genitals and gender, it would be a mistake to assume that the connection has been severed. For infants assigned female—virtually all infants with a 46,XX karyotype and some with a 46,XY karyotype—many surgeons continue to recommend early genital surgery. In the wake of a resurgence of biologically based theories of gender difference, many physicians I encountered do not view genital conformity as necessary for gender-identity formation in these girls. They still, however, view it as important for "psychosocial functioning," which has been reconceptualized to accommodate this body of research. No longer explicitly linked to gender-identity formation, "psychosocial well-being" is newly tied to the importance of looking "normal." Early genital surgery is assumed to give the girl a chance at an otherwise unattainable normal life—an assumption, widely shared by physicians and parents, that rests on the notion that surgery imparts normal-looking genitals and ameliorates, rather than contributes to, the child's sense of difference (e.g., Dayner, Lee, and Houk 2004). The notion that surgery will necessarily improve a child's well-being may explain why many surgeons and other clinical specialists continue to recommend the procedures—and parents continue to opt for them—despite evidence that surgery may harm. (It is also true that some parents opt for surgery for their children despite clinicians' recommendations that it *not* be performed.)

Surgeons' belief that genital surgery delivers psychosocial benefits remains strong. In consequence, those I have spoken to often overlook or downplay the fact that while adults' experiences and claims of poor outcomes may be scientifically unsubstantiated (despite some studies showing poor outcomes), so, too, is the clinical belief that surgery contributes to psychosocial well-being and preserves genital sensation. In an effort to reconcile these debates, medical professionals now routinely urge caution in surgical decision making and indicate that more data are needed to resolve competing claims. However, the data from outcome studies, while certainly important, are unlikely to resolve these debates. Rather, it seems more likely that they will contribute to them. As I discuss later in this chapter, outcome studies of genital surgery are difficult to carry out. To the extent to which they are possible, they will likely suffer the same methodological limitations as prior studies. In particular, small sample sizes and highly variable study methodologies and evaluation criteria will render the results open to reinterpretation and contestation.

When these debates first erupted, surgeons frequently dismissed the complaints of adults with intersex diagnoses as those of the disgruntled minority. They might have had poor outcomes, the argument went, but they represented the very vocal minority with the worst surgical results; a silent majority felt pleased with their surgical results (see chapter 8). Yet as increasing numbers of studies have begun to demonstrate poor surgical outcomes, some surgeons and other clinical specialists discount even these findings. Some also argue that research on surgical practices and outcomes are often irrelevant since measures of adult functioning assess surgical techniques almost two decades old. Surgical results may have been poor in the past, the reasoning goes, but since surgery has improved today, one can reasonably expect better surgical outcomes as well. This argument has a dual effect: it simultaneously acknowledges previous poor outcomes and effectively discounts any concerns about poor outcomes from today's surgical practices. (Of course, the logical conclusion to this argument is that future surgical techniques will be better still, so why not wait?) By charging that adequate studies are impossible because they will always assess old techniques, surgeons and others deflect current as well as future scientific and anecdotal evidence of poor surgical outcomes.

Some surgeons argue that surgery is better now because the specialty performing the procedures has changed: the pediatric urologists performing many of the surgeries today, they claim, are more sensitive to the need for preserving sexual sensation than other specialists were. Others cite advances in surgical techniques and instrumentation, including modern sutures and reconstructive surgery. With surprising frequency surgeons also mention the cessation of clitorectomy as an indication of improved surgery. Not removing the clitoris entirely as in the "old days" is read as a technical improvement and given as a reason to have faith in the excellence of today's surgical techniques. Surgeons also turn to medico-scientific rationales for early genital surgery, which include technical ease, more pliable tissue, supposedly more aesthetically appealing results, reduced scarring, improved healing, and better nerve regeneration. These arguments offer justifications for doing the surgery *early*; they do not provide any logical support or evidence for the idea that it is necessary in the first place. Nevertheless, the psychosocial and medical arguments can be persuasive, and hearing them outside the context of the debates, one could reasonably conclude that surgical practices result from a considered and deliberate reflection on scientific evidence. (One would also assume that the outcomes are good.)

The rhetoric of technological ease, expertise, and improvement masks the anxiety raised by gender-atypical bodies and avoids any moral discussion of the pathologization of such bodies, the conceptualization of genitals as malleable organs—rather than as healthy parts of people—and the cultural and medical imperative to make normatively gendered subjects. It also mutes any discussion about the cultural logic mandating these surgeries, which assumes specific connections among genitals, gender, and sexuality making these surgeries seem necessary, even inevitable. The valuation of sexual function over sexual pleasure, a lack of knowledge about female sexuality and human sexual development, and discomfiture with atypical genitals also play a role in making genital surgery seem essential to the child's future well-being. In short, surgery locates the problem in the child's genitals, not in social conceptualizations of what counts as sex. In analyzing and questioning the practices surrounding early genital surgery, I am not suggesting that genital surgery per se is bad, nor am I suggesting that an individual's desire for a smaller clitoris or for a vagina is somehow trivial or unwarranted. Rather, I aim to explore why practices around early genital surgery, especially on females, are so resistant to change.

The Heterosexual Paradigm

Each culture ranks different sexual practices and identities according to what is most normative and socially acceptable, resulting in what the anthropologist Gayle Rubin has called a sexual hierarchy (Rubin 1992). The hierarchy, which resembles a class system, reveals how a culture evaluates sexual behaviors and relationships, from those most highly valued to those most despised (Miller and Vance 2004). Many Western cultures in modernity have assumed an intimate connection between maleness or femaleness—a distinction based on physical organs, traits, and characteristics—and the correct form of erotic behavior (Weeks 1986). The cultural rules surrounding sexual conduct and ideas about appropriate sexual behavior have been tied to the belief that reproduction constitutes the natural and desired outcome of sexual activity. As a result, heterosexual, monogamous, reproductive sex has been privileged above all other erotic behaviors and relationships in the sexual hierarchy.

The greatest influence, and also the greatest harm, of the sexual hierarchy lies in the ways it both animates and is embodied in diverse fields and institutions, including medicine (Miller and Vance 2004). Over the past hundred years, medicine has had a unique role in monitoring bodies that violate sexual norms (and thus the hierarchy) through categorization, diagnosis, and treatment. Prevailing cultural attitudes about normative (i.e., proper) sexuality and its relationship to gender were codified in Money's early work (and hence the traditional intersex management protocol), with his classic definition of gender role as "all those things that a person says or does to disclose himself or herself as having the status of a boy or man, girl or woman. . . . *It includes, but is not restricted to, sexuality in the sense of eroticism*" (Money, Hampson, and Hampson 1955: 285; my emphasis). Money later modified this definition somewhat, but one aspect remained the same: sexuality by definition was classified as a part of gender role. In this framework, a male gender role assumes (even requires) sexual desire for females and vice versa. Males and females thus require each other, and their complementary parts, such that an infant properly assigned will be heterosexual in desire and will express this desire via penile-vaginal intercourse.

Genital surgery, then, seeks to produce genitals that not only conform to the gender of the infant but that stabilize particular forms of sexual desire and behavior. By reflecting and reinscribing social norms and expectations for appropriate female and male sexual desire and behavior, genital surgery enables and disenables certain sexual acts and thus certain ways of being. The assumption that infants need particular genital configurations to help them fulfill these expectations comes at the expense of leaving intact those body parts that enable sexual satisfaction. As a result, concerns about sexual pleasure have been secondary to ideas about normative bodies and the sexual practices in which they are expected to engage. Put another way, "penetration in the absence of pleasure takes precedence over pleasure in the absence of penetration" (Fausto-Sterling 1995b: 131).

The prevailing treatment paradigm for children with intersex diagnoses assumes that the connections among physical traits, psyche, and behavior (or sex, gender identity, and sexual orientation) are innately gender appropriate and that heterosexuality is the natural result of appropriate gender development. Intervention in children with intersex diagnoses is thus both meant to produce the expected, supposedly natural outcome and to prevent aberrations or deviations from that outcome.

While believing they are providing the best possible care to individual patients, clinicians (consciously or otherwise) also regulate social categories and the sexual hierarchy. The pediatric urologist Dr. J states the aim of treatment: "Intimacy is a big part of humanity, and we're talking about giving these children the tools to achieve intimacy." Others suggest that surgery to create more "normal" genitals is unquestionably an improvement over leaving a child with a "grossly" enlarged clitoris, a micropenis, or a "short" vagina. Yet embedded in this philosophy are numerous assumptions about what constitutes intimacy and sexuality. Typically, heterosexuality is seen as the natural sexuality and the successful sexual outcome for treated children; penile-vaginal intercourse as the exclusive or most important sexual act; and genital appearance as taking priority over sexual pleasure and sensation.

These assumptions, as my interviews with clinicians reveal, drive many decisions about not only genital surgery but also about gender assignment. Infants born with a small penis may be assigned as female because they do not meet a clinician's criterion for sufficient penis size—one long enough for vaginal intromission. A female child with a larger than average clitoris may undergo reduction surgery partly not to upset future male sexual partners—and to assuage her parents' fears that her atypical genitals might result in gender-atypical sexual desires and practices.

Assumptions about normative erotic desires and sexual practices also frequently reveal themselves as self-referential. Chapter 3 has discussed some of the ways in which behavior assumed to be distinctively male or female has become part of a circular reasoning process to determine the gender identity of individuals who practice such behaviors. When the urologist Hugh Hampton Young published his 1937 treatise on intersexuality, he included a chapter on individuals for whom "no positive diagnosis of sex could be made." Without any way to determine and thus stabilize the sex of these individuals, Young was seemingly at a loss to understand and categorize their sexual behavior: he referred to them as "practicing hermaphrodites." In these cases Young based his determination of sex on the sex of the patient's sexual partner and on an assumption of heterosexual behavior. If a "practicing hermaphrodite" had intercourse with a man, then Young understood the individual to be a woman, and vice versa (Kenen 1998).

A clinician I interviewed spoke of a patient assigned female in infancy and who transitioned to live as a male in his teens: "I watched this boy as he grew up, and he was clearly male in how he thought, the way he looked at the world, the way he dealt with people. Clearly male in a sexual orientation—he was oriented to women—but the struggles he had dealing with his sexuality were gigantic, compared to the typical teen. Ultimately, this boy moved in with a girl that he was in love with, and when I left, they had been living together for almost two years" (Dr. R). For this clinician, virtually every behavior can be read as gendered, and heterosexuality is the natural expression of desire for someone who identifies as male. Evidence of the inherent maleness of his patient is found in multiple behaviors (e.g., how the patient thinks), but the crowning proof of his masculinity (and thus maleness) is his coupling with a female. His patient's normative outcome provided proof of a biological origin for these observed gender differences.

Other examples of self-referentiality persist in contemporary clinicians' working definitions of genital morphology. Clitoral reduction is often performed on female infants with CAH in part because of the perception that a large clitoris will inhibit normal sexual activity and expression. One pediatric surgeon, for example, cautioned that it might be alright to forego surgery on "less virilized" girls with CAH, but that without surgery it was "unlikely that they would develop normal sexual attitudes and function" (Tovar 2003). Yet medical definitions of normal female genital morphology in textbooks tend to be artificially narrow, reflecting a prescriptive rather than an empirical standard.[1] Such circular logic has not gone entirely unnoticed by medical professionals. As one pediatric endocrinologist has remarked to me, "If we're saying genital surgery is for normalization purposes, I would ask the surgeons how many penises they've reduced to make them more normal in size" (Dr. A).

The categories homosexual and heterosexual presume a fixed and singular biological sex and gender identity. Yet for individuals who have changed gender, these labels have limited meaning. As Dr. F asks, "How do you define homosexuality in an XY individual who had testes and high testosterone levels [and] who is phenotypically a woman? When you stop and think about it, it becomes a real mind-boggling set of issues to deal with. In somebody who is truly ambiguous, what's homosexual versus heterosexual?" Karen Rollins, a woman with PAIS who was raised as a male and now lives as a female, sums up the flaws in this reasoning: "Based upon my chromosomes [XY], I'm in a 'homosexual' marriage if I marry a man. I'm in a heterosexual marriage if you base it on my penile type. Or, if I'm in a relationship with a woman, chromosomally I'm in a heterosexual marriage, but with my gender I'm in a homosexual marriage."

The presumption of heterosexuality persists in current practices of intersex management, both in clinicians' recommendations for treatment and in such limited assessments of outcomes as exist. In part, clinicians claim to be responding to the sometimes unstated fears of parents that babies with intersex diagnoses are more likely to grow up to be gay. Whatever the case, the widely accepted measure of successful treatment is that the individual marries and raises children.

This bias clearly emerges in a survey developed by a gynecologist in Great Britain assessing the outcomes of genital surgery for women with CAH (Minto 2000). In its section on "sexual functioning," this survey asks only about heterosexual unions. The entire project rests on the assumption that respondents are in a monogamous relationship with a male and primarily practice penile-vaginal intercourse. The questions are structured in a way that assumes orgasm to be possible only through vaginal intercourse with a male partner. (This researcher and her colleagues, however, went on to publish some of the most thoughtful studies of and commentaries on genital surgery to date.)[2] A survey by Milton Diamond and intended for individuals with CAIS or PAIS makes similar assumptions. The order of the interview questions assumes sexual activity as the province of married people: questions about marriage precede any about sexual activity, and the last question explicitly links the two by asking, "When do you think information about sexual relations and marriage should be offered?" (Diamond n.d.: 13).

Marriage and reproduction have been used as specific criteria in outcome studies evaluating genital surgery. One study of outcomes for clitoral reduction, for example, states: "A 21-year-old woman with congenital adrenal hyperplasia is married and, when examined last, was in the final month of her pregnancy" (Allen, Hardy, and Churchill 1982: 353). Dr. O, a pediatric urologist I interviewed, confirms this view when he says: "If they are able to get married, have babies, live a normal life, then I feel I have done my job," which assumes the patient will share the doctor's desires and values.

Conversely, individuals who express a preference for sexual partners of the gender to which they have been assigned (or with which they identify) are often seen as examples of treatment failures. Dr. B relays a comment made by another clinician at a medical conference: "A woman from NIH [the National Institutes of Health] said, 'You can take somebody and do everything so they come out fine, but they could wind up lesbian.' " Parents often also expect a heterosexual outcome, implicitly if not explicitly. Dr. Q, a researcher, notes: "Parents [of CAH girls] ask me that question, 'Will my child be lesbian?' all the time. I would like to have better information for them." One clinician I interviewed says, "Most parents are no more open-minded than doctors. I would never say to a parent, 'Have an open mind. Your daughter might be lesbian, but there's nothing wrong with that!' I'd probably get beat up'" (Dr. E). The view that homosexuality represents a failure of treatment may help explain the findings of a recent survey of pediatric endocrinologists and pediatric urologists that found predicting sexual orientation an important consideration in gender-assignment decisions (42 percent for pediatric endocrinologists, 57 percent for pediatric urologists) (Sandberg et al. 2004).

Though the topic is infrequently raised in public discussion (and even discouraged on some Web sites), many parents of girls with CAH wonder whether their daughters will be lesbian because of their exposure to high levels of androgens in utero. For those who view homosexuality as biologically determined, excess androgens raise concerns about supposedly male-typical behavior, including sexual desire for females.

For analytical purposes I have separated conceptualizations of feminine and masculine sexuality; however, these are inextricably linked. In the minds of many clinicians and parents, feminine sexuality is defined in relation to masculine sexuality and vice versa. From this formulation it follows that female parts must fit with male parts. This assumption can lead clinicians to take considerable measures to make a baby's genitals conform to social norms. Neither clinicians nor parents contemplate leaving a child with atypical anatomy to broaden his or her erotic possibilities and pleasures as an adult; indeed, as Suzanne Kessler ironically observes, making such a suggestion is so far out of bounds as to seem laughable: "You know, 'Wow, she's going to be in demand. She's going to be really'—Well, maybe not" (author's interview, SUNY Purchase, Purchase, NY, June 21, 2000).

Masculine Sexuality and Masculinizing Genital Surgery

Although physicians are increasingly reluctant to recommend female assignment and feminizing genital surgery for males with a small or malformed penis, or no penis, they have not ruled it out entirely. The recent "Consensus Statement on Management of Intersex Disorders" notes that the magnitude and complexity of phalloplasty should be taken into account during the initial consultation with parents "if successful gender assignment depends on this procedure" (Lee et al. 2006: e492). The language is obtuse, but it essentially means that if a penis is not big enough—or is not there at all—female assignment and feminizing genital surgery may be advisable, because there is little surgeons can do to create or significantly enlarge a penis (see chapter 4). (The implication that the creation of a vagina is more straightforward reflects what some have referred to as the "easier to dig a hole than build a pole" philosophy [see Hendricks 1993].) The statement, however, conveys no hesitation about performing so-called masculinizing surgery—what Anne Fausto-Sterling has called penis building (2000: 630)—to try to improve upon penises that are big enough, but otherwise imperfect.[3]

The consensus statement discusses masculinizing surgery primarily in the context of "DSD [disorders of sex development] associated with hypospadias" for which, it says, surgeons should perform standard techniques for surgical repair. Hypospadias occurs when the tube that allows urine and semen to exit the body opens somewhere other than at the center of the tip of the penis. It is classified according to the location of the opening. In mild forms, the urinary hole opens not far from the tip of the penis; in moderate to severe forms, the urinary hole opens anywhere along the shaft of the penis to its base (e.g., scrotum and perineum). In boys with severe hypospadias, the genitalia may appear ambiguous at birth raising questions about gender assignment.

In the classificatory scheme for hypospadias, more severe means more feminine; the further away from the glans the urinary hole opens the more likely it is the child will have to urinate sitting down. Surgeons then typically counsel parents to pursue surgery over concerns that the boy's urine stream will be abnormal and that he will not be able to urinate in a standing position or to penetrate and inseminate a partner. As one university medical center's Web site explains, "If the deformity will not impede normal standing urination, normal sexual function, or deposition of semen, surgical repair is not required" (http://pennhealth.com/ency/article/003000.htm). Since these activities are affected to some degree in all but the mildest forms of hypospadias, surgery is routine.

Physicians have not typically considered hypospadias an intersex condition, because it is rarely accompanied by atypical development of the reproductive or endocrine system and thus raises little question regarding gender assignment. Technically, the diagnosis of hypospadias can only be made if the infant has already been determined as male. Consequently, surgery for the mildest to most severe forms is performed not because the gender remains in doubt but because the genitals in some way fail to measure up.

The phrase "DSD associated with hypospadias" reflects a larger conceptual struggle about whether hypospadias should be considered an intersex condition. Most physicians are adamant that hypospadias is not an intersex condition. For those clinicians for whom DSD is simply a replacement term for intersex, hypospadias is not a DSD. For many others, however, DSD encompasses a broader range of conditions that includes all forms of hypospadias. As a clinician and researcher, Dr. T, remarked to me in an e-mail, "hypospadias is not intersex, but it is a DSD." The phrase "DSD associated with hypospadias," reflects this ambivalence. Read a certain way, it suggests that a DSD may be accompanied by hypospadias, but that hypospadias is not itself a DSD. The title of a recent conference, the Second World Congress on Hypospadias and Disorders of Sex Development, equally reflects this ambivalence.

This may seem like a fine point, but it is a hotly debated one with important consequences. In the past intersex activists, among others, have argued that hypospadias counts as intersex because it constitutes atypical sex development and because surgery to correct it is driven by the same motivation to normalize atypical genitals as other surgery for intersex conditions. Surgeons and other physicians have vehemently countered that activists incorrectly include hypospadias among intersex conditions because "it boosts their numbers," meaning that it allows activists to claim a higher frequency of intersex conditions and hence a larger number of affected people. Indeed, including hypospadias would increase the number of people with intersex conditions considerably, as it is one of the most common congenital anomalies in males, occurring in about one in three hundred births (Sheldon and Duckett 1987; Baskin and Ebbers 2006).[4] Activists and others additionally charge that pediatric urologists and surgeons resist viewing hypospadias as intersex because these surgeries are their "bread and butter," constituting a large and lucrative part of their surgical practice.[5] Though few clinicians, if any, would argue that hypospadias is intersex, many clinicians who specialize in treating DSD now view hypospadias as a DSD. By interpreting all forms of hypospadias as DSD, clinicians essentially have opened up hypospadias surgery, previously bracketed off as something other than intersex genital surgery, to the same criticisms garnered by other genital surgery—including the charges that surgeons are driven by deeply ingrained gender ideals and the desire to normalize atypical genitals, and that the surgeries are characterized by poor outcomes.

Concerns about complications and poor outcomes are not unfounded. Today, hundreds of surgical techniques for hypospadias have been described in the medical literature.[6] Despite the technical advancements implied by such a voluminous literature, no consensus exists about which techniques produce better outcomes and whether surgery on the whole is more helpful than harmful (Macedo and Srougi 2001). There are surprisingly few long-term follow-up studies demonstrating the efficacy of these procedures, which raises concerns about unsatisfactory results, complications, and the need for subsequent surgical repair. Even less understood are the psychological, psychosocial, and psychosexual effects of hypospadias and its surgical treatment.[7] This situation has led one surgeon to conclude that the factors contributing to success remain largely unknown and, although more research is needed to evaluate the various techniques and outcomes, that the subjectivity of clinical assessment will render evaluation difficult (Mouriquand 2004).

One of the most common complications, especially related to repair of more severe forms, is urethral fistula, characterized by an outflowing of urine at the site of the repair that can happen years after the operation and necessitate further surgery. A rarer complication is the shrinking (stenosis) of the urethra, which makes it difficult to completely empty the bladder, thus leading to urinary infections. Surgeons often treat this condition through dilation or further surgery. In some instances, complications result from an initial surgery for conditions that may accompany hypospadias, such as for chordee, where the penis bends downward due to a band of tissue connecting the tip of the penis to the base or gland.

Unsatisfactory aesthetic results due to excessive sutures, loose or excess skin, or a poorly placed or oddly shaped urethra, constitute other complications, as does hair growth in the urethra. Surgeons today try to avoid hair-bearing skin when using skin grafts, but if it is used or if surgeons use cortisone to promote healing, hair growth can ensue and result in urinary tract infections, requiring depilation using a laser or cautery device or, if severe, the excision of hair-bearing skin and repeat surgery. As with other surgery, the process of healing may lead to scarring, which can be thick, stiff, and have little sensation. Finally, some patients suffer from multiple failed surgeries, thus being referred to as hypospadias "disasters," or worse, "hypospadias cripples" (Paparel et al. 2001; Stecker et al. 1981).

Some suggest that surgery creates more problems than it corrects. Alice Dreger writes of a man who wishes his failed surgeries, which resulted in the need for life-long catheterization, had never been performed (1998a: 178–79). Sharon Preves also interviewed an individual with hypospadias who underwent over fifteen surgeries and who suffered considerable harm (2003). Given that a hypospadic penis enables pleasurable sexual sensation and orgasm, while surgery risks damaging erotic sensation and creating life-long complications, one may easily wonder why the risks and harms of such surgery may be outweighed by the harms of having a slightly different penis—which from what epidemiological studies tell us, is actually quite common.

Feminine Sexuality and Genitals

Among infants with atypical genitalia who are assigned female, two types of surgery are frequently recommended: clitoral reduction and the construction of a vagina or a vaginal opening (vaginoplasty). These two procedures are deemed necessary for different reasons, but both often have to do less with the well-being of the individual than with social ideas of what female sexual organs should be: capable of reproduction and of vaginal intercourse. A large clitoris is viewed as resembling a male penis too much to be left alone. Vaginoplasty is deemed necessary for sexual coupling with males via penile-vaginal intercourse. One surgeon I interviewed states this philosophy very succinctly: "If you took the case of a girl with a small degree of virilization, who has a relatively normal vagina so it's big enough for intercourse, then all they need is something done with their clitoris so it looks less like a penis" (Dr. O). In effect, a female infant who undergoes clitoral and vaginal surgery is primed to fulfill a particular version of sexuality, one in which her pleasure is secondary to that of future male partners.

READING GENITALS FOR CLITORAL EXCESS

The clinicians I interviewed often referred to an enlarged clitoris in highly subjective and pejorative terms, using expressions such as *grotesque*, *deformed*, or *abnormal*. In an oft-cited 1966 paper, surgeons recommend clitorectomy for an enlarged clitoris, arguing that the "disfiguring and embarrassing phallic structure" proves detrimental to a female gender assignment (Gross, Randolph, and Crigler 1966: 300). Dr. K, a nationally renowned pediatric urologist, echoes this negative sentiment: "Have you seen a baby with CAH? It's grotesque! You're the mother of a newborn with clitoromegaly [enlarged clitoris] and fused labia. Are you going to change the diaper every time? Are you going to hire somebody? Who's going to be the caretaker? Are you going to let any neighbor change that child? Absolutely, positively not. You are tied to that child every minute of every day of every week of every month! You're not going to go on vacation." Not only does a large clitoris appear grotesque to this clinician but it apparently also has the power to stop parents from working, going on vacation, or otherwise living a normal life. By his reckoning, surgery will not make the child's life normal so much as it will make the parent's life normal. A large clitoris also bothers Dr. P, another surgeon I interviewed: "Most of the ones we see have significant clitoral enlargement, and I can't leave that. These girls don't look right. It's unsettling. It's repulsive. You just cannot leave them looking like that!" Dr. G, a pediatric endocrinologist, tells me of another girl whose condition did not justify medical intervention but nevertheless clearly disturbed social norms: "The other day I saw a child who's about five. We cannot find anything genetically or biochemically wrong with this child, and yet she has a clitoris that's three times bigger than the average child her age. There's no other sign of virilization with her. No history of androgen exposure in utero. We've been over it and over it and just can't figure it out. Whatever it is, she's an outlier. And I was thinking that if she had CAH, a surgeon would *choose* to render her with less size" (original emphasis).

Why this dramatic response to an enlarged clitoris? What is disturbing about the size of an otherwise very small organ that is not on public view and whose size poses no medical risk? Notions of male-female difference rest on understandings of male and female bodies as not simply different but as opposite and incommensurate. Men have penises, women do not. The anthropologist David Valentine and the transgender activist Riki Wilchins have noted that although the genital area accounts for only 1 percent of the body's surface area, genitals "carry an enormous amount of cultural weight in the meanings that are attached to them" and constitute much of what individuals and society "come to understand and assume about the body's sex and gender" (Valentine and Wilchins 1997: 215). A large clitoris, one that visibly resembles a penis, upsets this primary understanding of sexual difference. No one would argue that genitals mark the only difference between males and females, but genitals that do not appropriately signal sex, or else missignal sex, throw into question fundamental understandings of males and females. A large clitoris with readily discernible erectile tissue raises anxiety about gender boundaries. The literal excess of a large clitoris places females outside the normative gender boundaries and is like a phantom male organ inhabiting female sexual anatomy. As a woman with CAH comments: "The last time I dated a boy, he went down my pants and felt a little hard-on and said, 'What are you? A he/she?' " (Lisa Todd). Gender-atypical genitals raise issues beyond the individual: if one cannot stabilize the sex of another person, one's own sexuality may become destabilized.

Concern with clitoral size in Western society dates back to the nineteenth century, when female sexual excess, sexual pathology, hysteria, and other mental states became associated with the female genitals (Foucault 1978; Scull and Favreau 1986; Groneman 1994; Odem 1995). During the Victorian era, orgasm was thought unhealthy for women. In this context Sigmund Freud published his *Three Essays on the Theory of Sexuality* (1905), in which he deemphasized the role of the clitoris in female sexual pleasure, noting that its role was secondary to that of the vagina. In Freud's formulation, "mature" female sexuality and hence orgasm were centered in the vagina, not in the clitoris (Freud 1962). This shift in the site of "erogenous excitability" foregrounded a central role for penile-vaginal intercourse in women's sexual satisfaction.[8] With the demotion of the role of the clitoris in female sexual pleasure and reproduction, it hardly seems surprising that Victorian doctors argued for its removal as therapeutic treatment.

Isaac Baker Brown, an obstetric surgeon, provided an early, vocal, and public endorsement of female circumcision in his 1866 book *On the Curability of Certain Forms of Insanity, Epilepsy, Catalepsy, and Hysteria in Females*, in which he argued that masturbation led to disorders of the nervous system in women that began with hysteria and progressed finally to epileptic fits, idiocy, mania, and death. He proposed the excision of the clitoris as a cure (Sheehan 1997). Although Brown was expelled from the London Obstetrical Society shortly after making this proposition, his premise that masturbation was intimately tied to the development of nervous disease went undisputed (Sheehan 1997: 327). As late as 1936, *Holt's Diseases of Infancy and Childhood* recommended the cauterization of the clitoris as a cure for masturbation in girls (Holt, Howland, and McIntosh 1936; Lightfoot-Klein 1989; Hall 2001).

This tendency to "cure" masturbation through clitoral surgery still appears to have some currency. Several clinicians I interviewed mentioned parents who originally decided against clitoral surgery for their daughters but changed their minds when they discovered their young daughters masturbating.

> I had a little two-year old girl who had a very, very enlarged phallus and they [the parents] didn't have surgery. The little girl was masturbating, and the parents just fell apart and were back in the office the next week for surgery! (Dr. P)

> Things like nocturnal erections or masturbation in little girls with clitoromegaly is a situation I've encountered quite a few times, and that's actually pushed many parents towards surgical intervention. (Dr. N)

> One little girl was having big problems with masturbation, which was also a big concern to the parents. . . . They decided to do clitoral reduction because the masturbation was becoming a big problem for the parents. (Dr. M)

Another fairly widespread concern among parents and clinicians is that a large clitoris will lead to a more masculinized sexuality, or to sexual desire for females. Surgery aimed at giving the child more gender-typical genitals will therefore provide her with an appropriate sexual preference (for males). One clinician comments: "A lot of this is visceral and intuitive. If you're a woman with a big clitoris, you're likely to turn into a dyke. People never say it in that kind of a cruel, inappropriate way, but that's the association people make" (Dr. B). Some individuals are less concerned about the size of an enlarged clitoris than with its erectile capability. One clinician I interviewed echoes this concern in discussing a patient who did not have clitoral surgery: "She was always very careful about it [her clitoris], and she said she's learned to control it, so that it didn't become erect, and wasn't as noticeable" (Dr. A).

Historically, an enlarged clitoris was reduced through amputation or clitorectomy, a method Hugh Hampton Young turned to after unsatisfactory results from trying to conserve clitoral tissues (1937). It is unclear to whom the results proved unsatisfactory (surgeon or patient), but clitorectomy may have been adopted because of its relative ease: instead of attempting to retain particular parts, the surgeon removes the clitoris entirely. Clitorectomy was still the primary method for clitoral reduction in 1955, when Money published his theory of gender-identity development—at a time when the vagina was still widely perceived as the site of female sexual orgasm. Money asserted that the removal of the clitoris would not damage erotic function, basing his assertion on evidence from twelve women: "There has been no evidence on the deleterious effect of clitoridectomy. None of the women experienced in genital practices reported a loss of orgasm after clitoridectomy" (Money, Hampson, and Hampson 1955b: 295). In a later study, Money reported the results of postoperative erotic sensitivity with (the same?) twelve women: five appear to have experienced orgasm, four did not, and three provided no information (Money 1961). With only half of the study participants reporting orgasm, Money drew the following conclusion: "The point of these data on orgasm and clitorectomy is not, however, that some clitorectomized patients did not experience orgasm. On the contrary, the point is that capacity for orgasm proved compatible with clitorectomy and surgical feminization of the genitalia in some, if not all, of these patients (1961: 294).

In a 1966 paper by Robert E. Gross, Judson Randolph, and John F. Crigler, the authors recommend clitorectomy for an enlarged clitoris and criticize those who "have been reluctant to advocate excision of even the most grotesquely enlarged clitoris" based on "the belief that the clitoris is necessary for normal sexual function" (1966: 307). They claim that the clitoris is unnecessary for "normal sexual function," arguing that a number of "sound studies" demonstrate "normal sexual response" in females who had undergone clitoral extirpation (307). These practices were not always, or even mostly, predicated on the promise of scientific findings. For the latter claim, the authors cite a paper by Joan Hampson (1955) and unpublished data from the anthropologist Philip H. Gulliver.[9]

During the second half of the twentieth century, clinicians held differing views about the importance and role of the clitoris in sexual function, and surgical procedures reflected that uncertainty. At roughly the same time that sex researchers such as Alfred Kinsey, William Masters, and Virginia Johnson were writing about the role of the clitoris in sexual satisfaction and orgasm,[10] surgeons were advocating the removal of the entire organ, arguing that this did not impair female sexual functioning, maybe partly because the vagina remained the perceived site of female sexual orgasm. With the emergence and circulation of new research to the contrary, surgical procedures began to shift from complete clitoral extirpation to other methods. By the early 1960s, the surgeon John Lattimer developed "clitoral recession," a procedure that aimed to hide the clitoris by in effect burying the shaft and the glans but also emphasizing the importance of preserving the sensitive tip of the glans and the erectile tissue for erotic stimulation (Lattimer 1961; Randolph and Hung 1970). Nevertheless, surgeons continued to debate whether to perform clitorectomies into the 1970s. Published within a few years of each other, one paper advocated clitorectomy (Jones 1974), whereas another invoked the work of Masters and Johnson to argue against the procedure, concluding that clitorectomy did indeed negatively affect sexual function (Fonkalsrud, Kaplan, and Lippe 1977). Today, surgeons aim to remove the erectile tissue of the clitoris while preserving the clitoris's dense innervation (called the neurovascular bundle), which they hope will preserve sexual sensitivity.

If there is a preoccupation with male phallic size—when is a penis too small?—there is an equal preoccupation with female phallic (clitoral) size: when is it too big? As in the case of penis size, guidelines have been lacking. The first table for female neonate clitoral size was published in 1980, twenty-five years after Money introduced his treatment paradigm (Riley and Rosenbloom 1980). It reported the average size of a female clitoris in infancy as 3.45 millimeters (about one-eighth of an inch). Later researchers established a mean of 4 millimeters, but they found a much broader range of clitoral size at birth, from 2 to 8.5 millimeters (or one-sixteenth to one-third of an inch) (Oberfield et al. 1989).[11] Neither study specified, however, at what point the clitoris of an infant could be considered enlarged.

In the medical literature, authors frequently refer to an enlarged clitoris without specifying its actual size. In a chapter on surgical feminization in a recent urology textbook, the authors refer to minor, mild, moderate, and severe clitoromegaly without providing any definition or assessment criteria (Hensle and Kennedy 1998). This lack of specificity is not unusual; unsurprisingly, it has carried over into clinical practice, where decisions about the need for and the extent of clitoral reduction are based primarily on the subjective judgments of the surgeon.

Although the physicians I interviewed all claimed to measure the penis of an infant with atypical genitals, only four knew that standards for infant clitoral size existed, and only three said they measured the clitoris. (All of the female clinicians I interviewed, who were pediatric endocrinologists, knew about and used these standards.) This does not mean that physicians are less concerned with clitoral size; rather, the mode of measurement differs. Most clinicians use visual assessment, not measurement, to judge clitoral enlargement. As one surgeon told me: "I know one [an enlarged clitoris] when I see it. . . . We're not talking about gray zones. It's usually very, very obvious." Dr. L, a pediatric urologist, comments, "It's an impression of how it looks. If you open the diaper and see a phallic structure then, clearly, that's going to be objectionable. I don't have an actual measurement. Maybe it's my sexism or something, but I've never measured clitoral size. It's by visual. As we're doing the surgery, we'll look at it stuffed back and say, 'That looks good, or it still really looks abnormal.' It's very much a judgment thing."

In these aesthetic assessments of infant genitalia, the clitoris is not linked to sexual functioning or pleasure. Rather, the goal of feminizing genitoplasty is to create "normal-looking" female anatomy. Despite this goal, normal female anatomy has been scarcely defined.[12] As Kessler has noted, an examination of any of the current sketches or photos of female genitalia in clinical writings on intersex conditions leaves the impression that the clitoris is merely a "nub," "button," or "pea," located above the urethra (1998). Such descriptions refer only to the outwardly visible portion of the clitoris—the glans (rounded head) and the hood—which comprises only one-tenth the volume of the entire clitoris (Dickinson 1949; Sherfey 1972;

O'Connell et al. 1998). Beneath the skin of the vulva are numerous hidden structures, including erectile tissue, nerves (the highest concentration of any part of the body and twice that of the penis), glands, and muscles which are rarely depicted (surgical articles, on occasion, showcase a completely extirpated clitoris); these have received scant attention in the literature.[13] The result, as Dr. N says, is that "you base your reconstructive surgery on what you think is normal."

THE CAPACIOUS VAGINA

For an infant with an intersex diagnosis who is assigned female, surgery will often be recommended to create, enlarge, or elongate a vagina. Over the years of debate, vaginal surgeries have garnered less critical attention than clitoral reduction surgery. This may be because clitoral reduction is more common, although that possibility is hard to confirm because no national statistics are collected. It is more likely that clitoral surgery causes more controversy because vaginal surgery is often deemed more necessary from a cultural standpoint. That is, a large clitoris may threaten femininity, but the lack of a vagina threatens to undermine an individual's status as a female. As one mother I interviewed said: "If you have a female child born with fused labia, that is a medically necessary surgery because, obviously, the child is going to have to have a vagina to be female" (Louise Rutherford).

Depending on the configuration of the genitals, vaginoplasty may involve creating an opening to the vagina, lengthening the vaginal canal, or creating a new vagina. Just as there are multiple procedures for clitoroplasty, there are numerous techniques for the surgical construction or alteration of the vagina.[14] The procedure used depends on the individual's anatomical configuration and the surgeon's preference.[15] Vaginoplasty procedures may involve skin grafts from another part of the person's body, but the tissue selected must not contain hair follicles: suitable areas include the upper rear and inner thigh, buttocks, inner abdomen, inner pelvis, and upper back.

While skin has commonly been used, other tissues are used as well, including human amnion (fetal sac), peritoneum (membrane lining the abdominal cavity), and bladder mucosal grafts, which are thought to reduce complications such as pain, infection, and scarring. A common complication of this method is that the body may reject the skin (or other tissue) or develop abnormal openings (fistulas) between the rectum and the vagina or between the urethra and the vagina. Skin-flap techniques may also lead to more scarring and inadequate lubrication, and multiple surgeries at the vaginal opening may numb the tissue.[16]

The technique used reflects the surgeon's skills, training, and experience. Skin-flap vaginoplasty was the predominant technique performed on individuals now over twenty years of age and is still favored by gynecologists. Skin grafts do not provide a moist vagina and tend to scar down over time, leading some surgeons to prefer other techniques. Pediatric surgeons and urologists tend to favor using the colon or bowel, which they feel is better lubricated and more pliable, enabling the vagina to "grow" with the individual and producing less stricture than skin. (Gynecologists do not have the training to perform reconstruction using bowel tissues.) However, using bowel tissue also incurs the risk of excessive mucous discharge, the sensations of bowel obstruction during arousal and penetration, and occasionally an unpleasant smell. Some clinicians feel vaginas made with colon tissue require less dilation to keep them open. Consequently, this surgery minimizes the thorny issue of performing dilation in young children, but it is more invasive, requiring laparotomy and major bowel surgery. Whether a surgeon chooses to use bowel or skin graft, depends on the depth of the vaginal cavity they want to reconstruct and the availability of other tissues. Some techniques being explored use grafts obtained from inside the mouth —a popular technique for reconstructing a scarred urethra in males—or engineered tissue for vaginal reconstruction.[17] Finally, there are methods that involve continuous pressure on the vagina achieved through surgical techniques (Vecchietti 1965, 1979; Fedele et al. 1994; Veronikis, McClure, and Nichols 1997).[18]

Unlike clitoroplasty, vaginal surgery often cannot be done in a single procedure in infancy (called a one-stage repair) because of the way the genitals heal (see, e.g., Alizai et al. 1999). The most common complication of healing is stenosis, a shrinking of the vaginal opening that requires either repeated dilation or further surgery. Other complications may also require additional surgery. These follow-up operations are most often done just before or at puberty, or later. Money's hope that early surgical intervention would leave the child with no memory of the event is rarely fulfilled with vaginoplasty because of the frequency of surgery at adolescence. As a result, the timing of vaginoplasty has proven controversial, with some arguing that all vaginal surgery should be delayed until the child can participate in the decision-making process. Members of the Androgen Insensitivity Syndrome Support Group (AISSG), for example, "totally decry the practice" of performing vaginoplasty in infants and children (www.aissg.org/33—surgery .htm). They feel that better results are obtained via pressure dilation (whether with a penis or vibrator) by motivated adolescents and adults, so that surgical vaginoplasty should be considered only after other techniques such as pressure dilation have been exhausted, after the tissues have stopped growing, and when the patient can have a say in the procedure.

Why does a five-month-old need a vagina? It may be helpful to ask first what the vagina is for. The answer depends on who one asks, as well as on the underlying diagnosis. Vaginoplasty in women with CAH, for example, will allow for menstruation and reproduction, as well as for vaginal penetration and intercourse. In individuals with a 46,XY karyotype (e.g., those with AIS or microphallus), vaginoplasty serves no reproductive purpose but can enable penetration and intercourse. Virtually all surgical articles state that vaginoplasty is needed to allow heterosexual intercourse, and a capacious vagina is widely regarded as an ideal surgical outcome. One of the surgeons I interviewed makes a representative comment on successful vaginoplasty: "It's how soft it is. A four-centimeter vagina that stretches versus a four-centimeter vagina that's rigid, that's a big difference. It's not absolute size; it's capaciousness, the elasticity of the tissue for intercourse. I like [using bowel segment to create a vagina] because the sigmoid's right there and you plop it in, and it's capacious enough, so even if it shrinks a little bit, it'll be adequate for intercourse" (Dr. K).

Many parents I interviewed share this surgeon's view that the vagina must be capacious enough to receive a male penis. Justine Good, for example, says, "Our daughter will need [genital] surgery to separate her labia and open her vagina eventually to have intercourse." And Sara Finney says that her daughter's "vagina was long, so that she didn't need any reconstructive surgery." One woman with CAIS notes the widely perceived importance of a vagina for intercourse: "I think the thinking is that if part B will not fit in part A, you're never going to get to the point where you're willing to have a sexual experience if you feel fundamentally inadequate to do what is, for most heterosexual couples, *the* sexual act" (Rosalind Bittle; original emphasis). Surgery is only one option for expanding the vagina, however, and some attest it is not the best one.

By analyzing the practices surrounding vaginal surgery, I am not suggesting that the desire for a vagina for intercourse is unimportant or frivolous, nor am I belittling any individual's desire for a vagina for intercourse or any other purpose. Rather, I am trying to illustrate the way in which contemporary surgical practices render female sexual pleasure as secondary to male sexual pleasure. The clitoris, overwhelmingly tied to female sexual pleasure, is reduced, whereas the vagina, more central to male sexual pleasure, is enlarged. A vagina perceived as abnormal or inadequate—and the consequent preoccupation with creating a capacious vagina, often in childhood—locates the problem in the female's genitalia, not in conceptualizations of what counts as sex. Penile-vaginal intercourse undoubtedly carries a great deal of cultural weight, but studies of current sexual practices, not to mention lay observation, tell us that people enjoy many other forms of sexual activity in addition or instead. One surgeon acknowledges this fact: "How does one become intimate with another person? It's not just with the penis and the vagina. There is much more to a relationship than that" (Dr. J). Nevertheless, early vaginoplasty seems generally accepted as necessary for female assignment in children with intersex diagnoses. The Web site for one department of pediatric urology explains that early vaginal and clitoral surgery is recommended "because female patients are able to undergo a more natural psychological and sexual development when they have a normal appearing vagina."[19] This comment, however, is tied more to commonsense cultural ideas about what females *are* than to any in-depth analysis of what females *need* to flourish.

Rationalizing Early Genital Surgery

When Money first laid out his treatment paradigm, he suggested genital surgery be performed early to enable the child to develop a gender identity in accord with the assigned gender and to remove any parental doubt about the sex of the child. Very few clinicians I spoke with gave reasons for genital surgery that echoed Money's rationale. Dr. P, a pediatric urologist, says: "It's my belief, which is a theory of John Money's, that it's necessary for the child to have a proper sense of oneself in either gender and that the genitalia be appropriate to that child's gender." Two other physicians advocated early genital surgery to remove parental doubt as to the sex of their child. Dr. D wonders if "with an enlarged clitoris, if the parents are viewing it and thinking it looks too much like a phallus, will they be able to raise the child as a girl when they're seeing this and getting confused?" Dr. L concurs: "I have always been very much in favor of early surgical repair. I think it creates a good deal of 'cognitive dissonance' to think 'this is your little girl,' but every time you open the diaper you look down there and say, 'Wow, this is not right.' It creates a lot of confusion in the family." But other clinicians' retreat from Money's rationale for early genital surgery does not mean that they no longer advocate the practice, but rather that they give other reasons for doing so. For some clinicians, early genital surgery is gender assignment. Conflating genital surgery and gender assignment, some clinicians feel a child with atypical genitals will be left in "sexual limbo," as one surgeon has said (qtd. in Guttman 2000). Another surgeon expresses a similar concern, "The child wouldn't really be assigned as male or female at birth if you're delaying surgery" (Dr. M). By making the genitalia conform to one sex, he is granting the child a gender. This reasoning reflects a kind of genital determinism, which rests on the assumption that both genitals and gender identity are naturally dimorphic, and that the former is the essential marker of the latter. Interestingly, however, some of these same clinicians believe that atypical genitals, for example an enlarged clitoris, would not inhibit the development of a female gender identity. Dr. N says, "Let's say a girl with an enlarged clitoris ends up not having surgery. I don't think that could be problematic in terms of her developing a gender identity as female." His colleague observes, "I think a female who has a clitoris, who has masturbated, who has had clitoral orgasms, would say whether you have a medium, small, large, short, wide clitoris, makes no difference in terms of your identity, so why do a surgery that has the possibility of, in some way, diminishing sensation or causing problems? I can see that because women relate differently to it than males" (Dr. O). These two clinicians express a more fluid understanding of the relationship between genitals and gender. The second surgeon, Dr. O, touches on an issue few other clinicians mentioned: perhaps males, who have performed most of these surgeries, have undervalued the sexual benefits of a larger clitoris. Nevertheless, both of these surgeons recommend early clitoral surgery (I discuss the reasons why below).

Today, both surgeons and pediatric endocrinologists frequently give a host of practical and medico-scientific reasons for performing genital surgery early rather than later in life (often before one year of age) based on the belief that results in children are much better than those in adults. Yet these arguments do not provide a rationale for doing surgery in the first place. Dr. M, for example, says: "I think our chances of doing a better job, surgically, are enhanced by the fact that we know that the healing process is much, much better the younger you do the surgery." Dr. O says that "children heal better, scar less, contract less," and another surgeon observes, "You'll find that pediatric urologists find that operating on a small baby is easier, much more aesthetically appealing, and you can get these children looking phenotypically as close to normal as possible at one time" (Dr. N).

For surgeons the mark of surgical success and outcome is deeply materialist, limited to factors such as healing time and aesthetic results. These medico-scientific elements are understandably important to surgeons because they affect the surgical outcome, but they do not address the effect of surgery per se on the individual's life and well-being. This broader framework addresses a different concern: not whether surgical results are better or worse, but to what degree surgery itself is helpful for those undergoing it. Yet very few consider the distinction between what *can* be done and what *should* be done; that is, even if we could guarantee "good" results, that assurance would not answer the question of whether a surgery should be performed.

A second set of arguments concerns the belief that surgery has improved dramatically in recent years. Surgical results may have been poor in the past, the reasoning goes, but since surgical techniques have improved, it is reasonable to expect that surgical outcomes will be better as well. Some surgeons believe that surgery is better because the specialist performing the procedures has changed; that pediatric urologists, who perform many of these surgeries today, are more sensitive to the need for preserving sexual sensation than other specialists were. Dr. N comments, "Pediatric urologists pioneered new and better techniques for clitoral surgery. [We] were not looking at this as, 'Well, let's deal with this problem and cut the damned thing off; it's in the way, and they're not going to need it anyway.' They were saying, 'We're reconstructive surgeons. Let's refine clitoral surgery so the results become uniformly good.' " He also cites advances in technical instrumentation, including more delicate surgical instruments and better optical magnifiers and sutures, as a reason surgery has improved in recent years. Although these advances have undoubtedly improved the surgeons' ability to perform procedures, and perhaps surgical outcomes as well, no evidence suggests that these shifts translate directly to improvements in the quality of life for those undergoing surgery.

As noted earlier, surgeons often point to the cessation of clitorectomy as a reason that surgery has improved. Dr. O says, "Surgery is certainly better, now. We don't remove the clitoris anymore." Echoing him, Dr. H observes, "If you go back just a bit, the resection of the clitoris would be considered a reasonable way to go, because what do you need it for anyway? We don't do that now." Dr. N adds, "Now when we do these operations, we go right in and we keep all of this stuff; we just take out the corporal bodies. So this whole thing [pointing to an anatomical drawing of the clitoris] is preserved and with that, the clitoris is almost luxuriant."

Clitoral reduction procedures designed to retain the more sexually sensitive parts of the clitoris would certainly appear an improvement over the complete removal of the organ. Many clinicians, however, erroneously read refinements in surgical procedures as ensuring the preservation of clitoral sensation. Although surgeons have incorporated an increased anatomical understanding into their techniques (e.g., Baskin et al. 1999), which they hypothesize *should* preserve sexual sensation and orgasmic capability, available data do not support the claim that current surgical techniques preserve sensation. Moreover, follow-up studies done today on adults would assess techniques that the surgeons feel are obsolete. Surgery, then, will always be one step ahead of anyone's knowledge of its efficacy, and the argument of obsolescence provides surgeons with a powerful argument to discount all activist charges and findings of poor outcomes now and in the future.

Clinicians also provide psychosocial reasons for early surgery such as trying to allay societal and parental discomfort, decrease the risk of a child's memories of traumatic surgery, and ease the transition to adulthood. Most clinicians link atypical genitals not to parental doubt about the sex of their child, but rather to the inability of parents and society to handle atypical genitals:

> Now a large percentage of women work, caregivers take care of the family, and the mother says, "I can't take my kid home looking like this, having nonmembers of the family changing this kid's diapers." There's much more pressure on doing this operation younger, not because we want to. (Dr. N)

> I have a preset concept that virilized females with congenital adrenal hyperplasia, ought to be fixed, surgically, early on. I think that gives the *family* the best chance to be normal. (Dr. K; original emphasis)

> There are few times where the parents have the "understanding" to be able to deal with it. That's the ideal world. No question about it. But the parents I meet cannot handle this. They want their child to be normal. If we don't do the surgery, I get this impression that we're sending them out into a field full of mines. (Dr. I)

One surgeon I interviewed says that many parents who initially refused clitoral surgery for their daughter changed their minds because of social pressures:

> When a nanny changes the child's diaper and sees that great big phallus, they're going to say, "This is a girl?" I've had that happen in my practice. I've had them go away and come back eight months later and say, "Aunt Sarah changed the baby's diaper the other day and she saw this and said, 'Are you sure this is a girl?' " This kind of thing, and they're right back in

the office. A significant proportion come back for clitoral surgery, and it usually is a traumatic event like that. (Dr. P)

Even a clinician who adamantly opposes genital surgery for males nonetheless stresses surgery for an enlarged clitoris to allay parental anxiety: "If they have a small, a mild clitoral enlargement, moderate clitoral enlargement, and if mom and dad want this, if it's important to them that it be corrected, I would recommend it" (Dr. R).

One problem with performing surgery to allay parental anxiety is that it is unlikely to work: the surgery cannot destigmatize the condition; it can only reshape the genitals. Clinicians and parents often assume that surgery will be a panacea for a variety of concerns—such as fears about homosexuality or a girl's XY chromosomal type—that have very little to do with the child's genital configuration and thus persist after genital surgery. Moreover, implicit in surgical decisions is a series of often unstated and unexamined assumptions by parents that surgeons too often do not dispel. The most important one is that surgery will create "normal-looking" genitals. What counts as normal, however, is subjective and rarely articulated. Moreover, it is elusive. The surgeon Jeffrey Marsh has noted that surgeons tend to say they will "correct" a certain feature when what they should really say is that they will *partially* correct it: "It is rarely the case that the operation is so perfect that there is no residue of the original difference" (Marsh 2006: 122). Indeed, this "conceit of cure," as Abraham Verghese has called it, can make surgeons feel that medicine can provide an answer to most things, even to social problems (Verghese 2006).

The surgeons I interviewed also recommend early genital surgery in hopes that the child will not remember the surgery as an adult. A typical comment is, "I like to do it when they're little, because they don't remember" (Dr. P). Clinicians and parents understandably want to spare the child any emotional distress associated with surgery. However, this view assumes that the emotional overlay of undergoing genital surgery in adolescence or later is worse than learning later in life that surgery was performed in infancy, before the individual could participate in the decision. It also fails to take into account the frequency of repeat surgeries, most of which occur later in life when the child will not only be aware of what is happening but will also require an explanation. Some clinicians (and parents) hope that by performing surgery early, there is a chance that the child may never find out about it. It seems a stretch, however, to assume that individuals will not find out—through self-examination, the inspection of others' genitals, an awareness of their own complications, a need for repeat surgeries, clinical records, or conversations with family members.

In response to criticisms of early genital surgery, clinicians suggest using hormone treatment to slow genital growth to see if a child with a larger clitoris will "grow into it." Dr. B explains this philosophy: "If you manage the kid well, the androgens come down and the kid continues to grow, but the clitoris only grows at a normal rate, so the discrepancy between the size of the kid's phallus relative to the kid's body size diminishes. And three months later, it's not that big" (Dr. B). Hormonal management gives pediatric endocrinologists a way to defer surgery, which may prove helpful especially if they disagree with a surgeon advocating early surgery. Yet although hormonal treatment has kept the clitoris in some individuals at what clinicians or parents feel is an acceptable size, if they deem a clitoris too large after treatment, clinicians or parents may still opt for surgery. Underlying this recommendation is the notion that some clitorises are just too big to remain on female bodies. One pediatric urologist who says he is "happy to see if the kid can grow into [her clitoris]," quickly adding, "but you don't understand the kind of patients we see. We see congenital adrenal hyperplasias who have a five-inch, six-inch long phallus in their abdomen. You can't leave that!" (Dr. P). It's safe to assume that Dr. P misspoke and meant to say five to six centimeters (2–2.5 inches). The slip is instructional in that it shows how large the clitoris looms in his imagination and likely in that of others.

Other clinicians argue that vaginoplasty should be done when the child is older, if at all, pointing to the frequent need for repeat surgeries if the procedure is done on infants. Claude Migeon, a prominent pediatric endocrinologist at Johns Hopkins, has said that vaginal operations on young girls leave them with "scar tissue from surgery. They experience difficult penetration. These girls suffer" (qtd. in Hendricks 1993). In some instances, surgeons suggest trying pressure dilation instead of surgery. Dr. J subscribes to this view, rejecting the technical reasons adduced for early surgery:

> The argument for early [vaginal] surgery is a fallacy. There is no such thing as one perfect operation and the child will never know. That's never going to happen. It's a big operation, and we're really creating an internal organ, using other tissues that are not meant to be in the genital tract. That's the rationale for doing it early—do the hard part early, so when the [patient] is an adult it'll be easier. But when you create an artificial opening with an extension internally, because it's manmade it will not stay open unless you do dilation regularly. If the child is five, they're not sexual

> beings yet, so they have no clue why they're being dilated. It hurts, and it is a terrible process for a child and the family. (Dr. J)

One recent study shows that women with complete AIS who did not undergo any surgery, nevertheless, on average had vaginal depths greater than in those who did, leading the authors to conclude that surgery is "by no means an absolute certainty" (Purves et al. 2007). Adolescents and adults will arguably be more motivated to dilate, thus increasing the likelihood that the vaginal opening or introitus will stay open. In addition, adolescents and adults can be told about the procedure and give informed consent—or not. Finally, waiting to perform surgery until adolescence enables the individual to take advantage of improvements in techniques made over those intervening years. Dr. J recently explained his rationale for advising a mother to delay vaginal surgery for her daughter, who is genetically XY: "You don't do anything that you're going to regret. You can always do surgery later. I told the mother, 'If the child grows up, and she's happy being a girl, we can do the surgery then. The difference is she is a part of the decision.' There's a whole magnitude of difference to dilate to a child who has no idea what's going on versus an adult who understands what's going on and making a decision for herself or himself" (Dr. J). Although his caution stems from his concern that this child may grow up to identify as a male, the advice seems apt when considering any vaginal surgery.

Surgical Results

Most professions—including, increasingly, the medical one—regularly assess the effectiveness of their practices in achieving predefined goals and refine their techniques based on these results. Despite its prevalence over the past five decades, genital surgery has been subjected to comparatively little systematic review. One reason why surgeons and others have been slow to evaluate surgery is that they have assumed both the necessity of treatment and the well-being of their patients. But another reason is the difficulty of carrying out outcome studies. The number of individuals with intersex diagnoses is relatively small, the cases are highly variable, surgical techniques have changed over the years, the population remains partially hidden, and it is difficult to keep track of patients from infancy into adulthood. Moreover, some surgeons have not felt comfortable approaching their patients because they have not wanted to make them uncomfortable either by "digging up the past," forcing them to a recall part of their life that may be emotionally fraught, or subjecting their genitals and sexual behavior to scrutiny.[20] Dr. K touches on this difficulty: "I want to test sexual and vaginal function in some of the ladies, and others, who I've made vaginas for over the years. If Jane had pelvic sarcoma and I made her a vagina, I feel very comfortable asking her to come in for sexual testing, blood flow, neurological response, clitoral response, and all the rest. I have *no* problem with that. But for me to delve in and ask Mary, who's mixed gonadal dysgenesis, raised female, but had testicular tissue, I feel uncomfortable asking her that" (Dr. K; original emphasis).

Why would it be easier to ask the individual without an intersex condition about her surgical outcome? One reason may be that the surgeon felt that individual had a more difficult life, thus making him understandably hesitant to probe what he imagines will be painful experience and memories. Another possibility is that the stigma associated with intersexuality makes him feel uncomfortable, and he assumes the individual has similar feelings. But it is also possible that the conversations feel different to him because he sees these individuals (and their vaginas) differently. The woman with pelvic sarcoma is an adult who presents with a horrible disease, who turns to him for help, and who may very well feel grateful that he has restored any function at risk from (or damaged by) the sarcoma. She was also born a physiologically typical female. His discomfort asking the woman with gonadal dysgenesis about her surgery may result from the fact he has difficulty thinking of her as a woman, which he hints at with his use of the term "others" to refer to individuals with intersex conditions for whom he has "made vaginas" who were not "born" female. Vaginal surgery in these individuals is not aimed at restoring function, but at erasing signs of gender ambiguity, placing the focus on how the genitals appear more than how they feel and function.

The net result for the treatment of those born with genital anomalies is that there is scant proof whether treatment intended to improve the psychosocial adjustment of the child fulfills this promise. Individual physicians continue to base their treatment recommendations and practices on their own subjective judgments. Moreover, their criteria for the success of the treatment often pay less attention to the psychological well-being of the patient than to cosmetic appearance, imprecise or misleading measures of sensation, and the efficiency and advancement of surgical techniques. As chapter 6 will show, this approach overlooks the considerable anguish and distress expressed by some individuals who underwent genital surgery and assumes that the silence of others implies satisfaction with the outcome.

CLINICAL STUDIES

Although surgeons have published numerous reports on operative techniques to reduce clitoral size, they have published relatively few corresponding studies looking at the outcomes of these surgical techniques. Existing outcome studies of genital surgeries are limited for several reasons, a number of which apply to other types of surgery as well. Studies have often utilized small sample sizes and surgeons' assessment of their own work, raising significant concerns about observer bias. Other factors like age at first surgery, age at evaluation (from immediately post-op to twenty or more years after surgery), the surgical method used, criteria for evaluation or success, and modes of assessing sexual sensation all vary enormously in these studies, making surgical outcomes challenging to assess and generalizations from these findings difficult to formulate. Nevertheless, the available outcome studies of reduction clitoroplasty and vaginoplasty paint an incomplete and often dismal picture of early genital surgeries.[21]

In the roughly twenty studies evaluating clitoral surgery available at the time debates began heating up, criteria for a positive outcome, when specified, are highly variable.[22] Some studies have none (e.g., Randolph and Hung 1970; Rajfer, Ehrlich, and Goodwin 1982; Joseph 1997), calling into question whether they can properly be considered outcome studies. Others purport to evaluate the cosmetic appearance of the genitalia, but do not specify what the ideal appearance should be (e.g., Sharp et al. 1987). Some studies seek to evaluate the preservation of clitoral function, but provide no criteria for assessment (Kumar et al. 1974). One study uses a "bi-electrode stimulator" (Allen, Hardy, and Churchhill 1982), while another study relies on self-reported orgasm to assess sexual function (claiming that "all reported orgasm"), but says nothing regarding the quality of and conditions necessary for orgasm (Costa et al. 1997). The language employed in these studies remains vague: for example, clitoroplasty "gave very good cosmetic results . . . and quite satisfactory functional results" (Pinter and Kosztolanyi 1990: 115). The results vary with some surgeons' determining that patients have a "pleasing cosmetic result" (Bellinger 1993: 652), while others simply note that results are "cosmetically satisfactory" (Mininberg 1982: 355). In one study, the authors report "disappointing results, even in the hands of specialists": the anatomical outcome was "unsatisfactory" in almost half of the study sample and "satisfactory" in the remaining girls (Alizai et al. 1999: 1588).[23] (They did not collect any information on sexual sensitivity, likely owing to the girls' young age.) In another study, the results were stated simply as "satisfactory," but the criteria for success remained unspecified, and four of the six women evaluated required a second surgery (Oesterling, Gearhart, and Jeffs 1987).

The most common complication of clitoral surgery is scar tissue, which may result after a single surgery (e.g., Randolph, Hung, and Rathlev 1981). Some women require additional surgery (Allen, Hardy, and Churchill 1982; Randolph, Hung, and Rathlev 1981; Van der Kamp et al. 1992), which may or may not be read by the clinicians as a complication and will likely result in more scar tissue and decreased sensitivity. Attempts to assess sensation or sexual pleasure are often absent, imprecise, and variable; the studies tend to focus on cosmetic results and rarely use patient's subjective assessments. Yet in some studies women reported painful erections after clitoral recession (Allen, Hardy, and Churchill 1982) and "painful stumps" after clitorectomy (e.g., Nihoul-Fékété 1981), which likely contributed to surgeons' abandoning these techniques.

A closer look at one report highlights the problems inherent in many outcome studies. In 1995, the pediatric urologist John P. Gearhart and his colleagues published an assessment of postsurgical clitoral sensation by measuring the conduction of nerves (using an electrical impulse) in the clitorises of six infants (Gearhart, Burnett, and Owen 1995). Concluding that nerve conduction was preserved in five of the infants, they proclaimed that this finding "clearly shows that modern techniques of genital reconstruction allow for preservation of nerve conduction in the dorsal neurovascular bundle and *may* permit normal sexual function in adulthood" (486; my emphasis).

Although the researchers are careful not to overstate their findings by arguing that sexual sensation is preserved, they nevertheless assume that innervation is tantamount to erotic sensation. Nerve conduction is a low bar; given the dense innervation of the entire area, some nerves may be damaged, thus causing pain when the woman is sexually aroused or touched. Moreover, these measurements were taken during surgery, before the wound was sutured, and so fail to account for any loss of sensation caused by skin scarring. Because the patients are infants, no data are available regarding the individuals' sexual sensation or sexual pleasure as adults. When I asked Dr. K whether he took sexual pleasure into account in surgical decisions for his patients, he responded: "Of course. Studies show they have nerve conductivity, so I feel confident they have sensation." Others have sought to document blood flow as a proximate determinant of sensation, but like nerve conductivity, this may or may not be correlated to one's subjective sense of sexual pleasure.

By the early 2000s, roughly thirty scientific studies had evaluated different types of vaginal surgery.[24] These studies are plagued by the same methodological difficulties as those for clitoral surgery, including variable surgical procedures, sample sizes, and ages at which the surgery was performed and evaluated, as well as differing criteria for evaluation, leading one pediatric urologist to refer to them as "show and tell" studies. Overall, follow-up studies of vaginoplasties have larger sample sizes than those of clitoroplasty. Some studies have as many as eighty participants (e.g., Mulaikal, Migeon, and Rock. 1987). Several have as few as two (Di Benedetto et al. 1997; Sakurai et al. 2000). Most, however, have between twenty and forty subjects (e.g., Azziz et al. 1986; Allen, Hardy, and Churchill. 1982; Fliegner 1996) and in total assess over one thousand women.[25]

The criteria used to evaluate outcomes for vaginoplasty vary. They include comfortable vaginal penetration, the frequency of coital activity, successful vaginal penetration, or penetration without pain or bleeding (Azziz et al. 1986; Hecker and McGuire 1977; Bailez et al. 1992; Costa et al. 1997). Several even use marital status as an indicator of successful outcome (Hensle and Reiley 1998; Mulaikal, Migeon, and Rock 1987). Another uses anatomical criteria, such as the position of the vaginal opening (Nihoul-Fékété 1981). Several studies provide no detailed evaluation criteria (Hendren and Crawford 1969; Allen, Hardy, and Churchill. 1982; Newman, Randolph, and Anderson 1992; Sotiropoulos et al. 1976). Very few studies explore psychological issues as an indicator of surgical success (e.g., Fliegner 1996).

What do these studies say about the success of vaginoplasty? If one takes coital activity as the marker of success (presumably the more frequent, the better the outcome), then fifteen women with frequent activity (defined as once per day to twice per week) had successful vaginoplasties out of twenty-three women total (Hecker and McGuire 1977). In another study of thirty-eight women, only twenty-three women were sexually active, and of those, only eighteen had "satisfactory intercourse" (Fliegner 1996). Another study found that of twenty-three women, seven had no sexual activity,[26] and of those who had no dilation after surgery (eight women), only half had "satisfactory intercourse," which was not defined. Seven of the eight who had postsurgery dilation had satisfactory intercourse, but the study gives no information about the women's feelings and experiences with dilation (Costa et al. 1997). In one study of eighty individuals, over half had an introitus that the researchers deemed "inadequate" and "disappointing" (Mulaikal, Migeon, and Rock 1987).

Although many of these studies evaluate the frequency of sexual activity, they do not evaluate its quality, for example by asking whether the women find their sexual experiences pleasurable. Moreover, the studies tell us very little about the sexual activities in which the participants engage. Finally, they provide us with almost no information about how these women feel about the surgery itself and their genitals, and how these feelings affect their ability to be sexually intimate.

A few studies have used the need for repeat surgeries as a criterion for surgical outcome. In one study conducted at Johns Hopkins, the researchers assessed the outcomes of vaginoplasty in twenty-eight women; of these, only six women needed no further surgery, and the rest had three or four surgeries each (Bailez et al. 1992). Another study (Creighton, Minto, and Steel 2001) found that of forty-four adolescent patients who underwent feminizing genital surgery, forty-three needed further surgery. Two of the authors, the gynecologists Sarah Creighton and her colleague Catherine Minto from University College London Hospitals, found these results so poor that they wrote an editorial arguing that most vaginal surgery should be deferred until adolescence (Creighton and Minto 2001a).

Why are the results for vaginoplasty so dismal? Dense scarring and the closing of the vaginal opening (stenosis) are common complications (e.g., Allen, Hardy, and Churchill 1982; Fausto-Sterling 2000: 85). Another study lists additional complications such as chronic pain during intercourse, excessive vaginal secretions, and a total closure of the vagina (Hensle and Reiley 1998).

In 2001, amid the debates over genital surgery, and recognizing the shortcomings of available data, Creighton and Minto began publishing outcome studies on genital surgery, along with other authors. This group, which includes the psychologist Lih-Mei Liao and the gynecologist Naomi Crouch, has been notable for its sustained attention to the issue. Taken as a whole, their studies paint a dismal picture of the cosmetic, functional, and anatomical outcomes for clitoral and vaginal surgeries. Among their findings are: that clitoral surgery can damage adult sexual function; that most children undergoing feminizing surgery require further treatment in puberty; that sexual dysfunction is common (Creighton and Minto 2001b; Creighton, Minto, and Steele 2001a, 2001b; Creighton, Minto, and Woodhouse 2001; Minto et al. 2003a; Minto et al. 2003b; Creighton 2004; Crouch et al. 2004). Reading their studies and editorials one gets the strong impression that there is little, if anything, to be gained from performing genital surgery on children.

Perhaps not surprisingly, their studies have received mixed reactions. Some adults who underwent surgery as children have been thrilled with their work. One group, Intersex Initiative, created a "fan page" for the researchers summarizing their findings, and ISNA created a similar Web page, proclaiming, "UK Researchers: a knack for asking good questions." Some surgeons, however, have been less enthusiastic. Several clinicians I spoke to are generally dismissive of their findings, arguing that the poor results are due to British surgeons' inferior surgical technique and capabilities. Some surgeons feel that the researchers are "antisurgery" and that their studies are thus biased. In an editorial accompanying one of their studies, a clinician suggests an alternative reason for the poor results: women's sexual functioning could be hindered by shyness or other problems related to the gender assignment and rearing (Slijper 2003).[27] Echoing an oft-repeated argument, the pediatric urologist John P. Gearhart suggested that the poor findings might be moot as "most of these patients were operated on years ago by children's surgeons" (Frimberger and Gearhart 2005: 294).

It is hard to overlook the gendered aspect of these debates: many highlighting poor surgical results and expressing the need to rethink early surgery are women, whereas the most vociferous defenders of early feminizing genital surgery are often men. In some ways, this is not surprising as the majority of those who have undergone surgery (hypospadias notwithstanding) are women, whereas men continue to dominate the surgical fields most likely to deal with intersex cases. Nevertheless, this demographic fact fails to address how male surgeons' status as men and surgeons (and the authority that accompanies that positioning) obfuscates the way gender threads through these debates. For example, surgeons counter women's complaints about poor surgical outcomes with charges of a lack of objectivity and representativeness (see chapter 8). Clinicians often make the same charge against female researchers investigating surgical outcomes who have found poor results. Cloaked in the rhetoric of scientific argument, surgeons discount women's complaints (or research) as biased or unscientific, sidestepping fundamental issues at the center of these debates. In this case, these are questions centering on what it means to be normatively human: Are atypical genitals a medical condition? What are genitals for? Who should make these determinations? Biomedicine has significant discursive and practical power in making these determinations, and in rejecting alternative views, but a central impetus for these debates is that many have argued that what constitutes the normal is a social, not a scientific, question.

One of the problems with many outcome studies is that they have not included any subjective assessment from patients. This deficiency results in part from a hierarchy of acceptable forms of evidence that prioritizes clinical judgment and can often exclude subjective perceptions of the actual recipients of care. A failure to include patient assessment often leads adults with intersex conditions to dismiss a study's (rosier) findings and to argue for the inclusion of their personal experiences, which many willingly share with the media. Clinicians often dismiss these narratives as anecdotal and unrepresentative of the majority of patients, who they assume are doing well. In this context, patient self-report and experiences often do not constitute appropriate evidence. Adults with intersex diagnoses and others who challenge the view of genital surgeries as successful are confronted with the dominance of the scientific hierarchy. If self-report or experience-based evidence is not gathered by clinicians or published in medical journals, clinicians do not count it as evidence. Here, then, it is the methods used to collect the evidence, not the evidence per se, that define whether it is acceptable. Paradoxically, some doctors see their own report in the clinic as acceptable evidence, but not adults' self-report outside the clinic.

Clinicians' underlying concern with patient narratives is that the anecdotal data offered by adults with intersex diagnoses may be biased toward those with poor outcomes, but in the absence of systematically collected data, many clinicians rely on anecdotal or clinical experience to support their own current treatment practices. Often adults' claims and experiences are dismissed as anecdotal, but the truth is that the only available evidence seems to contradict surgeons' and others' belief that early genital surgery both preserves sensation and provides cosmetically appealing and functional outcomes.

As long as these hierarchies of acceptable evidence persist, with clinicians citing scientific studies as the gold standard, on the one hand, and some adults with intersex diagnoses arguing for the legitimacy of their experiences, on the other, there will be no consensus over what counts as credible evidence, the standards and criteria used to decide valid and admissible evidence, and the truth claims that each side derives from these. This is the primary reason why outcome studies are unlikely to resolve these debates.

If one agrees that, at a minimum, patient assessment is critical to assess deeply subjective factors such as sexual and psychosocial well-being—as many adults with intersex diagnoses insist and as any sexuality researcher or psychologist would agree—such studies cannot be conducted until individuals are old enough to participate meaningfully, which means studies would assess techniques and instrumentation almost two decades old. Surgery, then, will always be a few steps ahead of anyone's knowledge of its efficacy.

CLINICIANS' ASSESSMENTS

The clinicians I interviewed often made speculative assessments of their patients' longer-term outcomes. One surgeon notes that he has very little knowledge of how his patients have done: "Their success is fairly good, although I haven't seen these patients at age twenty-five. I'm kind of curious to see how they do through puberty" (Dr. M). Most of the surgeons I interviewed, however, seemed highly pleased with the results of surgical procedures; very few admitted to poor outcomes in their own patients. Dr. N says of clitoral reduction: "The surgical techniques are pretty good at this point. You preserve the neurovascular bundle, you remove some of the erectile tissue and a little bit of the hood. I think that the results are actually quite decent. If it's done at a younger age, I must say that I've been quite happy, and the parents have been very happy with the appearance."

Surgical outcomes can be assessed with regard to sexual sensation and appearance. As do the studies discussed earlier, this surgeon assumes that surgical techniques designed to preserve clitoral nerves will not harm sensation, an assumption that has little empirical support. Read a certain way, this surgeon's comment could be construed as dealing only with cosmetic outcome. Although he says that results have been "decent," he is also discussing results assessed shortly after surgery. A child's genitals, however, will change over time, and what looks good in an infant now may not do so fifteen years on. Later in the interview Dr. N. acknowledges: "Several of my patients' clitorises have atrophied after surgery and shrunk down to something that I thought was unacceptably small. The kids didn't say anything, there were no complaints, but I felt there was not as luxuriant a vasculature. . . . People talk about fabulous results, everybody knows that nobody does as well as everybody says they do. When you get surgeons to talk candidly, they say, 'I still get fistulas, I have strictures, I have shrinkage.' This is one of those things that comes by infrequently, and so much is at stake that you want to make sure it goes right."

Every clinician I interviewed viewed the retention of sensation in the clitoris as important; but how important can they consider it when rigorous studies documenting the preservation of sensation have not been done, and yet surgery presses forward? Of the surgeons I interviewed, all but one said that they believed but could not be certain their patients still had sexual sensation after clitoral reduction surgery:

> We don't know that we're decreasing sensation. If you keep the dorsal neurovascular bundle, we think we're preserving most of it, but at the same time, we don't know. (Dr. O)

> Obviously, we plan surgical techniques around not damaging sensation, but I think the whole idea of sensation . . . I mean, how do you measure sensation, particularly in that area is so very, very difficult. . . . If I felt that we were de-innervating all the clitorises that we operated on, I would be very concerned, but I don't believe that's true. (Dr. L)

> You can't know what sensation will be like later on. We think we know, but we've never proven it. . . . Obviously, our techniques are better, our instrumentation is better, our understanding of the anatomical defects are better, our handling of the tissues is better, but, you know, we haven't proven it. (Dr. N)

One respondent draws a flawed analogy to male anatomy: "We know there are nerves all the way around the penis, and yet we do circumcisions all the time. It's extremely unusual for males to complain about problems with sensitivity from their circumcision. There're millions and millions of men who've had circumcisions, and very few complain of any problems from it, so it does not worry me" (Dr. L).

Despite efforts to preserve nerves, surgeons may occasionally excise them. During a panel on clitoral innervation at a medical meeting held in 2004, a pediatric urologist presented slides of recent clitoral surgeries. In one, which he deemed a "pretty good result," the clitoris was not visible. In another picture of tissue that had been taken from a female infant, he commented, "Here is a situation where the pathologist ended up with the nerves. We'd prefer the patient have them."

Although most clinicians agree that the effects of surgery on clitoral sensation remain largely unknown, it seems highly plausible that surgery can (severely) impair sensation. Clitoral surgery removes erectile tissues that are anatomically intricately connected to clitoral innervation—large bundles of nerves course along the corporeal bodies that enable erection—thus reducing clitoral erection and likely sensation. This may explain why some women who have had external portions of their clitoris amputated report the ability to experience orgasm with the clitoral "stump," while others who have only had erectile tissue removed report no capacity for orgasm.

In 1999, the pediatric urologist Laurence Baskin and his associates published an anatomical study of the clitoris using three-dimensional imaging techniques that offered detailed information about the anatomy of the clitoris (Baskin et al. 1999). The study showed that clitoral nerves extend around almost the entire tunica (a dense connective tissue layer covering the corporal bodies that enables erection), that several bundles innervate the tip of the clitoris (the glans), and that the densest concentration of nerves occurs along the surface of the corporal bodies (the primary erectile bodies of the clitoris). Baskin recommended a modification of current techniques of reduction clitoroplasty to minimize damage to the neurovascular bundle. The findings themselves are of interest; but more interesting still was their reception: in surgical circles, some found Baskin's work passé, while it caused alarm for others.

When I asked one of the top surgeons in the United States whether Baskin's work has caused him to rethink clitoral surgery, he had the following to say: "I *don't* know [about clitoral sensation]. The papers by Baskin are *very* disturbing to me because it has been my long-held presumption and prejudicial view that it [clitoral sensation] was preserved! See, it's not big news that the nerves wrap around, what's big news is his perhaps understated presumption that surgery injures sensory ability later on. Maybe he doesn't come out and say that in writing, but any time he presents he sure says it" (Dr. K; original emphasis). I asked this surgeon and another, both of whom have attempted to assess postsurgical sexual sensation, what they would do if they found that their current techniques led to nerve damage. Both said they would unquestionably develop another surgical procedure, but this response overlooks the other, simpler possibility: not to perform surgery at all.

The split over the necessity of early clitoral surgery was reflected at the 2004 meeting of the American Academy of Pediatrics Section on Urology. During the panel discussion "Implications of Female Genital Innervation," a pediatric urologist discussed the potential problems with clitoral surgery in the areas of cosmetic appearance and functionality, stating that these problems persisted even when the procedure was clearly the wish of the patient. He argued that the "most important" consideration was "how to preserve function," and added: "I think there is a very important question: no one has proved it is a problem to have a large glans or a large clitoris, so should we be doing this?"

Another surgeon, however, defended early clitoral surgery by saying, "We remove just a little erectile tissue, which is different from the past," and showed a short film in which he demonstrated how he "attacks" the clitoris. However, he also stressed that the nerve pathways of the clitoris, labia minora, and surrounding tissues differed markedly from what surgeons had been assuming. Many surgeons have been cutting exactly the places research suggests are the most sensitive and those central to sexual response. He repeatedly stressed the anatomical complexity of the genitals and pointed to the analogous structures of the penis and the clitoris, which raised the question for me and, likely, others in the audience of how sensation could not be harmed by these procedures.

Nevertheless, surgeons tend (as many parents do) to presume that it is enough for the child to have the proper genital configuration for heterosexual intercourse. Many clinicians I interviewed found it difficult to talk about sexual pleasure: the topic almost never came up unless I asked. The few clinicians who discussed pleasure often limited it to a remark about believing that clitoral surgery did not harm sexual sensation. This almost universal lack of attention to sexual pleasure by clinicians struck me as odd, given that the clitoris is the only organ of the body—male or female—which has no known purpose except pleasure.

Judgments of successful surgery tend to emphasize the improved appearance of previously "grotesque" genitals. Dr. K equates progress in surgical techniques with improved cosmetic results: "The way we do surgery today, they have a pretty good cosmetic result. In years past, some of the pictures, the girls wind up with kind of bullet holes. One here and one here, and it was really awful. It was *terrible.* But that sort of thing we don't do anymore!" (original emphasis). Those "bullet holes," however, were likely considered evidence of a good surgical result years ago. Other clinicians I interviewed were less positive about recent progress, often pointing out that clitoral surgery frequently resulted in very small clitorises and was likely to affect sensation, and that surgically created vaginas did not look normal.

Pediatric endocrinologists, who often have less of a vested interest in surgical success, expressed greater doubts about the outcomes of these surgeries. One says: "I saw slides last week that showed several of horrendous surgical outcomes. It was a real shocker. Most of those surgeries were done by a very highly regarded surgeon. These were males and females who, postsurgery, had genitalia that were unrecognizable as relating to anything normal. . . . [We] all came away from that thinking to ourselves, 'My God, it can really be that bad' " (Dr. B).

Dr. D, another pediatric endocrinologist, says he inherited several cases "where I just couldn't find a clitoris and it begged the question, did the clitoral reduction end up being a clitorectomy. . . . I don't think they know." A colleague, Dr. E, adds that posturgery clitorises do not "look normal" to her, adding, "A lot of the time it's a scarred-down blob of tissue that looks awful." Dr. B opines that what he was about to tell me would "seem outrageous, but it is hard for me to think of patients I've taken care of with these problems who've had what I would consider good outcomes. There just aren't very many. The girls who've had surgery? They don't look good at all. I mean, really not good." Dr. G offers an explanation: "The concern I have is that initially it may look like a good result, but both cosmetics and function have the capacity to change over time, and even more so over a longer time. Many normal-appearing structures [clitorises] get bound down over time, and there isn't much there by the time twenty years go by. My question is how do I know that the surgery is good enough, and will be good enough?"

Some surgeons acknowledge similarly poor results with vaginoplasties:

> To be candid with you, if you look at the literature, many vaginoplasties, despite the quality of the people that are doing them, have to be revised. We do pretty good vaginoplasties, they look quite good, but there is scar tissue around the introitus, and the question is, will that genital skin become supple and stretch with time? (Dr. N)

> If you look between their legs, it does not look like a girl at all. All you see is a flat surface with a hole. And then, basically, internally, there is a tubelike extension, which is what we call a vagina. (Dr. J)

Not only do several clinicians acknowledge that a surgically constructed vagina does not look normal but there are also other troubling complications, the most frequent of which is stenosis. Dr. M also notes that using bowel segment to construct the vagina brings the risk of perforation and fistular formation. When I ask Dr. L if most people who have had some kind of vaginoplasty need a procedure again at puberty, he replies: "Nobody knows, but I wouldn't be surprised if 90 percent needed further reconstruction." A recent study by Creighton and her colleagues of subjects who had undergone vaginoplasty, which revealed that 98 percent needed further surgical treatment, would appear to confirm this (Creighton, Minto, and Steele 2001b).

A common tendency among clinicians, particularly among surgeons, is to conflate improvements in medical techniques with improvements in outcome. A procedure that is easier and more efficient for the surgeon is assumed to produce a better result for the patient. Dr. O says, "In the past three years, vaginoplasty has changed tremendously. I wasn't satisfied with how we did vaginoplasties before, because the vagina was inadequate, so you couldn't have normal penetration. Now we can make an adequate vagina." Dr. K concurs: "When I first started it was a horrible job! A lot of very difficult, unsatisfying surgery. Now, it's really fun because you can fix it!" (Dr. K).

When clinicians acknowledge failure, they may ascribe it not to limitations of technique or skill, or to the inadvisability of the procedure, but to the patient. Dr. P tells me: "If you look at the literature, many of the very masculinized congenital adrenals will have to undergo touch-up surgery at puberty to become sexually active. They may have a little stenosis, which is not uncommon, because they're terribly masculinized, they look like boys at birth." This prominent pediatric urologist acknowledges that some individuals will have multiple vaginoplasties; however, in his view, the cause is the excessive masculinization of their genitals, not the practitioner or the procedure. Surgeons frequently echo this view when they remark that a patient was "very masculinized and needed a lot of work." Presumably, patients would need fewer surgeries if they were less masculinized. However, stenosis is a problem in any vaginoplasty, regardless of the degree of masculinization of the infant's genitals.[28]

PARENTS' ASSESSMENTS

Parents of very young girls frequently state that their daughters have excellent postsurgical results; these assessments, too, are based on external appearance often only a few years after the surgery. Nicole Day, whose daughter with CAH had clitoral and vaginal surgery at two and a half months, says, "She was only in the hospital for two days, and she recovered very well at home. She looks very good now. You can barely see any scars." Others have commented:

> At the age of two, Dr. [X] recommended that we have her genitalia surgically repaired. The surgeon was wonderful, and the surgery was very successful. He performed a clitoral reduction and also made sure her bladder and other internal organs were in the correct places. She now looks like a perfect little girl! (Mother posting on a Web site for those with what are called genetic mosaics, a condition in which different cells in the body have different genetic sequences or genotypes)

> My 7-month old daughter just had her surgery done on Monday, and it looks beautiful. She now looks like the little girl that she is supposed to. (Mother posting on an Internet discussion board for parents of children with CAH and adults with the condition)

Yet parents who feel reassured when their daughters come out of surgery looking "fine" or "normal" may be surprised and upset by changes that take place as their children grow. One mother I interviewed said the surgery to reduce her daughter's clitoris and open her fused labia left her with "abnormal-looking genitals." Rebecca Davis, who has two daughters with CAH, says that the surgical procedure to reduce her oldest daughter's clitoris has created "endless problems, abscesses, infection, massive scarring, she has constant pain."

Two mothers I interviewed were disturbed by the results of their daughters' vaginoplasties:

> From the top portion up, she looks like a little girl, but the bottom outer labia looks like scar tissue from a burn. I neglected once to tell a babysitter, and the babysitter called me in a panic. I said, "For the love of God, I forgot to tell you she had reconstructive surgery, and that's how she healed." (Gloria Jackson)

> She has tremendous scarring. They did the surgery at eighteen to try to remove the scar tissue and open the vagina up, because it had pretty much scarred over. See, they wanted her to have sex. But I could've told them at puberty she wasn't going to be heterosexual. (Rebecca Davis)

Medicine is often driven by hopes, desires, and even fantasies of abolishing disease and ameliorating life. Physicians and patients (or their surrogates) often do what technically can be done simply *because* it can be done, sometimes at the expense of any proven benefit because they are caught up in what Mary-Jo Del Vecchio Good (2001) has called the "biotechnical embrace." (Indeed, one mother I interviewed is hoping "to find out more about corrective surgery to fix the corrective surgery.") Supposedly normalizing genital surgery has failed to incite much critical commentary from physicians and the parents of young children despite the degree of intervention, which some have argued transgresses myriad ethical boundaries, because it is the body's transgressions of gender norms that cause even more concern.

Of course there is no guarantee that people on whom surgery is not performed will fare any better, since living with atypical genitals will certainly have challenges as well. But early surgery adds another layer of complexity to the issue because it is irreversible, may cause harm, involves bodily intervention, and because the patient does not participate in the decision. Surgery is not the only or best way to help these children, and in the rush to surgery it is too easy to forget we have other tools at hand to help children and their parents adjust to their condition, and that people with these conditions may require ongoing, lifelong support regardless of surgery.

The current medical protocol for the treatment of intersexuality requires congruence between the child's assigned gender and his or her genitals. It also presumes heterosexual desire and behavior and thus surgically produces genitals believed to facilitate these sexual activities. Once people are fitted with appropriately gendered parts, it is expected that they will use them regularly and heterosexually. One of the more surprising and sad findings in my interviews with adults, recounted in chapter 6, is the degree to which they avoid all forms of intimacy, not just sexual forms. Treatment for intersex diagnoses (which invariably involves genital surgery) may be intended to help these children enjoy a more normal sexual life as an adult, but there are stark contrasts between how clinicians conceptualize "normal" sexuality and how their patients conceptualize and live out their own sexuality.

PART 3

Living the Medicalized Body

SIX

Wanting and Deciding What Is Best:

Parents' Experiences

As previous chapters have shown, intersexuality has been a flashpoint of social controversy about the relationships among sex, sexuality, and gender, as well as the subject of much recent clinical controversy and speculation about what constitutes good treatment. The following two chapters focus on the experiences of those who must live most intimately with the consequences of these social and clinical approaches to the issue: the parents of children with intersex diagnoses and adults who were themselves treated as children and adolescents. I discuss the parents' experiences first not to privilege them over the experiences of their children but because parents' concerns, hopes, fears, and decisions regarding their babies largely shape the conditions of that individual's life. Moreover, it is my sense that parental adaptation to the condition may be the most important factor for determining the child's quality of life; the question remains to what extent encounters with medical caregivers and current treatment foster their adaptation or hinder it.

For the parents of children with intersex diagnoses, all the usual anxieties of child care and rearing are compounded by the diagnosis. They experience not only protective anxieties about how family, friends, and strangers will treat a child who may be perceived as fundamentally different but also conflicting ideas about sex and gender as they are embodied by the child. Although all children experience some health problems during infancy and childhood, for most these remain relatively mild and intermittent and do not interfere with their daily life and development. In the case of some intersex conditions, parents can face the typical pressures and stresses of caring for a child with a chronic medical condition such as frequent medical visits and hospitalizations, home health care, the administration of medication, multiple surgeries, and the attendant pain and discomfort for the child. Parents and child-care providers may often deal with the child's health issues on a daily basis, for example, by administering and keeping records of the child's medications. Much like children with chronic illnesses who may be ill or well at any given time, they are always living with their condition.

At a deeper level, the stigmatized nature of gender-atypical bodies separates intersex conditions from other conditions and may raise fears about how the features of intersexuality may be perceived by others and about the extent to which they should be concealed to protect the child. Although public awareness of intersexuality is relatively new, the fears it invokes about gender difference and deviance from the norm are old and embedded in our culture and social structures. Intersexuality is thus not just a medical issue; it is a complex sociocultural issue that embodies a variety of perceived threats to individual well-being and social stability. Parents may feel tremendous guilt that they did "something wrong" to cause the condition and that their child will suffer as a consequence, but they may also fear that their child's condition may make it harder for them to accept and love their child.[1] These combined pressures, which often meld sadness, phobia, shame, worry, frustration, anger, and fear, create unique stressors for parents: for several parents I interviewed the marital stress associated with the diagnosis led them to seek counseling and, in some cases, divorce.

Because of the emphasis in the traditional protocol on rapid gender assignment and early surgery, many parents come under pressure to make treatment decisions quickly; and, indeed, many are anxious to embark on a course of action that they believe will protect the child from being perceived as freakish or unable to live a "normal" life. In the absence of rigorous long-term studies regarding treatment outcomes for genital surgery, parents face complex moral decisions about what is best for their child. Inextricably tied to ideas about the child's best interest are parents' views about what bodily parts and capabilities are required to be male or female. Parents are thus put in the position of assessing whether their baby is appropriately and sufficiently gendered, effectively making them gatekeepers, along with clinicians, responsible for making irreversible and embodied decisions about the standards of maleness or femaleness. In some instances, parents' dilemmas about care result in an internal debate in which they must weigh what they feel is best for the child against what physicians are recommending or what they feel society will tolerate. For some parents, surgery may appear to offer the remedy to all the physical, emotional, and social difficulties they fear. But because gender assignment and genital surgery can be fraught with problems, they must also confront the possibility that the medical intervention intended to help their children instead may harm them physically and psychologically. Some parents defend their decisions as the best possible choice under very difficult circumstances; others are haunted by guilt and their children's resentment. In interviews and on the Internet discussion boards I participated in, parents frequently expressed anxiety, anger, and guilt about the difficulties of making choices about and for their children.

Parental adjustment and early treatment decisions have the potential to influence dramatically children's quality of life, affecting their emotions, thoughts, and behaviors in everyday life and, perhaps most poignantly, their relationships—those that are public as well as those most intimate. All too often, however, parents are neither encouraged to take the time nor given the support to explore their varied emotions as they arise and before they must make critical and often irreversible medical decisions. Before pursuing remedies that reshape the body in an effort to help children whose quality of life is often believed diminished because of their atypical anatomy, it seems fruitful to examine the context in which treatment decisions are made.

Birth: The Hospital Experience

The birth of an infant with an intersex diagnosis takes place in a well-established context of medically managed, hospital-based childbirth. Doctors exert considerable control over the birth process, thus mooring birth in medical cultures and institutions. One of the most significant consequences of medicalizing intersexuality, both conceptually and structurally, has been the redefinition of atypical bodies from normal and natural, albeit rare, to pathological configurations requiring medico-scientific monitoring and intervention. Beginning at (or even before) a child's birth, intersex diagnoses are subtly—and sometimes not so subtly—framed as abnormal in a way that strongly influences parents' emotions, concerns, and, ultimately, their decisions. Because gender atypicality is medically managed, most parents learn almost from the moment of birth to understand and describe their children's bodies through the language and lens of medicine.

Chapter 4 described some of the anxieties faced by parents of a newborn baby with gender-atypical genitalia or anatomy; these are partly created by the clinicians' own doubts and uncertainties in encountering such anomalies. These doubts, in turn, may arise from a perceived failure of highly technologized medicine to produce the desired outcome: a completely "normal" newborn baby.

Over the past century, birth and women's reproductive capacities have been increasingly configured as requiring management. Construing childbirth as a "potentially pathological" event opens a space for medical science to intervene and assist or even improve on nature (Arney 1982; Franklin 1991). As a consequence, technological interpretations of reproductive events frequently override women's own knowledge and experience of their bodies (Young 1990; Rapp, Heath, and Taussig 2001). In this context, an anomalous birth is framed as a medical "problem" with a solution that emphasizes modifying the individual medically and surgically to improve the child's health and adjustment, but also, ultimately, to fit social norms.

The medicalization of reproduction shapes social relationships and practices surrounding reproductive events. Framing conception, pregnancy, and birth as medical issues requiring expert management has consequences not only for the birth of an infant with an intersex diagnosis but also for the cultural meanings of these events. In some respects, of course, advances in gynecological, obstetric, and neonatal medicine have expanded women's reproductive choices and improved the experiences and health of newborns, including infants with intersex conditions. Clinical interventions have proven critical for managing the life-threatening aspects of CAH, for example. Concomitantly, however, such technologies and interventions can be alienating and can leave little room for parents to understand their newborn outside of the context of the diagnosis.

At the birth of an infant with an intersex diagnosis, physicians and technologies define the birth as a "disorder of sex development." Medical instruments, tests, and technologies objectify the newborn as the physician overwhelmingly determines how parents view and respond to the birth. Construing such a birth as a rare physiological abnormality not only paints the event as one that requires medical intervention but also works to isolate the parents and their child socially by identifying them as anomalous. Shirley Friedman recalls that "for three hours, they left us in the room with no support and family asking, 'What did you have?' and we didn't know *what* to say. It was just very, very shocking, confusing, and very, very alone, because there was no one there for us" (original emphasis).

Although a physician may claim to have treated a number of similar cases, parents of a newborn rarely have the opportunity to meet or seek advice from others who have been through the same experience. Indeed, they may be actively discouraged from doing so. Tina Haywood says that she and her husband "asked [the doctor] if she could put us in touch with other people, and she gave us the same pat answer that the other two pediatric endocrinologists did: 'Other people probably don't want to talk about it.' I almost think that she feels like she has more control over the situation the less knowledge you have, and the less likely you are to question her."

This isolation is exacerbated by the general lack of social, emotional, or psychological support or counseling offered to parents. Rather than helping them through the understandable mourning of the loss of the "perfect child," clinicians instruct parents early on to view their children as abnormal or pathological and requiring intervention, intervention that may help to "restore" the child. Although many other chronic conditions have well established networks for referral to psychosocial care and peer support, the relative rarity of intersex diagnoses and the attendant shame and stigma associated with them has meant that medical caregivers often have either not known about or presumed no specialists were available to provide support, or else assumed that parents would rather not discuss their child's condition, thus failing to refer them to available support. Some parents report learning only much later that they should have been offered counseling. Shirley Friedman observes that the doctors "told us they didn't know anybody, so there wasn't even an option. Eight years later we found out accidentally that there was a circle of people that had been set up: a psychiatrist, a child psychiatrist, a social worker who had been working in that capacity for about five years, in the same hospital, and we were never referred to them."

The lack of referral in this instance might have been an unfortunate oversight. But in some cases the lack of referral to support, especially to peer support, could also be intentional. Clinicians may not want parents to meet one another for many reasons: from viewing the condition as stigmatized and not believing parents would want to connect with one another to thinking secrecy about the condition would foster a better outcome. Monica Cole recounts that "the weakest link in the medical care we received was from the hospital social worker who met with us before we left the hospital and gave us the phone number for the March of Dimes." If given time, however, parents can move past the initial fear and discomfort to see another reality: that they have given birth to a beautiful baby. Too often, however, parents do not have the opportunity to talk honestly, intimately, and at length about their feelings and experiences, including the freedom to explore the full range and meaning of their experience, which might include the sadness and difficulties engendered by their child, but also the strength, joy, love, and hope.

Most of the parents I interviewed described extremely awkward moments as delivering doctors and nurses struggled with what to say at the birth of their children. Few clinicians are as outspoken as the nurse who told Sara Finney, "What you have here is not right" (see chapter 4); yet parents often described the chaos in the delivery room and the astonishment and disbelief at not knowing their newborn's sex: "The doctor delivered her, and the resident mumbled, 'You have a boy.' The doctor immediately turned around and shot the resident a look and said, 'Shh. We don't know, yet.' After that, they took her away, and I had no idea what was going on. I'd never given birth before, so I was thinking, 'They're taking her away to clean her up.' The doctors and nurses came back and said, 'We don't know if you had a boy or a girl.' We were stunned. How could you not know?" (Tamara Dawson).

Interestingly, a number of parents I interviewed challenged the clinicians' gender announcement at birth, illustrating how bodily signs of gender are open to interpretation. The parents ascribed their own interpretation of the baby's sex either to commonsense observation or maternal intuition; doctors responded by ordering more clinical tests. Shirley Friedman, the mother of a girl with PAIS, recalls: "The doctor held her in his arms and said, 'Congratulations, you've had a girl.' He passed her to me, and I looked at her and said, 'I don't think so.' I immediately questioned the sex of my child, which was a complete shock. I felt she looked like a boy. . . . Her genitals, to me, looked very, very male."

Sara Finney, whose baby was eventually assigned female after much deliberation, recounts: "When she first came out, they said she was a girl, but I had seen her come out, and I'm thinking, 'Either they called it wrong, or that's a bit different for a girl.' I looked at my husband and said, 'I don't think so. That's not a girl.' " And Gloria Jackson, who claims she knew the sex of her daughter (who has CAH) before birth despite an ultrasound suggesting the baby was a boy, reports her experience at birth: "I had a midwife, and she said, 'Oh, you have a little girl.' And my husband, who had helped pull her out, said, 'Are you sure? Because that looks like a scrotum and a little penis.' Her clitoris was very enlarged and she had almost completely fused labia. My midwife . . . went to go call the obstetrician who was on call. I looked, and I said, 'Wow, that does look like a little boy part, but I also know that this is a little girl that I am holding.' I don't know how I knew, I just knew."

The medicalization of birth can turn the birth of a baby from a joyful celebration into a profoundly alienating clinical experience. With some frequency, the birth of a baby with gender-atypical anatomy involves acute, life-threatening complications, but these can usually be treated.[2] Even so, because these births are relatively rare, there is usually tremendous medical curiosity about the child. In medical parlance, these children represent "fascinomas." News of these births travels quickly in the hospital, and doctors, nurses, medical residents, and students may flock in to see the newborn, all too often without consulting or in some cases even acknowledging the parents. The physician in attendance may whisk the baby away for further examination or call for immediate consultation with a specialist, at times without telling the parents why. These acts are upsetting, particularly when parents do not understand what is happening.

> They [clinicians] were gone three hours, and I still had not seen [her baby]. At that point I knew something was wrong, and I began to panic, so I crawled down to the nursery, and there's the physician and the nurse who had attended the birth and about ten other nurses, all around my daughter. I said, "Hello, I'm the mom. Who are all these people?" They said, "Well, I'm Dr. so and so." And I said, "Who are all these other nurses?" And they said, "Oh, we're from downstairs." And I said, "What are you doing up here?" They replied, "We've never seen a baby with CAH." And I said, "You'll have to wait." It was horrible. I was in shock. I know they were there to learn about CAH, but it wasn't the time to do it. (Gloria Jackson)

> There was a busload of medical students finding their way around us to get a peek at our baby. He [the attending clinician] didn't really ask our permission, but he talked to us in such a way that he was soliciting our agreement without giving us a whole lot of leeway to say no, like, "This is a teaching hospital, and we generally keep these students with us, and this is so and so, and so and so, and. . . ." He skipped right over the "Do you mind?" part! (Tina Haywood)

As chapter 4 shows, bewildered parents may be further confused and distressed by conflicting possible diagnoses offered by different physicians and nurses in the hospital; and they may be unable to grasp hasty, incomplete explanations or those couched in complex medical terms. The net result of such experiences is to instill in the parents of infants with an intersex diagnosis the idea that their child's condition or atypical genitals will dominate the child's life, as well as theirs, from now on. And for many this indeed holds true. The entire family often adjusts how it lives to accommodate the child's diagnosis.

What to Tell Friends and Family

Parents' confusion and fear about their child's condition can strain families and relationships, starting with decisions about how to announce the birth to friends and relatives. Parents often feel confused, vulnerable, and shocked. Many do not want anyone to know that the birth was not routine, and in a culture in which the first question from well-wishers is overwhelmingly "Is it a girl or a boy?" they certainly do not want to admit they do not know the sex of their newborn.[3] It hardly seems surprising that many parents desire not to disclose their child's condition to others. They worry that others will not understand or may view their children as abnormal or freakish. As a result, some hide from family and friends by avoiding the phone and e-mail while they try to sort out their own feelings and understanding of their child's biology, health, and future. Monica Cole, whose daughter has CAH, initially told people she had a son: "We delivered in a small hospital which gave us the privacy to not have news of our baby's birth go out to friends. We needed time to take it in, and we didn't want to explain what we didn't understand."

Even when parents feel emotionally prepared to discuss their child's condition with family and friends, they can feel they lack the vocabulary and knowledge to do so. Tamara Dawson, whose daughter also has CAH, says she and her husband initially told their family and friends only that "we have some problems, and the doctors are pretty sure it's a girl, but we're not going to know for sure for a few days." She adds that her emotions "were totally out of whack. We were screening phone calls all week because you knew it was going to be someone wondering if you'd had the baby, and we couldn't tell anyone. During that week, we were in contact with my family and two friends. Months later, we found out that we were supposed to be offered a social worker to help us deal with this, but we never were."

How much information each family will choose to share with relatives and friends while they wait for further diagnostic information and sex assignment is highly variable and may depend to some degree on how the situation is presented to them by clinicians, as well as on their own understanding of gender development and their tolerance for its variability. Justine Good, whose young daughter has CAH, recalls that "after she was born, the nurse came in and said, 'Do you still want visitors?' I thought that was callous and said, '*Of course*, I still want visitors! I have a child here, I have family in town, I have friends in town. I'm going to tell them what's going on.' I just went 'Ugh,' but I guess they had to ask that" (original emphasis). She did not, however, let her visitors know that anything was wrong until she saw them in the hospital, and even then she felt reluctant to disclose everything she had been told: "I couldn't admit to myself that they didn't know whether she was a boy or a girl. That was too much, so I just told them that her labia were fused and it didn't look like she had a vagina. I sugarcoated it. I didn't lay it straight that they weren't sure if she was a boy or a girl."

Rebecca Davis, who also has a daughter with CAH, describes telling people, almost two months after announcing that she had had a son, that she actually had a girl: "It's a total nightmare, and people don't understand! Worst of all, our families couldn't deal with it. My husband couldn't deal with it. There was just me. I told the people that I thought needed to know and pretended that the other people had just misunderstood, and went on about my business and took care of my baby."

Some parents are also faced with the question of what to tell the baby's siblings, who have been confidently expecting a new brother or sister. Sara Finney, the mother of a daughter with mixed gonadal dysgenesis who was initially assigned as a boy, explains what she told her ten-year-old daughter:

> We called Julie [their oldest daughter] from the hospital and said, "You have a sister." But later that night I told her, "The doctors made a mistake. She was deformed down there. You have a brother." My mom came down a few days later and took Julie home with her to take care of her for a few days. By then my mom knew we had a girl, but we didn't know what to tell Julie, who still thought she had a brother. My mom went out and bought an outfit, and Julie said, "Gram, boys don't wear pink!" My mom told her, "It doesn't matter with babies." My mom then brought Julie to visit [her sister] and the nurses were in the room saying, "She, she, she," and Julie goes, "Okay, is it a boy? Or is it a girl?" And we said, "You have a sister. She was very deformed down there, we had to get her fixed."

By contrast, Monica Cole says that their older daughter Lara did not think it odd that they did not know if she had a brother or a sister; she was just happy to have a sibling.

> We started talking to her [Lara] soon after Cindy was born and kept it simple: "Cindy is missing a hormone her body needs, so we give her medicine. Everyone has different body parts, and her clitoris is large because of the hormone her body needed." Now I talk to Cindy when she's taking a bath with Lara. "Your body looks different than Lara's. Lara has dark straight hair, and you have light curly hair, etc., and your clitoris looks different also. It grew very large when you were a baby in my

tummy, and you didn't have some stuff your body needed. Now we give you medicine that your body needs." It'd be wonderful if they grew up comfortable about it, because it has not been a secret.

Rather than hiding her daughter's condition, Monica Cole made efforts to share information with both of her daughters and to normalize difference and variety. Her choice, however, is far from typical.

WHAT TO TELL THE CHILD

Whether the decision is due to fear, shame, a desire to protect the child, or simply a reluctance to explain the situation to others, parents may conceal an intersex diagnosis from the child, his or her siblings, family members, and friends. Despite John Money's recommendation that children be given age-appropriate information about their condition, and the trend toward honesty and full disclosure in medicine, some physicians still advise against telling the child anything about their diagnosis, anatomy, or treatment.

When children do ask questions, parents and doctors may withhold information or tell the child partial truths—for example, that they have a hormone imbalance or that they were born without a uterus—which are defensible because medical subtleties are bound to be lost on small children. Parents are also often reluctant to tell their children too much about their diagnosis for fear that the child will share the information with others at school and will suffer teasing and ridicule as a result. But in some instances these silences and gaps in knowledge persist into adolescence and adulthood. Parents weigh what they feel will be best for the long-term psychological well-being and adjustment of the child and may lean toward the prospect of a happy child who does not know or knows very little. They may justify their own reluctance to discuss the subject in terms of the child's welfare. In choosing this option, however, they minimize the long-term consequences of keeping a secret from a child about her or his own body and medical treatment. Honesty is one way to foster trust between the physician and the patient; without it, individuals may feel betrayed by both their parents and physicians. Moreover, giving patients complete information helps them become informed participants in important health care decisions. Many parents and physicians worry about the harmful effects of disclosing too much information to patients. But assuming that such disclosure is done with sensitivity, compassion, and tact, there is little empirical evidence to support such a fear. Moreover, it is very likely that the child will eventually become aware of the secrecy and may conclude that the truth about his or her body is so bad that no one dared speak of it. This experience may leave the child overwhelmed with feelings of shame about being different and isolated because of her or his difference, and destroy the trust between the child and parents: the child may grow up to feel betrayed.

Most parents I interviewed whose children are older said they had not shared much information, if any, with the child about her or his condition. Ramona Diaz, whose daughter is twenty-eight and has PAIS, says that "we never really talked about it. All of her doctors said, 'Don't ever discuss this with her.' Had they not said that, I probably would have talked about it a little more with her, but I probably would have also said, 'Let's just keep this within the family.' " Tamara Dawson, whose daughter is now seven, describes what she told her child when she was two years old about going to the hospital for genital surgery: "We didn't do much as far as telling her what would happen. It was just, 'This is what we're doing today.' "

Many parents said their children never brought up the subject or asked questions; they kept things to themselves. One mother says, "I think she really tried to not let it bother her." However, as I show in chapter 7, this may be because many children, sensing the topic is not a comfortable one for their parents, avoid it.

Another option is to present a difficult experience honestly but also optimistically and evenly tempered as a way to bolster children's self-esteem and their sense that they are accepted and loved as they are. Rebecca Davis explains that she was "really straight" with her two daughters about their CAH diagnosis and medical history to make sure they could talk to the doctors about their own bodies and health: "There weren't any secrets, and whatever they asked me, I answered honestly. I'd bring things up, talk a lot about the surgeries." She drew pictures of their anatomies and explained what was different and what had happened with surgery. "The only thing that I didn't tell them, because the doctors were so adamant about it, is that they had had male names. I did not tell them that until they were grown." Of course, telling the child the truth in no way guarantees that life will be easier. Moreover, by the time children are given more detailed information about their condition and medical treatment, it is likely that some irreversible decisions have already been made. This was the case with Rebecca Davis and one of her daughters. Although her honesty did not diminish her daughter's anger about having undergone clitoral surgery, it did enable them to talk about her reasons for choosing it.

Parents may persist in telling a child very little despite the fact that the child must often take daily medication, go for frequent medical visits, and undergo surgery or other procedures, such as vaginal dilation. These experiences can be distressing for both parents and children. It goes without saying that many parents are concerned about the general physical well-being of their children, particularly in cases of potentially fatal conditions such as salt-wasting CAH. In some respects, these experiences may not differ much from those of children with other chronic conditions for which treatment is monitored with regular tests and checkups. Many practitioners see individuals with CAH, for example, every three to four months to assess physical changes and monitor hormone levels. These visits, which involve blood and urine tests, may be frightening and painful for the children. One mother described an endless series of blood tests and heel sticks and bruises.

Children who undergo this kind of frequent medical monitoring may also be included in grand rounds, educational sessions that involve medical students, residents, and specialists. For children with intersex diagnoses, these visits might involve full-body or genital exams, photographs, and a discussion of the case in front of the child. Such experiences strike many of the parents I spoke to as intrusive and dehumanizing. Justine Good says of her daughter: "If she's helping other doctors to learn about it, that's fine, but when she gets older I don't want everyone poking and prodding her and every doctor experience to be, 'I've got to look between your legs.'" Shirley Friedman observes that such procedures were all right when her daughter was a baby, but that she "put a stop to it when she got older."

Some parents see these practices as coercive. Rebecca Davis, who has two daughters with CAH, says she felt she could not object to genital exams for fear her daughters would be denied care: "I never let anything happen to them that I didn't explain first, but I couldn't prevent grand rounds. I wish I could have. I know in my heart we would not have had the financial help and we couldn't have been a part of the program if I didn't agree to these."

Tina Haywood, whose twelve-year-old daughter has CAIS, is uncomfortable with the fact that at each visit her daughter's endocrinologist conducts a full physical and examines her genitals. "My daughter knows where the focus [is], and she doesn't like it much." Despite her concerns, she has been reluctant to try to stop the exams because the doctor "has made a big point about how she *has* to examine her. She *has* to. I accepted that the first couple of times, but things have been tense enough between us about the fact that we aren't going do the surgery. I didn't want to have yet another battleground with her" (emphasis original).

Gloria Jackson, whose four-year-old daughter with CAH had clitoral surgery at ten months, says that her daughter's endocrinologist examines her daughter's genitals every three months. She adds: "I constantly ask him, 'I know you have to monitor her clitoris, but at what point can this stop?' At what point can my daughter say, 'No, I'm not comfortable with you looking at my bottom like that'? He hasn't given an answer yet, but I want to raise my daughter to be able to tell someone 'No' when someone is doing something that doesn't feel comfortable. I'm cringing to think about this happening when she's thirteen. This is an awful enough period for a young girl, how does that affect her confidence that every time she goes in the doctor's office, someone's pulling her pants down and looking at her?" The constant medical surveillance through genital exams and the monitoring of hormone levels, as well as the administration of daily medication, can make these children feel they differ from others. Parents often struggle with their own feelings about the child's condition while trying to keep up a brave front for the sake of the child. Many feel a sense of grief or loss for the way they imagined their child's life would be: the Finneys, Pamela Healy, and Monica Cole's husband, having settled on a female assignment for their infants, all reported grieving for the loss of the son they thought they had had. But they also grieve for their imagined life in which they pictured a "perfect" baby with no health concerns.

PARENTS' FEARS

Parental concerns do not end at health issues. Added to the worries engendered by medical visits, medications, and surgeries are fears that the child may have been assigned the wrong gender, anxieties about how the child will survive all the typical experiences of childhood, puberty, and adult life with the perceived handicap of their diagnosis or biology, or of having atypical genitals. How will a girl with CAIS, for example, handle the fact that she will not menstruate or be able to give birth? Pamela Healy says her concerns about her daughter's well-being come in "waves" and are "like a big cloud hanging over your head, and it crosses my mind every day in some way. Because she has AIS, everything is magnified." Or, as another mother says, typical parental concerns are "kicked up a notch."

Implicit in parents' comments about their fears are ideas about what makes boys or girls. As chapter 5 has shown, these ideas may be reinforced during consultations with clinicians about the diagnosis, gender assignment, and genital surgery. Although many parents couch these anxieties in terms of the intolerance or insensitivity of other people toward bodily difference, their comments also hint at the parents' own discomfort in dealing with a child's atypical biology or anatomy. Tina Haywood describes her reaction to the birth of her daughter with AIS: "All I could think of, and this is such a terrible thing to say, but it's the truth, were people that I've seen where you really had to stare at them for a long time to decide if they were male or female. Not because I have any intolerance for people like that, but because I have a lot of empathy. I so much didn't want my child to be like that because I imagine that it's a really, really painful existence." Monica Cole acknowledges similar feelings about her daughter: "I'd like to say I'm totally fine with it, but every once in a while seeing her genitals surprises me. I am not concerned about the people who love her; we accept her. I am worried about others. My instinct is to cloister her away from society to spare her."

Such concerns are commonly expressed as anxieties about how the child will fare in such settings as the gym locker room, the public swimming pool, or in sex education classes. Jim Finney favored genital surgery for his baby on the grounds that "I don't think you can subject her to ridicule amongst other kids. You can't raise her in the center. She has to be raised as a male or female, period." One mother, posting on a Web site devoted to CAH, writes: "We are very happy we decided to have our daughter's severe case of verilization [*sic*] corrected. . . . We were never worried about others thinking anything about how she was formed. We were thinking of her and her alone. She has cousins her age she has been able to bathe with them and when she has a sleepover she doesn't have to hide in the bathroom when its time to change clothes. She can stand proud and be a normal little girl! Just do it for her and you'll not look back." This mother chose genital surgery so her daughter would feel normal, not, she says, because of a fear of what others might think about her daughter's genitalia. Paradoxically, however, her concern that her daughter would have to hide her atypical genitals if surgery were not performed centers precisely on what others might think.

One mother of a girl with CAIS expressed concern about the "health talk" that takes place in fifth grade. Although her daughter was not yet in school, she decided to attend to find out what information was presented about menstruation and other physical developments that her daughter would never experience, so that she could prepare her child beforehand, but also so that the teachers could normalize her experience by explaining to the class that some women are infertile.

Many parents also wondered whether an intersex diagnosis will affect the child's quality of life and ability to develop intimate relationships as an adult:

> I worry like any parent about so many things. Some of this has to do with CAH. I am worried about when she becomes sexually active. What if she

becomes involved with some young man, or maybe a young woman, and it doesn't work out, and it gets around school she's some Vagina Frankenstein? . . . Any time you add gender and vaginas or penis, it really causes discomfort for some people. I'm trying to think about the best way to prepare my daughter for dealing with that, because she's going to have to tell people she's intimate with. She's going to have to realize that some people are going to make assessments of her based upon that. (Gloria Jackson)

Pamela Healy expresses similar concerns: "Is she going to get married? Is she going to have boyfriends? Is she going to be able to tell them? Is she going to tell somebody, and then they'll tell everybody else or make fun of her? I said to a friend, 'Who's going to want to marry her? She can't have kids.' She said, 'It would take somebody special to marry her, but who else would you want to marry her?' " Healy found herself subject to fears not only about her daughter's life but even about her death: "What keeps me up at night is if something were to happen to her and they found a body, and it was so badly decomposed that they had to do a chromosome test, and they said, 'It's a male body, obviously it's not her.' Does she have a male skeleton? I just envision telling the police [about her diagnosis] and, this is morbid, and then seeing this headline splashed across the news saying, 'Looking for a Woman with Male Skeleton.' "

For some parents, other implications of the diagnosis add to their concerns for their children. Pamela Healy says: "She may not be dying from this [AIS], but is she going to kill herself because of it? I read somewhere that's likely." There have been surprisingly few studies addressing levels of psychological distress in people with intersex diagnoses. Those that exist have been limited by small and heterogeneous samples and diverse measures. A recent meta-analysis of these studies, however, suggests that people with intersex conditions do experience higher levels of psychological distress, but that it is difficult to isolate the specific determinants of that distress (Schützmann et al. 2007). Nevertheless, these rumors of an increased risk of suicide persist, especially concerning genital surgery. If you do not surgically "fix" a child with atypical genitals, the thinking goes, he or she will be more likely to commit suicide. I have spoken to several adults who have considered suicide. Their desperation did not result from their diagnosis per se, but from how their family, physicians, and society had perceived and treated them throughout their lives. Some claimed to have suicidal thoughts not because surgery had not been performed but precisely because it had. This is not necessarily because the results were poor, though for many they were, but because the surgical decision formed part of a series of decisions and actions that led the child to feel there was something horribly wrong about her or his body and condition and hence self.

Anxieties about an infant with an intersex condition may also lead parents to worry about the sex and gender identity of their other children. Pamela Healy who has two other daughters, learned that the chances of AIS affecting other siblings was one in four. As a result, she says, she is "looking at all my girls differently, like, 'Could she have it?' "

Because parents know their children have gender-atypical anatomies and biologies, they often worry that their children will look "odd" as they develop. They may also be concerned about gender-atypical behavior. Some parents frequently look for bodily and behavioral signs marking their children as atypical for their gender. Any deviation from gender stereotypes can become a cause for concern. Because parents are highly attuned to the way their child's body or biology differs from that of a typical boy or girl, this knowledge can lead them to read a wide range of behaviors and ways of being as suspicious—behaviors that likely would not raise concern without the diagnosis. Pamela Healy says: "Certain things trigger my fear. She's a bigger girl. She's not overweight, but she's big boned, and she eats a lot. She's got a big belly, and she's sloppy, and she's really active. A lot of people see her and say, 'That's just how my boys were.' That bothers me, and I wonder, is she like that because of her chromosomes? You read too much into stuff like that because it can have a different meaning for your kid. I still love her the same, but I look at her differently, and I don't want to." When Healy attended a support group for women with AIS, she reported that the most reassuring thing about the meeting was that she "couldn't tell who had AIS and who didn't because there's no sign. This woman came up to me, and it could have been somebody sitting next to you in church. They weren't strange-looking, you know?"

Because of their heightened awareness, parents may read a surprisingly wide range of behaviors as "cross-sex," in effect creating a rigid behavioral gender binary and attributing these differences to the child's "mixed" biology. Interestingly, when parents have more than one child of the same gender, but only one is born with an intersex condition, they can interpret the same behavior differently in the child with the intersex condition. One mother often compares the behavior of her daughter with CAH to that of her other daughter, who does not have CAH: "She [the daughter with CAH] is more interested in sports than Lara and plays less typical girl activities. I would not think about this if she did not have CAH, but knowing she was exposed to all of those androgens makes me wonder what that might have done to her" (Monica Cole).

Parents of girls with CAH often focus on behavior that is understood as overly masculine, which they attribute to prenatal androgen exposure. Their assessments necessarily depend on commonsense (and stereotypical) notions of "typical" female and male behavior. Although parents of children without intersex conditions may have similar concerns, for parents of children with intersex conditions the stakes are much higher: any atypical behavior may signal the child's confusion about gender identity. Everything is magnified and viewed through the diagnosis.

Another mother of a daughter with CAH mentions her concern regarding her daughter's behavior as it related to an exposure to high levels of androgens in utero: "One thing that really worried me was thinking that she would not want to play with little-girl toys, since she had so many male hormones. Although she loves to play ball with her daddy, she also loves to play with her dolls and dress up in fancy clothes. She is truly a typical little girl!" (Mother posting on a Web site for those with genetic mosaics).

Concern about masculine behavior is not exclusive to parents of girls with CAH. The mother of a girl with PAIS spoke about her concerns over what she perceived as her daughter's tomboyishness: "She [her daughter] has always showed signs of being a tomboy. In mean name-calling, my son would call her a boy. She's always had boy friends, always liked to do boy things. It's something to watch for and a concern. But my sister and my best girlfriend like to do all the guy things. I don't think I've overreacted to that because in this world, you can do whatever you want. If you enjoy baseball as opposed to curling your hair, there's nothing wrong with that. But it's been hurtful to me, because my son would call her a boy and that stirred up so, so much" (Shirley Friedman).

Another mother of a girl with PAIS discusses how she and her husband hypothesized the condition might affect their daughter's behavior and how their expectations were proven wrong: "We had a lot of questions about androgen exposure and the brain. We got John Money's books and speculated about how it might affect her personality and abilities. We thought she would be a passive child, sweet, demure, and she's not! She's really feminine, but in a very flamboyant way. She reminds me of the really out-there gay male friends that I've had. She *loves* to dress up, and she really states her mind, and she has very feminine mannerisms. She cries easily and takes things really hard. She's not good in math" (Tina Haywood; original emphasis). For this mom, every behavior can be read as distinctively masculine or feminine. She uses commonsense (and commonly held) ideas about male-female difference, including stereotypical notions about mathematical ability, and understands them firmly in a biological framework. One also gets the sense that her daughter's female-typical behavior is a comfort to her, an indication that her daughter is "really" female. She adds: "We're at a point now where it would be fine for her to be somewhat masculine, but a few years ago it wouldn't have been OK for her to be masculine in any way, because that was my biggest fear, of her being a frightening hermaphrodite-type person. She's very, very feminine now, but we didn't know that about her when she was two weeks old." She locates evidence of a successful gender assignment not in her daughter's sense of herself as female, but in her gender-typical behavior.

With their heightened attunement to gender-typical behavior, some parents can take on the role of reinforcing behaviors they feel are appropriately masculine or feminine and discouraging those that are not. Managing gender-atypical behavior is a way for parents to manage their anxiety and fears about gender difference and deviance. Other parents, however, are unconcerned about gender-atypical behavior: "I was a tomboy, and when I see someone stereotype girls, I think, 'Don't pay attention to it.' I do see a little difference. Around fourteen, not that she *has* to wear makeup, but she *ran* from makeup. Some of her attitudes have been a little extreme about not wanting to do things that are feminine. I don't know if this is because she's a teen and would be this way anyhow. She has hair down to her waist, so I think, 'Well, she doesn't want to look like a boy' " (Dorothy Powers; original emphasis).

Although this mother has noticed gender-atypical behavior in her daughter, her daughter's feminine appearance has lessened her concerns. She offers a clue about why gender-atypical behavior preoccupies many parents of girls with CAH; not acting female may be a signal that they do not feel female. While she is not bothered by her daughter's behavior, Powers observes:

> Many parents of girls with CAH are terribly concerned about their kid's behavior, though they won't say it publicly. At the same time I don't want to say that there isn't an [androgen] influence there [on her daughter's behavior], because I'm not sure. She's got model cars, an Apache helicopter; she's very interested in weapons. The only thing I could think was that the Gulf War had a big influence on her, because we watched it. She wants a Samurai sword for Christmas, but so does one of her girlfriends, so that

means nothing. If her CAH has anything to do with it, it's a bonus, to make a very well-rounded person, with *many* interests. (original emphasis)

Another mother of a girl with CAH is not concerned about her daughter's apparently masculine behavior because she herself was a tomboy as a child: "I was a tomboy; I still am, so that doesn't concern me at all. I hope whatever she enjoys to do, she can. I hated dolls. I'm not frilly, and I don't need her to be" (Justine Good).

In these parents' accounts, feminine traits include playing with girls' (and not boys') toys, acting passive, wearing makeup, not doing sports, not being good at math, and having long hair. Masculine traits include an interest in cars, sports, and weapons. Even parents who are not bothered by their daughter's supposedly masculine behavior still refer to it as masculine. Male and female behaviors are often viewed as distinct and mutually exclusive and tied to biology.

Ensuring Happiness: Parents' Treatment Decisions

As chapter 5 has shown, many parents of infants with atypical genitals find themselves contemplating reconstructive surgery for their babies at a very early age. The arguments parents, physicians, and surgeons marshal for performing what is essentially elective surgery on a child reveal a great deal about the underlying anxieties of parents, as well as the public belief in the power of clinical medicine to cure all ills, even fundamentally social ones. If the pregnancy has gone well and parents have no indication that their child may be born with an intersex diagnosis, they can face confusion and emotional upheaval when they learn that rather than going home with the "perfect baby," as they imagined, they now face a future that requires they take in enormous amounts of medical information and make a number of significant medical decisions. The seemingly limitless possibilities of modern surgical medicine, coupled with the parents' motivation to normalize the child's experience, often leads parents to ask surgeons to "normalize" their children's genitals.

THE SURGICAL CONSULTATION

Parents' decisions about genital surgery are heavily influenced by the consultation with the surgeon, during which they are told what types of surgery may be involved, along with the potential benefits, expected results, and potential complications or harms. Dr. K sums up what he feels is important to cover in a surgical consultation: "The three areas that you talk about are, one, cosmetic results, how it's going to look. Number two, sexual capabilities, how are they going to function? And three is fertility. Parents' main concern is: 'Is she going to be able to be a normally functioning female?' They can't verbalize penetration. They just say 'sex,' and we know what they mean." He adds that most parents "don't focus on clitoral function. You have to lead them toward that and sensation. They don't mention it; I have to." Given that the benefits of surgery as well as its cosmetic and functional results are in dispute, parents should also be told about current controversies over early genital surgery. Dr. L, a pediatric urologist, echoes Dr. K when he suggests that future sexual sensation is hard for some parents to discuss: "Parents don't mention sexual stuff on their own. I make an effort to discuss it because it must be on their minds, but very few parents say anything about it. I make sure we at least mention it. Some parents, after we mention it, they don't want to talk about it anymore."

Although these surgeons said they discussed sexuality and future sexual sensation, other physicians and surgeons said they did not discuss the future sex life of the infant or sexual sensitivity after surgery. The pediatric urologist Dr. M says he does not mention these issues because they are uncomfortable for parents:

> Sexual functioning is not a comfortable issue. A lot of parents are focused on what they see; looking twenty years down the road and the issue of sexuality is something that they do not bring up. I have to admit that I don't bring it up because I have not encountered any patients with an enlarged clitoris who come to me and say, "This is a problem," or, "This is okay." I don't think there's much literature either regarding that, or at least I don't know enough to be able to counsel the parents very well and say, "This is the type of sexuality these patients should expect." Parents don't ask that question either. They're more concerned about, "Is this a boy or is this a girl?"

This clinician feels it is acceptable not to speculate about a child's future sexual functioning because so little is known. The fact that so little is known, however, is precisely the kind of information a clinician must share with parents. It also seems likely that if parents sense that a clinician is uncomfortable talking about sexual capability and function they will not feel comfortable bringing it up themselves.

Several parents I interviewed said that clinicians gave them no information about these issues. Ramona Diaz had the surgical consultation for her child over ten years ago: "When the surgeon spoke to me about clitoral surgery, he didn't mention anything about sensitivity. No mention whatsoever. This is something I learned later. To be honest, I just knew they were doing surgery to remove the growth [enlarged clitoris], but I didn't even think it would take away her sensation." Two other mothers who had their surgical consultations within the past five years reported different experiences: "I learned about it [potential loss of sensation] later. I guess I didn't think about it because here you have a mere baby. You don't think about their sex life when you've got an infant. There were so many other issues to face besides will she have any feeling later, but it does not help that the surgeon did not mention it to us," says Nicole Day. In Tamara Dawson's words, "He was telling us, 'This is what's going to happen.' We asked a few questions, but we really didn't know the questions to ask. We were going on what he was telling us. . . . Looking back, I would have asked, 'Is she going to be able to have a normal sex life? Is she going to have sensation? What are the long-term effects?' I would have wanted more information because you're messing with nerves."

Of those surgeons who raise the issue of sensation with parents, some are either unaware that surgery could impair it (as parents report), or they choose not to discuss that possibility with parents. Shirley Friedman says that her surgeon "very much reassured us that surgery would not affect her sensation, even when we had the second clitoral reduction, because that certainly was our concern. But when he spoke about the clitoral reduction, he felt that it was very cosmetically important; there was less emphasis on our concern about sensation." Sara Finney comments: "I got in contact with ISNA and Dr. [X], and they both urged me not to have the surgery. All I heard was 'genital mutilation.' I called our doctor and said, 'I just heard genital mutilation. What the heck am I doing to my child?' Immediately he calmed me down and said, 'ISNA are all failures, people who grew up with shame.' But I wasn't convinced. I was really concerned about the surgery. My pediatrician said, 'Look, she needs to be complete. You need to complete her. You can't raise her as both. There is not an option, male/female both. There's no option.' "

Many parents of young children that I interviewed said they were not told about the current debates or that some clinicians and adults with intersex diagnoses consider genital surgery optional, unnecessary, or harmful. Nicole Day recalls that her daughter's surgeon "didn't say anything about 'Some people are questioning this now.' Nothing like that." Tamara Dawson had a similar experience: "I didn't realize that you could do the surgeries in different stages or leave the clitoris alone. They never told me that."

Some parents are told about the potential loss of sensation, but only after bringing it up themselves. Others learn about it through their own research. One mother who chose clitoral surgery for her daughter explains her consultation, which began with the surgeon explaining his background, medical training, and previous experience in this area. She continues:

> He also said in some cultures children with CAH are abandoned because of stigmas in those cultures. He then drew what would happen: they're going to create a vaginal opening and canal; she would have to have her clitoris reduced. I got really alarmed and said, "What do you mean? How is this going to affect her sexual response and appetite? My daughter's getting a plateful; she doesn't need that on top of it." He explained, "I can't tell you it's going to be the same as you, because every person is different, but it will not be reduced or diminished, and she will have some scarring on her vaginal opening," which she still has, which is another story, and that "hooking up the plumbing should go well." He told us some adults who've had this procedure have picketed a couple of conventions and said, "I will tell you that maybe twenty years from now, she might tell you, 'You made a decision for me that I wouldn't have made for myself.' I've dealt with adults—not any patients of mine—who resent the hell out of it. It's your call." We decided she would have the surgery, per their advice, and at ten months, she had the surgery. (Gloria Jackson)

Another mother, who decided against surgery, discusses her consultation and says the surgeon gave them a description of the "best-case scenario": "He said that a second surgery at puberty may be required, but that it was a simpler surgery. He did not mention that it may require a third surgery, or that the clitoris could wither, or that dilation may be needed, or that there could be pain when the clitoris became erect because of the surgery, or even that the surgery could take hours and a long recovery time. He also said the preserved nerves would provide sensation. He said they partially based this on information from men who had sex-change operations and were able to keep their orgasmic capabilities" (Monica Cole). Their pediatrician also gave the parents copies of articles published in *Newsweek*, the *New York Times*, and the *Utne Reader* that covered current controversies and, as Cole says, "presented all the viewpoints." Cole says that after she read the articles, she discounted some of the information in them because she felt the adults were angry about the secrecy surrounding the surgery and that it "seemed that surgery was performed because of parents' and doctors' fears that these children would be confused about their sexual identity." She says she discounted the information because "I knew we would tell [their daughter] what was happening and why, and because we did not have those same fears." She told the surgeon she appreciated the articles, but that "I wasn't sure if we would be willing to let [their daughter] be one of the first kids to test out whether not doing surgery was right."

When they met with the urologist a second time, she said that two residents also sat in on the discussion. She and her husband asked about not doing surgery and felt that "they seemed to discount antisurgery voices as the extremist fringe. Our doctor was uncomfortable about the recent outpouring of upset patients' stories and defensive about his stand on performing surgery. I can imagine it would be very difficult to acknowledge that what you had been doing for decades had made patients furious. He told us that surgery had changed and that the nerves to the clitoris would be preserved."

CHOOSING SURGERY

Money recommended early genital surgery to normalize atypical genitals to ensure the development of a gender identity in accord with the assigned gender of an infant. Very few physicians I have encountered still view this as the primary rationale for early genital surgery. Indeed, as noted earlier, some do not view the child as having a gender without genital surgery. Many clinicians, however, cite other reasons for early genital surgery such as the best interests of the child and technical ease. Parents express similar reasons for choosing the procedures, though at times these may veil other concerns, such as their own discomfort with their children's genitals.

Money also argued that early genital surgery reduced the likelihood of traumatic memories of the operation. Even though some surgeons now dispute this idea, a good number continue to recommend and perform early genital surgeries. In some cases, genital surgery appears to take priority over other medical treatment for the child. One newborn underwent genital surgery before her life-threatening congenital heart defect was surgically repaired. A surgeon describes the cases of two patients born with cloacal exstrophy—where abdominal viscera are outside of the abdominal cavity—who had early vaginal surgery despite highly complex reconstructive surgery to allow the child to survive and the prospect of lifelong disabilities: "In two cases, one was twelve and the other one was nine. . . . They're both genetic males with testes, but they have no phallus. Along with that, they were born with truly life-threatening problems at birth. Now that they've been reconstructed, they have no colon, so they have an ileostomy bag. They have no bladder, so they have the urostomy bag. They don't have any intestines, so they have a feeding tube. It's a horrible way to live. But somewhere at the beginning, surgeons at this institution created a vagina for them made out of a throw-away structure that was supposed to be colon, but never quite formed."

One argument adduced in favor of early surgery, as chapter 5 has shown, is that it will free the child from the burden of dealing with other people's reactions to a perceived abnormality. But even if discomfort or repulsion is the motivating factor for genital surgery, is it ethical to surgically alter the child's genitals because of the fear they will make others uncomfortable? In many other areas of medicine this would seem like an extreme proposal, and parents would likely be shuffled out the door, but intersex conditions have often been bracketed off as unique because they touch on deeply held ideas about what males and females are (or are not).[4] Moreover, such fears are often exaggerated. One mother who did not choose surgery for her daughter has explained her daughter's genitals to child-care providers and has found they are not bothered by her revelation: "It hasn't been difficult to explain her ambiguous genitalia to child-care providers. By the time I needed to explain things, I had already narrowed down who I liked, and they were people who were fine with it. It just hasn't been a big deal with any of them" (Monica Cole). Not only can parents screen for care providers who are more accepting of genital difference, but they can also help their children accept and appreciate their unique bodily configurations. It would be naive to assume that screening babysitters and promoting self-acceptance will protect the child from all embarrassment, shame, or discomfort about their genitals (or bodies), but all too frequently parents and clinicians quickly conclude that the surgical fix is the only way to bolster's the child's self-image and to earn social approval. A physician further suggests that children themselves learn to steer clear of risky situations: "This whole rigmarole about, 'If they're seen by the day care worker, they're going to be treated weirdly, or people are going to flip out,' is all crap! None of those kids with abnormalities take their clothes off in front of anybody, even normal kids don't want to take their clothes off in front of anybody!" (Dr. R).

Other parents argue that early surgery is the kindest choice because it will spare the child the pain and embarrassment of discovering that his or her genitals are atypical, and that an adolescent should not be burdened with having to make a decision about surgery. Nicole Day comments: "As virilized as she was, I can't imagine that as a teenager she'd be real happy with mom and dad if we had left her as she was." A mother posting on an Internet discussion board for parents of children who have CAH explains her decision: "I have a six-month old daughter who also has a severe case of virilization. She looks like she has a small penis. Next week we will set up a date for her surgery. I believe in my heart that this is the best thing for her. I don't want her to have to make that decision when she is older. She will have enough to deal with when she is a teenager."

Some parents believe, as Money did, that surgery is an important step in sealing the child's gender identity: that eliminating genital ambiguity will resolve any gender ambiguity. Jim Finney, discussing the decisions he and his wife Sara made on behalf of their baby, explains his reasoning: "If you're going to raise her female, give her the physical characteristics of a female." A number of mothers posting on a CAH Internet board express similar concerns:

> We chose surgery for my daughter mainly because we did not want her to grow up questioning her sexual identity. We felt that she should look like a female, so we chose the clitoroplasty and the vaginoplasty. We felt that she would have a better self-image if she did not have a "phallic structure" and "scrotum."

> I with all my heart believe there would be psychological problems in a child that is female and has to grow up with a clitoris that looks like a penis. If I had an enlarged clitoris, I would be looking at this "penal" [sic] structure saying I look like a boy even with no doubt to my parents I am a girl. We are told in society girls have vaginas boys have penises. I wish to God we could change that to we all have unique bodies, but to totally eliminate cosmetic surgery would be just as unfair to the females that felt this would have given them a "normal" life, as to the females that had it done and turned out bad.

> There was no way I was going to have my little girl grow up thinking she had a penis; that is why we did the surgery.

Interestingly, all of these children are female, yet parents refer to their daughters' clitorises as penises. These comments suggest that at some point a clitoris stops being a clitoris and becomes a penis, a distinction based exclusively on the size of the clitoris, raising the question, can a woman have a penis? The boundary between a clitoris and a penis is indeterminate, but in cultural and medical terms the clitoris and the penis are considered so discrete that an organ of ambivalent length must be rectified because it raises questions about the gender of its owner (Morland 2004). In this formulation, the "penis" only becomes a clitoris through surgery.

Parents and surgeons also stress that providing the infant with "normal" genitalia will enable them to live a normal life: "Basically, it amounts to a cosmetic revision of the external genitalia so they resemble normal females and can get married, have children" (Dr. P). This view assumes, of course, that surgically altered genitals both appear normal and function normally, that atypical genitals alone are what would make a child with an intersex diagnosis feel different, and that children with atypical genitals are unmarriageable. By focusing their energy on the genitals, parents and clinicians downplay other important considerations that may create equally or more profound and long-lasting difficulties, such as parental adjustment to the diagnosis, frequent medical visits, the need for daily medication, the inability to menstruate and bear children, reduced or painful sexual sensitivity, or knowledge that one's chromosomes are gender-atypical, among others.

Because the medical rationale for early surgery is variable and often not supported by data, it seems probable that surgery is also done for the benefit of the parents. Some parents in fact acknowledge that surgery helps them psychologically. Although Sara Finney was reluctant to agree to vaginoplasty for her baby, she admits: "I'm glad we did it. It's mentally easier to look at her perfect." And Gloria Jackson confesses that the stresses associated with her child's diagnosis were just too much to withstand: "You reach a point where you think, 'Enough with all this intense medical crap! Let's just get this over with for all of us.' I could wait, but I will admit I thought, 'My gosh, that's years of worry for me, of thinking, "What if?" ' Two to four o'clock in the morning are the longest hours of the day; you worry about ridiculous things in those hours: 'What if she goes on a field trip and she falls and they somehow see her genitals? Are they going to know?' "

In performing early genital surgery, clinicians and parents often make several assumptions: that surgery creates genitals that look normal; that the children would choose genital surgery for themselves; that most people would be bothered by atypical genitals; that surgery will solve their anxieties and fears about intersexuality; that surgery will allow the child to lead a normal life; that surgery is the best or most important way to help the child; that atypical genitals are the most difficult aspect of intersexuality. Most of these assumptions turn out to be questionable at best.

As noted in chapter 5, there has been minimal follow-up data on the cosmetic results of genital surgery. What data exist are overwhelmingly based on clinician, not patient, assessment. Hannah Samson, a woman in her thirties with CAH who has had clitoral reduction and vaginoplasty, says: "They couldn't make my outward appearance be something that was at least acceptable to me and that led to so much pain inside of me." Several clinicians acknowledge that genitals that look more gender typical postsurgery may end up looking abnormal after several years: "Any time you operate on something, a good result has cosmetics and function, and both have a capacity to change over time, and even greater, over a longer time. Many very normal-appearing structures, over time, get bound down, and there isn't much there after twenty years. How do I know that the surgery is good enough, and will be good enough?" (Dr. G).

Surgery, then, is often viewed as a viable solution, but it does not always relieve parental anxiety, and it can cause other problems. Some children may have traumatic memories of the surgery and some, as adults, may be furious that their parents chose surgery for them. A child who has had surgery will still require help coping with her condition. It is quite possible that genital surgery can make some people feel different and abnormal, the exact opposite of its intended effect. Hannah Samson says, "It's all about outward appearances for them, but the help I needed was inside, because I just felt so different." Dr. B also suggests that "normalizing" surgeries do not necessarily make people feel normal: "The rationale behind calling it medically necessary is that the patient can't be raised as a male or a female if they're ambiguous looking. I don't mean because of parental doubt. The idea is that the kids themselves will look down and see if they look different than their brother or sister. But the point that I've made to a lot of people is it's hard for a kid to think they're normal when they're having multiple surgeries on their genitals, and every time they go to a doctor, the doctors are all staring at them."

REFUSING SURGERY

In part because of what they have heard or read about poor surgical outcomes from adults sharing their experiences, a small but significant number of parents and clinicians are now opting to decline or postpone genital surgery for children with gender-atypical genitals. I interviewed the mothers of several young girls who did not choose surgery even though their daughters had what doctors considered virilized genitalia. Their reasons for refusing surgery varied. Justine Good and her husband were opposed to anything they considered unnecessary surgery on principle: "If we had a boy, we weren't even going to get him circumcised." She was also aware of the risk of reduced clitoral sensation after surgery. However, when their daughter was five months old, they felt tremendous pressure from the surgeon to go ahead with the surgery: "He was really, really guilting me into having the clitoroplasty and the vaginoplasty done, saying she'll be socially, psychologically scarred from not doing it. So we went back and forth, and I actually made an appointment to have it done."

She also expressed the fear that if she took her daughter to the same surgeon to open her labia, he might perform the clitoral reduction without consulting her. This fear is not unfounded. Two mothers told me that a surgeon had reduced their daughter's clitoris a second time, without their approval, while performing other genital surgery. Three pediatric endocrinologists who spoke off the record said they have worked with surgeons who have done this. All had stopped referring patients to these surgeons. After agreeing to the surgery, however, Good canceled the appointment. "After I made that appointment, I cried every day. . . . I talked to pediatricians who said, 'There's a wide range of normal in women, and even women who don't have CAH, some have bigger clitoral hoods than others.' . . . The only reason our surgeon gave to have the clitoral recession was, 'What will the babysitter think?' That's not good enough! Who cares what anyone says except our daughter. If she wants to make that decision she can, but it's not my body. It's her body, and I can't do that to her. If she never can have sensation, I wouldn't be able to forgive myself."

She added that when she canceled the surgery, she knew "deep within this is her decision when she can fully understand the consequences." She explains, however, how easy it is for surgery to appear as a panacea: "At first, you just want to hear what you want to hear. You want to be told they'll make her look normal and that she'll feel wonderful. And so that's what you believe at first. When he said that there'd be reduced sensation, that's when I went 'Oh, God!' " The reaction of her family members has been mixed. Her sister, who is a nurse, is "one hundred percent behind our decision." However, this mother has concealed the decision from other family members who she feels will not be as accepting.

Another mother explains how she eventually decided against genital surgery for her daughter:

> Almost everyone thought we had to do the surgery. "How could we not, when she would get teased?" Even people whom we admired for their respect of diversity thought we should not leave that alone. Our next step was to research what we could on the surgical procedures. The gruesome details and photos of the surgeries in the medical journals were so different from the urologist's drawing. We were struck by the near universal references to "cosmetic" improvements. We also read studies on the

> social and psychological issues of CAH patients, but the articles were not very helpful in making a decision regarding surgery. It dawned on us that what was proposed for our daughter was chiefly to "improve" her appearance. Our daughter was two months old at the time, so small and vulnerable, it made me panic to think of her being under anesthesia and having hours of surgery to make others more comfortable with her. We decided to put off surgery. At our next appointment with our endocrinologist, we brought up our reservations, and he talked with us at length about the changes happening in the management of intersex patients and further supported our decision to delay surgery. (Monica Cole)

Justine Good sums up the difficulties this way: "It [no surgery] is the lesser of two evils. It's not easy to grow up like this [with atypical genitals], but it's not easy to grow up anyway. Differences just aren't allowed, but everybody has their differences. They just want her to not be ridiculed or look different. And we don't want that either, but we are very sensitive to it all, and we'll be good at having heart-to-hearts so she doesn't feel weird."

Dr. A. is a pediatric endocrinologist who supports considering alternatives to surgery. He tells me:

> As [the surgeon Justine] Schober says, "The intent of surgery is to make people feel better about their genitalia. Psychotherapy does the same thing, but isn't irreversible." Will the child be bothered by scars in the genital area? You're not going to have something that looks normal, that doesn't have any residual trace. . . . You have a difference, regardless of whether you do surgery or not. To a certain extent, that same argument can apply to the parent's problems dealing with it too. . . . There's not one follow-up study showing that the kids who have had surgery had a better psychological response, or parental psychological response, than kids who didn't. We're doing this all on a theoretical basis, not on a pragmatic basis.

Living with the Consequences

Of course, parents who opt for genital surgery for a young child do so on the assumption that surgery will, in fact, solve the perceived problem with few adverse consequences and improve the child's quality of life. Some express great faith in current surgical techniques, like one mother posting on an Internet discussion board for parents of children with CAH and for adults with the condition: "I honestly believe that due to complications, physical and emotional, I would be unethical and doing her a great injustice if I did not have the surgery done now while she is still young. It is fortunate that today we do have the medical science and expertise available to ensure a positive outcome for this surgery. Years ago this was often experimental and the surgeons were not experienced nor aware of the damage that could be done."

However, as Tamara Dawson discovered, surgery may not resolve parental anxieties: "Someone who took care of her before her surgeries had no problem with her, and that was a concern for us, but it still is, because she just doesn't look normal." The same holds true for Dorothy Powers: "Before her surgery, I wasn't going to have someone else changing her diaper and talking about her, but even after surgery she did not look right, which made me just as nervous."

Parents may also learn that the surgery was unsuccessful in more than cosmetic ways. All the parents I interviewed whose daughters are adults (ages seventeen to thirty-eight) have learned that their daughters have diminished or no clitoral sensation following surgery. This outcome has been enormously painful for these families and in some cases has caused vast emotional rifts between parents and children. One mother whose teen daughter has PAIS and who regrets having chosen surgery gives her original rationale for choosing it for her daughter: "In retrospect, I think we were very concerned others might notice. As she's gotten older, nobody makes you take a shower in high school anymore. The emphasis was too much on the cosmetic appearance when she was tiny, but also more when she was in public, or even later in her life, in an intimate relationship. . . . I worried that if we left stuff until she was older, it would be very difficult on her, and we wanted to get all that over with. I don't agree with that anymore; it's a much more open world" (Shirley Friedman). Her comment suggests that with time parents' fears and concerns about their children's genitals can become manageable.

Because surgery and its complications are irreversible, and because some parents have come to regret choosing surgery for their children, it seems wise to give parents ample time and support to explore their feelings and fears about their child's condition and body before making any surgical decisions; they may find that what seems critical at the birth of their child may not seem so several months or even years down the line. This shift will not happen quickly through one conversation, but rather it will take place over time as parents adjust to the reality of having a child with an intersex diagnosis. Also, as parents meet other parents who have faced similar fears and concerns, including those who may regret their surgical decisions, they may develop more realistic ideas of what they can and cannot expect from surgery, which will aid them in their decision. One mother says of learning that her teenage daughter with CAH has no clitoral sensation, "I made believe everything was fine throughout her childhood, only to discover this as a teenager. I'm almost glad that I didn't know this until she was fourteen, because at least I didn't have to think about it most of her childhood. I'm just sorry for those who have the younger daughters now thinking everything is going to be fine" (Dorothy Powers).

When we spoke, this mother was eager to know the experiences of women I interviewed who have had clitoral surgery. She has since written to me inquiring about surgeries to restore clitoral sensation. She feels tremendous guilt, anger, and confusion: "I get angry because I think, didn't they *consider* that [the potential loss of sensation]? . . . Don't they know what causes sensation in a female body? Or do they really not know this? Or did they think they were really preserving the nerves? This was 1983; that wasn't ancient history, that was the modern times. It's just so hard to accept."

Another mother whose daughter has PAIS and no clitoral sensation regrets that she was given so little information about possible adverse effects of surgery: "If I had known, I would definitely have thought about it a whole lot more. . . . I don't know that we would have made the decision we did, but it's too late to worry about it. We just didn't have information, so all we could do was rely on a few professionals to guide us. I can't beat myself up about it because we had nothing, nothing to read about it, no other parents. Now I give my opinion to lots of parents; all I can do is let them know about our experience" (Shirley Friedman).

Ramona Diaz whose daughter has PAIS and who wishes she had not chosen genital surgery for her daughter has found another way to deal with her disappointment and sadness: she has become an advocate for delaying surgery until the child is old enough to decide: "I feel very bad. I hurt because she hurts. I just want to spare people everything that she's been through. I feel the same way that a lot of intersexed people do now: Let that person make the decision when they are ready. If they are ever ready to say, 'Yes, I want to do this,' or, 'No, I don't want to do this.' Let them have the say in the matter. Not the medical profession. Not the parents." And for Rebecca Davis, whose two daughters have CAH, the stresses associated with their reaching puberty resulted in a broken marriage. A psychiatrist explained to her that having a chronic illness in a family creates an enormous amount of pressure, but she says, "I didn't have a great marriage to start with, but even those with really good marriages, most of them didn't make it." She is still struggling to cope with the anger of one of her daughters: "I can't make it right. Part of what she's so upset about is that she feels raped, and in a way she really was, and I couldn't help it. She can't see that I couldn't help it; all she can see is that I let it happen and ask, why didn't I protect her? Oh, God, that hurts."[5]

Mothers versus Daughters

The tensions and conflicts between mothers who have chosen "feminizing" genital surgery and the daughters on whom surgery was performed are today played out not only in support groups and therapists' offices but also in public forums such as Internet discussion boards. A common source for such debates is a new mother of a girl with CAH posting a query to a board about whether or not to choose genital surgery for her daughter. The topic generates such vehement exchanges that the most prominent board for parents of infants with CAH has created a separate "controversy board" for these messages.[6]

Although there now exists a group of very vocal parents who oppose genital surgery, there still appears to be a huge generational and emotional gap between mothers who have decided on surgery for their young daughters and adult women who have experienced it. Women with what they feel are bad surgical outcomes want to share their experiences with new mothers, making sure the latter are aware of the possible adverse effects. Jean Butler, a woman now in her fifties whose clitoris was removed at age five, regularly participates on these boards. She says: "I have been trying to convince these parents that doctors who say 'We can do it without changing sensitivity' is a frigging lie. There have been no follow-ups; any follow-ups there are say that it hasn't worked. I want them to know this, and many do not." Kristin Rupp, a woman with CAH who is in her thirties and who has had five genital surgeries, says, "All the parents say, 'I had my child's surgery when she was two months, and she's just fine now.' But these girls, they're six. You won't know until they are older like us. My mom thought everything was fine too, because I never told her until very recently because I looked at my mom as somebody who was in with the doctors, a kind of a coconspirator."

A woman who had clitoral-reduction surgery at age twelve, against her wishes, adds, "I want parents to know that genital mutilation is not far removed from our culture and that what is being done is irreparable. The individual should have that choice. Part of my message is that parents are capable of dealing with the birth of an intersex child without these drastic interventions, which exacerbate the problems and damage the relationships between parents and children instead of fixing them" (Teresa Diaz).

The desire for this kind of insider knowledge that doctors either do not know or fail to share has led to the rise of disease-specific support and advocacy groups over the past several decades, but in the case of intersexuality their comments are not always appreciated by mothers who have chosen genital surgery for their daughters. Echoing clinicians about improvements in surgical methods, one comments, "You should not give out such negative feedback to a new parent with a daughter with severe CAH who is deciding whether to have the surgery done. It just makes the decision that much harder to make. Medicine has come a long way in the past 30 years. I am sorry you had a bad experience with your surgery, but that doesn't mean everyone will" (posting on an Internet discussion board for parents of children with CAH).

Some mothers become extremely defensive about their decisions in favor of surgery: "I take GREAT offense to say I made this choice for my daughter because I was SELFISH! . . . It literally brings tears to my eyes that someone thinks this choice for my daughter was in the end to 'please' myself. It was a choice that I made for my daughter because I felt at that time and still do it was the best choice for her" (posting on Internet discussion board for parents of children with CAH).

Parents are concerned, rightly, with making treatment decisions that they believe will provide their children with the best quality of life. Although they hope surgery will fulfill this desire, the answer is not immediate or obvious, and they remain hopeful that they have made a good decision (one woman with CAH complained that parents only see with "rose-colored glasses"). Parents may think the risks and concerns of surgery end when the stitches come out, but the real effects are not known until years later, when their children are older and become sexually active. Parents whose children are still young often remain more focused on their immediate concerns and either are not worried about surgery's future effects on well-being and sexual sensation, assuming it will be fine, or value attempts at genital normalization above all else and assume that their child will also. For parents, the surgery is over by the time the child has physically healed. For their daughters, the surgery and its effects linger for the rest of their lives. This difference in perspective is illustrated by the differences between discussion boards for individuals with CAH and those for their parents. As one adult woman with CAH explains, "On the parents' board, you see the [hormone] levels; they don't want to discuss the surgery or sexuality. The parents, they're told if the child has the surgery they'll be happy, healthy, sexually functional, and not have any problems. On the adult board, there's a bunch more on fertility, sexuality, homosexuality—because for the child, what's important is totally different" (Kristin Rupp).

The Limits of the Current Surgical Decision-Making Model

In the wake of the ongoing controversy over early genital surgery, a recent consensus statement has urged caution in surgical decision making (Lee et al. 2006). Despite the call for caution, there is no suggestion of how caution per se may help physicians and parents arrive at a decision that lies in the child's best interest, one that will minimize pain and suffering and improve the child's quality of life. The call for caution alone is inadequate because it assumes that the risks and benefits of the procedures are knowable. But in fact outcomes of genital surgery are highly contested and largely unknown. Moreover, there are wide variations in what parents are told regarding the risks and benefits of surgery, and all too often surgeons either foster or else do not temper parents' unrealistic expectations about how surgery may improve the lives of their children. Thus, although proceeding with caution is important, this alone will do little to improve decision making as long as outcome studies remain few, the surgical consultation fails to fully inform parents, and the gap between expectations and actual outcomes remains wide.

The usual decision-making process for surgical treatment for infants with intersex diagnoses is governed by a number of assumptions about how decisions are made in the clinical context, assuming much of the communication and factors involved in decision making takes place during the surgical consultation, in which surgeons discuss the details of surgical options with parents. The surgeon is expected to cover the risks and benefits of the proposed treatment, as well as the alternatives including the option of no treatment at all. It is generally assumed that surgeons will lay out all the possible options clearly and objectively, that it will be possible to identify and rank the risks, benefits, and likely outcome of each option, and that parents will be able to weigh those considerations rationally, in the context of their family circumstances, to reach a decision that will promote the child's greatest welfare. As my conversations with both clinicians and parents attest, all of these assumptions are flawed.

Discussions of decision making about early genital surgery tend to focus on improving the surgical consultation if they focus on anything at all. Medical consultations about procedures often rely implicitly on the transmission model of communication, one that assumes "stable senders and receivers and unambiguous messages"(Sankar 2004: 431). In this model, "successful communication occurs when the message sent is the message received. . . . Any other result is a failure and is attributed to 'noise' or 'distortion' " (Sankar 2004: 431). This model of communication emphasizes the role of the doctor to make the relevant knowledge available to the parents, who then use this knowledge and apply it to their own circumstances to make an informed decision. The surgeon is expected to identify and present relevant information about the probable consequences of different courses of action. In this model, the communication of knowledge is viewed as relatively straightforward (Alaszewski 2005). One commentator characterized the surgeon's role as that of an "unbiased educator," suggesting that once parents have "obtained sufficient information" about surgery, they can decide whether to consent (Marsh 2006: 120).

It is further assumed that once provided with information, parents weigh all the options rationally to arrive at the decision that is in the child's best interest. They are expected to be told or to ask: What are the options? What are the possible consequences of each? How likely are they? How desirable are they? But this model assumes that physicians have precise information about exactly what will result from any choice made, that they will convey this information in such a way that parents are fully aware of all possible choices and their likely outcomes, and that parents have the time and ability to weigh every choice against every other choice. It also assumes that the parents' decision will be based on a desire to choose the best possible course of action for their child, unclouded by subjective or emotional judgments of their own or of surgeons.

This model underestimates the power of the clinician to define the situation through scientific or biomedical language, which is authoritative because it is perceived as objective and straightforward. It also elides the way in which what is communicated and what is heard will be shaped by the clinicians' and parents' understanding, perceptions, and valuations of benefits and harms. Moreover, it discounts the fact that clinicians consulting with the same family may disagree and that families often consult other parties when making such decisions (Lashley et al. 2000).

The production of knowledge about risk and the subsequent communication of this knowledge are complex tasks. Information about outcomes, however carefully researched and presented, is often incomplete, uncertain, and contested. Some fundamental questions about whether genital surgery harms sexual sensation, produces acceptable cosmetic and functional results, or improves well-being can never be resolved definitively because the answer will always be dependent on too many variables, including the surgeon's skills and judgment, the techniques used, the child's individual anatomy and physiology, the quality of postoperative care, the psychosocial resources available, relationships within the family, and the deeply subjective nature of sexual sensation and of perceptions of cosmetic appearance and function.

Another problem with this model, which applies in some degree to all medical procedures, is the difficulty of applying clinical evidence to the treatment of individuals. Knowledge about risks derives from studies of the effects of interventions in populations and cannot be directly applied to the clinical management of individual patients (Sullivan and MacNaughton 1996; Parker 2002; Schattner and Fletcher 2003). That is, outcome studies generate knowledge about the probability of harmful events occurring within *populations*. Parents need information on the risks and possible benefits for their own child. Evidence from research studies may suggest that some patients are harmed while others are not, but we cannot know which patients will fall into each group. Moreover, there is no way to know whether, as adults, individuals who undergo treatment will agree that the decisions made on their behalf were the right ones. In this respect, more evidence and outcome studies will not simplify or necessarily lead to better decision making. Parents are making quality-of-life judgments for their children: parents holding a tiny baby are being asked to imagine what will be best not only for that child in the short term but also for the adult he or she will become. For these reasons, surgical decisions are and will continue to be complex and demanding. With genital surgery in particular, because so little is known about the outcomes of these procedures, the rational-transmission model of communication underestimates the ways in which norms, cultural influences, and emotions play into surgical decisions. To a significant degree, parents' decisions about surgery are shaped by early birth experiences, by how intersex conditions are understood and framed both culturally and by all the parties involved, and by their own concerns, hopes, expectations, values, and fears.

Too often, the context in which surgical decisions are made is drawn too narrowly, usually to the surgical consultation alone. It is important to draw attention to the complexities of surgical decision making by discussing the other factors that influence surgical decisions, including parents' experiences immediately after the birth, in which intersex diagnoses are subtly—and sometimes not so subtly—framed as abnormal, deeply influencing parents' emotions and concerns and thus providing an important context for their decisions.

Fears of social stigma and a lack of support leave many parents feeling isolated and anxious, forced to make choices about treatment for their children on the basis of inadequate information. If the outcome is not a positive one for the child—and in a significant number of cases it appears that this is true—parents may later have to cope with strong feelings of distress and guilt. Although the medical profession generally acknowledges the effectiveness of counseling in treating a variety of conditions that have both physical and psychological manifestations, few parents of children with intersex conditions reported being offered or actively seeking counseling for themselves or for their children as either an alternative or an adjunct to surgery.

The power of the binary model of sex, often expressed in the well-meaning opinions of clinicians, relatives, and friends, as well as in their own convictions, virtually compels many parents to choose surgical options to help their children fit into a culture that expects everyone to be unambiguously male or female. For some parents, this concern and the concomitant anxiety to which it gives rise can overwhelm such that they override questions about the other consequences of intervention for children including physical pain, difficulties with sexual activity later in life, and the possibility of lasting mental trauma. (Some parents, of course, are unaware of the possibility of these negative consequences, in part because their child's clinicians may also be unaware of or downplay them.) The next chapter explores these experiences from the perspectives of individuals born with atypical genitals and other intersex conditions.

SEVEN

Growing Up under the Medical Gaze:

Adults' Experiences

A woman with AIS recently posted a query to a Web site open to the general public on which individuals ask questions and members provide answers. She writes, "[I look] like a normal girl on the outside, but my vaginal canal is considerably smaller than most women's," which makes penetrative sex with men "very difficult." She explains that this has not been an issue until recently because she previously slept only with women. Wanting to pursue sexual relationships with men, she was now seeking advice on how to broach the subject of her diagnosis and anatomy with potential male partners:

> I could: (*a*) Avoid clothes-off situations until I know a man well enough for the fact that I have AIS to be stated naturally. Not tempting. (*b*) Before any intimacy, tell the guy in question about the fact that I have AIS. This would almost certainly necessitate mentioning it well before I would with any new friend. (*c*) Tell the man that I simply don't want penetrative sex, and not say why. This is OK, but it is reasonably obvious to all but the most inexperienced that my vaginal canal is different from other girls'. So, my question: should I do (*a*), (*b*) or (*c*) (or a (*d*) that I haven't thought of). If (*a*) or (*b*), how should I go about doing the telling? With friends, it's easy, but I can imagine that a few straight men might have a problem with being intimate with someone who has XY chromosomes, and I should think that most will need to process it. (Woman on ask.metafilter.com/mefi/30147)

Readers posted many compassionate responses that suggested ways of being sexually intimate other than penetrative sex, and many thought she should choose option *c*. Several respondents, however, said that not telling a potential sexual partner was unethical:

> Genetically, *you're male*. You don't just have a smaller vaginal canal; you have (or had until they were surgically removed) *testes*. . . . Many heterosexual men have a strong distaste for having sex, or even "messing around," with someone who has balls. That may be, in your opinion, less than open-minded, less than sensitive and accepting, or whatever. But it's not—it shouldn't be—your decision, it needs to be theirs. Ethically, it should be your partners' *choice*. By letting them assume you're a genetic female, you're lying by omission. You're tricking them into a form of sex that they would not have had, given a free choice. (original emphasis)

Another respondent concurred that, ethically, the questioner had to tell a partner about her diagnosis before *any* sexual contact because it would be upsetting for most men to have sex with someone who had "balls." My point in sharing this exchange is to highlight the complications of living with an intersex diagnosis, which inevitably burden those with these conditions. Although coming to terms with a gender-atypical anatomy presents its own set of difficulties for those with intersex diagnoses, disclosure about one's anatomy appears to be especially fraught. Disclosure about gender-atypical anatomy, even if it remains otherwise hidden from view (e.g., testicles) or unconfirmed for most individuals (e.g., chromosomal type), can elicit confusion, misunderstandings, and deep discomfort, even repulsion. Such disclosure could even prove dangerous for individuals given essentialized notions of maleness (e.g., only men have XY chromosomes or testes) and homophobia.

Cultural understandings of what make us male or female are tied to medical understandings of what intersexuality is and how to treat it, thus shaping in significant ways the experiences of those treated. Children and adults with intersex diagnoses face a continuous negotiation between a medicalized understanding of their bodies and an experiential one: their perceptions of themselves are inflected by the perceptions of their doctors, their families, their friends and lovers, and society at large, and also by their own assumptions about how other people see them. The same debates about appropriate care for persons with intersex conditions that center on gender assignment, gender identity, surgery, and honesty emerge in discussions with all the constituencies. As this chapter demonstrates, a trickle down occurs from researchers' theories to clinicians' practices to parents' concerns to children's experiences to adults' reflections.

As previous chapters have discussed, few outcome studies on the adaptation to and the well-being of individuals with intersex diagnoses have been conducted, and it is difficult to draw useful conclusions from those studies because of the small sample sizes and wide variations in methodology. Moreover, existing outcome studies have overwhelmingly focused on gender identity, surgical outcomes, and sexual orientation, rather than on broader quality-of-life issues, partially because physicians and researchers have viewed "correct" gender assignment and genital surgery as the most important aspects of treatment. As I noted in chapter 2, heightened interest in intersexuality in the 1950s emerged not because researchers and clinicians were concerned with improving care per se, but because treatment decisions were tied to long-standing debates about what makes us feel male or female and what makes us attracted to men or women (the real interest lay in the origins of same-sex attraction). This legacy of interest in intersex for what it says about gender development more generally continues to frame our understandings of good care today. Overwhelmingly, the focus has been on making correct treatment decisions, assuming that gender issues, broadly construed, are the most important aspect of care and well-being. As a result, those conducting outcome studies have been particularly interested in trying to ascertain whether individuals with a particular diagnosis are more likely to identify as male or female and why. This focus also derives from the fact that researchers are not exclusively interested in outcomes for these individuals; many hope their research will shed light on gender-identity development more generally.

An individual who accepts his or her gender identity and is married (i.e., heterosexual) would, by the criteria of most studies, be said to have had a successful outcome; but these criteria tell us very little about how individuals adapt to their condition and its treatment, shame or stigma they or their parents feel about their condition, the strain or challenges the condition poses for family and personal relationships, and the experience of being different. This is not to suggest that life with an intersex diagnosis is exclusively characterized by pain and suffering, but it seems reasonable to assume that most individuals will have some degree of difficulty adjusting to their condition and their medical care. All would agree that the goal is for the child to be well adjusted, healthy, and to have good self-esteem. Yet we know surprisingly little from research studies about what individuals feel has helped or hindered their well-being, and thus about how to achieve these goals for children born today.

It should be said that there does not and cannot exist a typical intersex experience. Diagnoses, bodily types, treatment practices, resources, and mental resilience all vary among individuals and families, and all these factors interact in complex ways. Nevertheless, some common themes and experiences emerge from the accounts of individuals gleaned from interviews, correspondence, and online discussion boards. If one generalization can be drawn, it is that there remains ample room for improvement in the treatment and care of those born with intersex diagnoses. Although a few of the adults I contacted expressed satisfaction with their treatment and quality of life, the majority reported past and continuing suffering, both physical and psychological.

The disease-model approach to intersexuality, which broadly includes conceptualizations of, treatment practices for, and anxieties about gender-atypical bodies and creates a profound insecurity about the body and being, and one's right to ownership of both, affects lives in meaningful ways. The psychiatrist R. D. Laing coined the phrase "ontological insecurity" to refer to individuals experiencing a compromised sense of self where the ordinary circumstances of living threaten one's security and being, making self-preservation a major preoccupation (Laing 1960: 42).[1] Here I investigate the factors that contribute to ontological insecurity for those with intersex conditions as they attempt to contend with their condition and society's and medicine's responses to it.

Childhood and Adolescent Experiences

Children born with intersex diagnoses are not, of course, aware of their diagnosis or the uniqueness of their bodies at birth. It is only through experience that they come to understand their bodies as different. Whether this difference is experienced as ordinary or extraordinary, admirable or shameful, depends largely on how the child is instructed—via society, parents, and clinicians—to view such difference. Many clinicians long believed that the less the children knew, the better, even when some, such as John Money, advocated giving children age-appropriate information about their conditions. Consequently, children may go to the hospital and have multiple checkups, undergo repeated surgery, and take medications daily, all without knowing why.

As discussed in the previous chapter, parents are often reluctant to tell children too much about their diagnosis and treatment. Understandably, many parents feel their child is unable to handle the information, or they fear that others may treat their child differently as a result of knowing the diagnosis. It is difficult for parents to know what is in the child's best long-term interests: a child whose parents have sought to protect her or him from the curiosity and insensitivity of others may be traumatized to learn that important and potentially discomfiting information about her or his body has been kept a secret. Most adults I spoke to who discussed their childhood experiences confirmed some of their parents' deepest fears on their behalf: they recalled an atmosphere of shame and stigma about their own bodies, in large part stemming from the secrecy, confusion, fear, pain, and humiliation associated with medical visits and procedures. Ironically, the attempt by some parents to spare their children anxiety by keeping information from them and not discussing their condition may in fact have exacerbated these feelings, creating the impression that the child's body was unspeakably abnormal. Daniel Johnson, who has PAIS, reflected: "People are ashamed of that space in between, everybody wants to keep it a secret. They don't know what to do with intersex kids because they're a mixture, in limbo. It's purgatory for the adult, but limbo for the kid."

In many instances children get the signal that their questions are off limits, things they should refrain from inquiring about. All too often, these early silences and parents' unwillingness to discuss the child's condition or treatment persist into adolescence and adulthood and result in large gaps in the person's knowledge about her or his diagnosis, bodily characteristics and capabilities, and treatment. Jean Butler, who is in her fifties and has CAH, learned early on that certain aspects of her body remained off limits for discussion: "The cue I got early on from my mother was, 'Don't talk about this.' When I was about six and my older sister was eight, she and I had just had our baths, and we were playing naked on the bed and giving ourselves a genital examination. I counted that she had three holes and I only had two. We thought this was uproariously funny, and I went running over to my mother saying, 'Susan has three holes, and I only have two,' and my mother just said to me sternly, 'Get to bed.' That was it. We never discussed it again." Butler explains that this episode curtailed all openness in her family about her diagnosis, even between her and her sister, who also has CAH: "I had a couple of other experiences with my mom like this—the silence, the never talking about it, not even acknowledging there was anything to talk about—and finally I just closed up. I didn't talk to anybody about it, not even my sister. You are the first person. I had the feeling that there was no recognition, that there wasn't an issue because I tried once to bring up my surgery with my mother and she said, 'Your doctors assure me that you're just perfectly normal after your surgery, there's nothing wrong at all.' That was basically her attitude: surgery was fine and there's no problem. I think she didn't feel comfortable going into it, and as a result I didn't either." Lisa

Todd's parents told her nothing about her CAH diagnosis. She says it was not until she had the first vaginoplasty at age twelve when things started to "come together" for her: "At that point it was just, 'Oh, well, you were born without a vagina, so they need to make one.' This is the first we ever spoke of it, and this was while I was being examined for the first vaginoplasty."

Of course, telling the child the truth in no way guarantees that anything will be easier for him or her. Indeed, by the time many children get the details, irreversible decisions have already been made, ones over which they had no say. Secrecy seemed especially prevalent when it came to genital surgeries. Most of the individuals I spoke to who had genital surgery were, like Butler and Todd, told very little in preparation for the procedures. Others were told a few details about some surgeries and nothing about others. Teresa Diaz, who has PAIS, was taken in for vaginal surgery at age twelve. Her mother told her about this procedure and had prepared her for it, but what her mother did not tell her was that her clitoris would also be surgically reduced. She recalls: "I had some fear that they'd do surgery on my clitoris [after her mother saw it and became concerned about its size], but since they hadn't said anything, and they seemed to be very explicit about everything else, I started to breathe a little easier, thinking, 'Well, if that were going to happen, they would say something.' But after the [vaginal] surgery I was feeling around my pelvis, and my clitoris was not there, just a crusty area that felt awful. I was horrified, but I didn't express it. Removing my clitoris was confirmation of the fear that I might be freakish."

Indeed, many echo the sentiment that the treatment itself contributes to a feeling of otherness, of being different. Diaz notes that she had recently "discovered" her clitoris just prior to surgery, and rather than finding it disturbing, found it to be a pleasurable asset. Butler recalled that she "wasn't told anything about my surgeries. Never. Not in preparation for them and not after they happened." The same held true for Todd, who had a vaginoplasty at age twelve and a subsequent "repair" at age fourteen: "I was 'going on vacation,' my parents said. It was done in the summertime. I remember a lot of pain, a lot of blood, a lot of confusion. My parents didn't talk to me about it. The endocrinologist, I remember, did send up a psychologist the night before to tell me what they were going to do, but it was all about hormones. I really did not have a clue what they were talking about."

Kristin Rupp, who has CAH, observes that parents generally say that their children are fine because they are focused on the immediate health issues. She notes that this focus obscures all of the other issues the child experiences about feeling different, sensing something is wrong but not knowing what, and being told it is not alright to raise certain questions. She adds, “the kids don’t want to talk about it because they’re busy just surviving and living with the pain.” The silence, some argue, fuels the shame they feel about their bodies.

In addition to the trauma from these unexplained surgeries, some individuals recall a sense of discomfort, vulnerability, and humiliation due to repeated medical and genital examinations. Louise Rutherford, who has CAH, observes, “I remember feeling like a freak show because every doctor had to do a genital examination. It didn’t matter how much you cried or that you were cold. No one cared.” Several others I interviewed complained about being objectified by clinicians and of being spoken about as if not present:

> My sister and I had every endocrinologist looking at us. Then the urologist would look at you. And then if you were in the hospital, the nurses would look at you. And then, when you go to the pediatrician, he’d look at you. It wasn’t just once a year to see how everything’s going; it was so many times a year. It could be thirty people over a year who would actually see you. They’d poke and prod and say, “Well, the vaginal opening isn’t big enough,” and of course, they’d insert their finger to prove that. It was painful, but you never could really say anything. They wouldn’t look you in the face or the eyes. They’d come in and talk about you like you weren’t even there, and you didn’t have any say in the matter. (Kristin Rupp)

Parents were often present but were either unwilling or unable to stop these exams. Their perceived passiveness in some cases eroded the child’s trust of parents who did not speak up for or protect them.

Puberty can prove especially difficult for individuals with an intersex diagnosis. This is often the time when they begin to understand and gather information about their bodies, diagnoses, and treatment history, in addition to experiencing common difficulties of adolescence such as insecurity, conflicts with parents, and acute self-consciousness about their bodies and appearance. A number of individuals on reaching this age refused to submit to further medical exams or, in several instances, to attend any medical appointments.

> I had genital inspections every single time I went to the doctor. And not only that, I was used for training. It’d be me, my mother, the doctor, and then there’d be about a dozen others—nurses, residents, interns, medical students—and they would all be standing around looking at my genitals, and the doctor would be making comments, and I’d be lying there pas-

> sively while this was going on. I finally rebelled. I just refused to go to the hospital. (Jean Butler)

> They would lay me on this table, and the first thing out of my doctor's mouth is, "Pull down your pants." And I never understood why. And then all these other people would walk in, and I'd lie there, completely exposed, while these strangers are probing and looking at me and writing notes. My mom would be in the room, but she would never say anything. Later I would tell her, "I don't understand why they're doing this." I asked her several times as a little girl, "Why do they do this?" And she would just say, "Because it's what they have to do." That would happen every visit until I was probably fifteen or sixteen years old. And then I just quit going. (Hannah Samson)

These experiences had the unintentional effect of making these girls feel that something about their body was wrong or shameful and of giving them a sense of deep bodily vulnerability and of a loss of control over their bodies. This loss of control can lead to depression, anger, a sense of disconnection from their bodies and a disinterest in taking care of them, and a retreat from family members and friends. Many individuals I spoke to have learned strategies to manage their pain and vulnerability which can include trying to numb out their feelings or else withholding painful feelings in an attempt to master overwhelming or unmanageable experiences.

In addition, follow-up treatment and medications sometimes have adverse side effects or cause outright pain. As one example, after vaginal surgery dilation of the new vagina may be recommended to avoid scarring and the closing of the vaginal opening. Today, many parents and surgeons are reluctant to consider dilation for small children because of the potential for physical and psychological trauma. As Suzanne Kessler observes, "What meaning does the intervention have for inserter and insertee? Does the body part lose all its sexualized connotation or is it experienced by the girl as a violation by her parents?" (1998: 59). None of the parents I interviewed had performed dilation on their daughters. When dilation was considered necessary, it was done under general anesthesia. Many of the clinicians I spoke with no longer recommend dilation in infants or children, but many feel it is necessary at puberty, often as a way to avoid further surgery. A number of adults I spoke with had experienced dilation beginning at puberty. They recall the experience as painful, sometimes agonizing. Ramona Diaz recalls her daughter's attempts at dilation: "They wanted to try dilators to enlarge her vagina. She said, 'I can't use these. These are way too painful.' She tried them once or twice, and that was it." Kristin Rupp describes her own experience:

> When I had my final surgery at seventeen, they gave me a set of four dilators, which progressively got bigger to the size of a man's penis. The first two were uncomfortable, but I could insert them. It took a week or two per size. I was doing this with stitches, and it hurt like hell! By the time I got to the third one, I couldn't insert it, so there was no way for the fourth one. When I went to the urologist's office, I figured he'd realized that every time he was inserting his finger, the opening wasn't getting any bigger. He gets to the fourth one and manages to insert it. What I wanted to do was scream out in pain, but I didn't, because I had always been told you just don't do that.

Those aware of their condition also face the difficult decision about disclosing it to their friends and potential sexual partners. The desire not to be perceived as different and the mercurial quality of teen romances make it hard for many of these individuals to even enter into a relationship. At an age when the pressure to conform to social norms can seem overpowering, they fear becoming the subject of gossip among their peers. (Of those who underwent surgery ostensibly to normalize their genitals, most I spoke to still feel they look abnormal.) For some young adults, it is easier not to disclose their condition to anyone and to avoid relationships that might lead to someone learning about their condition or seeing their genitals. Rebecca Davis, who has two adult daughters with CAH, says puberty was when "all the stress began." Her daughters "didn't fit in with the girls, and they didn't fit in with the boys, so they just withdrew from the world."

It is perhaps not surprising that the combined stresses of adolescence, a chronic condition, and living with what they often perceive as a shameful condition and atypical genitals can lead to distress, depression, and even suicidal feelings in some young adults with intersex diagnoses. Elissa Ford, who has PAIS, says: "I was really isolated, extremely lonely, and couldn't talk to anyone about how I was different, my diagnosis or my body, or the surgeries. I was suicidal for most of my teenage years and a lot of my twenties." She stopped taking her hormone medication "because it was indicative that I was different." Noting that her mother told her that she needed it to live, she said, "I was suicidal at the time, so that didn't impress me much."

For about six years in her late teens and early twenties, Rupp, too, refused to take her medication for CAH. She had always been told that she might die if she did not take it; but, she said, she accepted this risk because "I grew up wanting to be dead because of all the stuff that was happening to me." She describes the ways in which she dumped the medication down the drain or otherwise avoided taking it, always making sure to get rid of the pills so that her mother would not find out: "Every six months we'd go to see the endocrinologist, and my blood work would be screwed up, and they couldn't figure out why. I don't think they realized that I wasn't taking my medicine, because they had beat it into my head that if I didn't take it I was going to die, but that's what I wanted." When I spoke to her, Rupp was again suicidal after her recent breakup with a partner of many years, who told her she simply could not handle Rupp's struggle with her condition, and her pain and discomfort during sex, anymore.

At sixteen, Rosalind Bittle, who has CAIS, had surgery that she had been told would be a hysterectomy. The truth was that she was going to have her testicles removed, a fact she did not learn until many years later. She relays that the surgery had traumatic consequences for her: "I became depressed immediately after my orchiectomy. I struggled terribly with depression and attempted suicide. . . . I wouldn't look at my face in a mirror for a couple years after the surgery. I was really suffering terribly." She adds that after the surgery the doctor told her mother that they had to decide what to tell her and who would do so. Her mother took on the task, but as Bittle says: "She took it on with the intent of telling me what she understood about it, but she didn't understand much, and my mom could never find the right time, in part because I was struggling for years over this. But the problem was that no one gave her the tools or taught her or helped her to understand and deal with this, so she was unable to help me. She was swamped beyond her abilities to cope. And she was terrified that I would find out about my diagnosis and that I would kill myself." Just as parents are often not offered psychosocial support at the birth of their children, it is also rare for them to get this support as their children age. Yet each developmental transition for the child raises new issues for both the individuals themselves and their families. These transitions are most often navigated without any help about how to broach sensitive topics concerning atypical anatomy, disclosure to others, fertility, and sexuality.

Adulthood: Asking Questions, Dealing with Consequences

Individuals with intersex conditions may be too fully occupied with enduring adolescence to ask questions about their condition until later. Those I interviewed said it was often not until their late teens or twenties, or sometimes even later, that they first began to talk to their parents and others about their bodies, diagnoses, surgeries, and feelings. Some found out about their condition through their parents, but many found out through other means, such as reading books and online materials and covertly asking family members or requesting their medical records. Kristin Rupp observes that it took her "most of my teens and twenties to figure out what happened to me, and then, at the end of my thirties, I was able to start dealing with it. I've noticed on the message boards you don't have a lot of teenagers; you really don't have a lot of people in their twenties. People start talking when they're in their thirties."

Daniel Johnson was born in a rural farming town in the southern United States. At birth, he was announced as a girl, with no indication that he had PAIS. He lived as a girl throughout his youth, but at puberty he unexpectedly began to masculinize: his voice lowered, his "clitoris" grew, and people frequently took him for a male. It was not until he was in his early twenties that he found out that he had been born with PAIS. Working at his university library, he frequently read the titles on the shelves and stumbled across John Money's *Man and Woman, Boy and Girl.* When he opened the book, what caught his eye were the pictures of atypical genitals and the full-body nude photographs of children and teens with intersex diagnoses. This was the first time he had seen any information that reflected his embodied experience, and seeing the pictures gave him a sense of recognition. Although he had never been told his diagnosis, he was able to figure it out by reading the book. This discovery marked the beginning of his transition to living as a male, which he had felt himself to be since puberty. He said it was a "sad situation because there was somebody that I liked, and I did not dare tell her the situation."

Those who learn of their diagnosis in adulthood may feel nervous discussing it with their parents or with other family members. For some, talking to their parents about their feelings, and even expressing anger, can improve these relationships. Ramona Diaz says that while conversations about her daughter's diagnosis have been painful, "It is much better for us to talk about this. It has brought us closer." Sometimes, however, the issue both reveals old fault lines in family relationships and creates new resentment and anger. Individuals may find their parents in an apparent denial that medical treatment could be responsible for their child's later suffering and unwilling to discuss the subject. Alison Sawyer, a woman in her forties with salt-wasting CAH, says: "One night I could not take it anymore and called my mom in a crying fit, wanting to know how old I was when they'd done the surgery and what was done. For her it was 'out of sight, out of mind,' that when they were done, that's the last she had to think of it." Hannah Samson, a woman in her thirties with the same diagnosis, says that her mother was very upset by the question, "What happened to my clitoris?" "It was like bringing up the past. In my mom's mind, once they did the surgery, it was over. I was fixed, and all they had to do was make sure I got my medicine. I think that's how the doctors see it too. I don't think it even occurred to them that it would affect the women, especially, so profoundly later in life."

Elissa Ford reports her mother's dismissiveness: "My mother says, 'Are you still talking about this? Your life is fine! You have a great job, you have a house.' She doesn't see anything else. I think most parents can focus on that and not see the rest of it." Jean Butler, whose mother so abruptly deflected her daughter's curiosity about her genitals when she was a child, feels tension with both her mother and her sister, who also has CAH but does not want to talk about it. She is concerned that parents today adopt a similar attitude that treatment or "surgery was fine, and there's no longer any problem." Referring to an Internet discussion board for parents of children with CAH and for adults with the condition, she notes that "they never talk about psychosexual aspects. It was exactly like my parents. They complain any time anybody tries to talk about surgery, sexuality, gender identity, behavior, and say that we're overfocusing on that, like that's all there is to CAH. Well, I talk about it because so many parents refuse to acknowledge that there is anything beyond the biochemistry."

Those who learn in adulthood what was done to them as children are often understandably angry with their parents. This anger has destroyed some relationships and severely strained others. Samson says she now realizes that her mother "did what she thought she had to do to save me. I understand that now, but I was so bitter towards her for a long time." Rebecca Davis says of one of her daughters: "It has just cost me more than I'll ever be able to verbalize to listen to her anger. I understand why she's angry with me, and I prayed for it to come because she'll never come to terms with herself until she gets past the anger, but it is hard because so much of my adult life has gone into this child who has now turned on me. She still loves me and I love her, but she's railing against me for doing the best I could do. She can't say to me, 'Mom, I know you did the best you could do, but I'm really angry.' She just *screams* at me" (original emphasis). The anger these individuals feel toward their parents has a basis in a violation of trust: they believe their parents betrayed them by withholding the truth about their condition or failing to protect them from pain and humiliation. Louise

Rutherford recalls how she felt about her medical visits: "My mom was always in the room with me. She never said anything, and I'm very angry with her for that. I'm having to learn how to forgive her for that. I now know there's a lot of circumstances surrounding that, but it's hard when you're a child, thinking your mom is going to stick up for you and she doesn't."

For many I spoke to, this sense of eroded trust extends to the clinicians who treated them. Lisa Todd says: "In *my* perfect world, they would *not* have cut [my clitoris] off. That's where my anger comes from, that it was not my choice to make. A doctor and my parents made this choice for their own comfort, for the comfort of babysitters, siblings, and relatives. Well, what about *me*?" (original emphasis). Others remain angry with their clinicians not only because they directly inflicted discomfort or even pain but also because they failed to observe confidentiality. Rupp recalls that she "never felt like I could trust the therapist because I knew if I said anything, that it would eventually go back to my mom. Like if I said anything to an endocrinologist or a urologist, if my mom wasn't in the room, it'd be written down on the chart, and the next time he talked to her, he'd say something." And Claire Halstead, who has PAIS and is now in her forties, says her clinicians took photographs of her, which she later saw reprinted in a medical journal article. She observes: "At the time they said they needed the photos to show other doctors, but they didn't say they were going to publish them in a journal. If he had told the complete truth, he would have said, 'I'm trying to make a name for myself in endocrinology, so I'd like some photos of you so that I can write an article about your condition and establish myself as one of the experts on this condition.' That would have been truthful."

As a consequence of these medical experiences, many of the adults I spoke to have developed a lasting distrust and fear of doctors. Some avoid any medical visits, even for non-intersex health issues such as a case of flu. But perhaps more frequent is avoiding routine gynecological checkups and pelvic exams. Rutherford says she is terrified of any suggestion of a physical exam. Butler concurs: "It became like, 'Don't allow anybody to know that there's anything wrong with you, because they're just going to go in there and fix it for your own good.' "

In addition, individuals may carry a lasting sense of discomfort and shame about their bodies. Bittle, who became suicidal after genital surgery, struggled with the implications of her diagnosis for her gender identity: "I've always considered myself a feminist, but I could not find a way to be a woman and not be able to have children. I felt degendered. I called myself an 'it' for a long time." Others are acutely uncomfortable with their atypical genitals, especially following genital surgery. Despite the importance surgeons place on a good cosmetic result, a number of the individuals I spoke to considered their bodies ugly and disfigured. Rupp notes of her surgically constructed vagina: "My girlfriend said it looks very different. It's not repulsive or anything, but it is different." Others had more negative reactions, including Samson: "When they separated my labia, they left too much skin. I've been able to compare myself to other women, and it looks like they just ran the scalpel down and sewed them. I'm very self-conscious and feel freakish because of this." And Todd, who is now in her thirties but had clitoral surgery at four months, describes her clitoris as looking like "a guy's penis when he's erect and cut it off right where it meets the body, that's what it looks like. Oh, it just looks so bad, awful really."

This self-consciousness is frequently accompanied by other feelings of being abnormal and alienated from one's own body. Teresa Diaz says that she "felt like my body was unruly, that it had betrayed me and that [surgery] was the price to pay." Rupp observes that when "I had surgery on my vagina and to reduce my clitoris, the doctors said 'We're doing this to make you normal,' but you already know that you're not because of the surgeries they've already performed on you. To them I was abnormal, but in my mind I was normal until they started doing all the surgeries." The adults I spoke with described bodily and mental symptoms that included severe depression, eating disorders, fainting spells, and panic attacks.

The combination of eroded trust, poor self-image, and a sense of feeling freakish has made it difficult for some adults to form close friendships, much less sexual relationships. Many feel unbearably vulnerable. Daniel Johnson, Jean Butler, and Karen Rollins, all now in middle age, have never been sexual with anyone. Butler explains: "Shame is the problem. I panicked at the first closeness. The first touching, just holding, just showing a physical interest where, immediately, there was this seizing up, and I could not bear it."

Johnson's feeling of shame about his body and genitals was compounded and complicated by his encounters with clinicians in his early twenties. After learning of his diagnosis in the library, and reading that some individuals with PAIS live as males, he visited clinicians in a nearby city to begin the process of transitioning to male. The doctors felt that his phallus, at two inches, was too small for a male gender assignment and strongly recommended he remain in a female gender role and undergo feminizing genital surgery. Johnson says: "The doctors wanted to make me more female, but I didn't understand that at all. They're telling me I've got an XY chromosome pattern, I am fertile, and all of these other things are either normal or just below the male level. I was thinking, 'What would I gain in feminization when I don't feel that way? And if I can have erections and orgasm and reproduce?' So I was like, 'Hell no.' But they still wanted to do the feminization surgery." Against their advice he began transitioning to living as a male at great personal cost, because it meant leaving his hometown and family, who he felt would not accept his decision. Even though he rejected the clinicians' recommendation to continue living as a female, he describes how he believed and internalized what they said about his penis size: "They said, 'Well, you'll be a lot less desirable as far as penetration of a female loved one.' They were telling me, 'Who would *ever* love you with this [small penis]? What could you have to offer?' And I *bought* it. I bought it because they just seemed so sure. But what was I going to do? I did not feel female and hadn't for years. I said, 'I'm not going to let you cut it [his penis]. If you can enhance it, help me get *more* of that [penile length], this is the kind of surgery that I want.' And one doctor, he got very angry with me, and he said, 'Well, why don't you do it yourself? You're college-educated!' "

Realizing that Johnson would not follow their advice, the clinicians denied him further medical care. He moved thousands of miles away from his hometown, found a surgeon who would perform surgery to make his genitals more typical for a male, and began living as a man. He says, "I was happy for a while because I was hugging girls, I was kissing, I was making out. Before I transitioned to a male, I had no dating, no kissing, no nothing." But the comments of the clinicians still haunt him, and, at forty-four years, feeling enormous shame about his penis size, he has never shared his naked body with anyone because he fears no one could ever accept his genitalia: "I could always attract females, but kissing was as far as I could go. Some girls would be like, 'Maybe he's not interested,' but that wasn't the thing! I would go out on two dates, and in my mind, on the third date, you must do something. But I would say, 'No.' After five years of this I began to get depressed. I had not been close with anybody. I don't know why I was always afraid. I guess it was just because I was ashamed of my body, my genitals." He gave up dating several years ago and now says, "I've never had sex because you have to prepare someone concerning a situation like this. I completely gave up on the whole situation. I just got fed up because I wanted to take it a step further, into intimacy, and I couldn't."

The burden of what clinicians said about Johnson's "inadequate" penis has stayed with him. Perhaps it is true that he would have felt inadequate even if the physicians had never made disparaging comments, but it seems obvious their comments scarred him. He now understands that penis size might not be as important as his doctors implied because women might prefer a range of sexual acts over penile-vaginal intercourse, but this knowledge is of little comfort to him: "What they [the clinicians] said was very harsh. I've now come to know that a lot of women don't like penetration, but I found this out much later on, when it was too late." In recent years, Johnson has struggled with depression and suicidal thoughts. He has given up on the idea of ever having a relationship with someone.

Others I spoke to began sexual relationships relatively late in life. Todd says she was in her late twenties the first time she ever had sex: "Previous to that I had boyfriends, but I just was not comfortable with sex because I just knew I was different." Alison Sawyer comments: "I have felt so much shame about myself and my body, and I think it came early on when it wasn't being discussed, where nobody knew, not even my brother and sister. It wasn't talked about to me or in open terms; it was a 'medical issue,' a salt issue, a growth issue. It's not necessarily the surgery that created the shame, it's the hiding and not being honest with me."

Although there are diverse forms of sexual pleasure and activity, many of which do not involve penile-vaginal intercourse or penetration, few adults I contacted reported satisfying sexual experiences. Not only did they find it difficult to become sexually intimate with anybody but they also found it immensely difficult and stressful to discuss their bodies and desires with sexual partners. One woman reports an encounter with a new partner: "Of course, I had to explain it to her. I wouldn't let her touch me, and I explained to her, briefly, 'I had surgery, and this is what happened.' I tried to explain it to her a few days later, and I never heard from her again" (Hannah Samson).

Like the woman who posted the query that opened this chapter, a number of individuals reported deeply unsatisfying or painful sexual activity.

> I have a very, very, very small amount of clitoral sensation. It's not completely dead, but it doesn't do anything for me. God help me if I should ever find the urologist who did this to me. I would call him a butcher. (Lisa Todd)

> I have some pleasure, but it isn't consistent, and the intensity isn't there. After the surgery I didn't have an orgasm for a very long time. (Teresa Diaz)

> It hurts every time. When I would have sex with [girlfriend], it would be uncomfortable if she penetrated me too far. I'd make her stop because it

> hurt too much. She'd say, "I'm a horrible lover," and blame herself because she couldn't give me the same pleasure that I could give her. What she said was in order for her to give me pleasure, she has to hurt me. It was this vicious cycle where because I had the pain and I'd make her stop, she believed she was a bad lover, so I backed off even more, and then she would say the things like, "Well, this would turn on most people," which was kind of like saying, "You are a freak and abnormal." (Kristin Rupp)

Even those who have been able to establish lasting relationships are afraid their condition will prove repellent to their partners. Rupp says of her girlfriend, "There was always a part of me that would say, 'She's going to leave because of this.' Maybe not now, but later. She's going to want somebody normal, someone where she doesn't have to try to figure out what's going to be pleasurable, as if this is all that matters." Eventually Rupp's partner of six years did leave for the exact reasons Rupp feared, an experience she feels has scarred her further: "I am really sad, because this has been my secret, and now I've shared it, and because of sharing it, I lost my marriage. And now I'll have to keep it a secret forever because no one will love me because of that. I don't know if I could take the chance of exposing myself to somebody else. I necessarily can't meet basic human needs. What kind of relationship can it be where there's no sex, ever?"

As Rupp's experience demonstrates, even couples who have been together for a long period of time have found their partners still uncomfortable with the issue or else tiring of it. Sawyer, who has CAH, was told by her husband of fourteen years that she "thinks too much" about her condition; he suggested she "stop looking back." Bittle, who has AIS and whose diagnosis was first suggested to her by her husband, was shocked to discover that her karyotype was XY. She recounts that the two of them reacted in different ways: "I just melted, and my husband would not talk about it. He said, 'It's medically interesting, but it doesn't mean a thing emotionally, and it doesn't bother me. I don't know why you're upset about it.' " She says she "put up with that for about two weeks" before she finally said, "I am really suffering. I am happy it doesn't have emotional meaning for you, but it has a great deal for me, and you're making this harder by shutting yourself off from me about it." He started to cry and told her that he had been afraid that if he acknowledged any significance in her diagnosis and XY chromosomes, she would think he felt differently about her. She said, "I can handle that. I just couldn't handle the silence that he placed around me, because at that point, I needed to talk sixteen hours a day about how I was feeling, and how do women have testes, and did this make me a lesbian, or did it make me a hermaphrodite, or did it mean that our marriage was illegal. It just had these massive implications for me."

Getting Help and Finding Others

Some of the individuals I spoke with described how they came to terms with their situation as they grew older. Many spoke of having lived in hiding, ashamed of their condition, until quite recently:

> I was in therapy for thirty years, and the first twenty years I just couldn't talk about it. I've just gotten to the point of being able to function as an apparently normal person, but there's so much that's still in there. (Jean Butler)

> Bringing voice to things, and saying things out loud for the first time is intimidating and scary. Part of me feels shameful that I'm forty-three and I'm just doing this, but I also know I wasn't ready before. (Alison Sawyer)

> I'm just beginning, as I speak with you and a few other people, not to be so ashamed. I've lived in shame; I've hidden all these years. What a waste. Each birthday brings such pain because I felt that by age thirty-five I would be well-adjusted, and one of my problems was that I believed that medical science would come up with a cure. (Daniel Johnson)

As Sharon Preves notes in her research on the life experiences of adults with intersex conditions, an important factor in adjustment for some was psychotherapy; another was simply becoming aware of other people with similar diagnoses, through the recent proliferation of online or face-to-face support groups (Preves 2003). Preves found that identifying with others who have had similar experiences coping with their differences increased individuals' sense of empowerment and acceptance, describing the process as one of "coming out." This process, she argues, involves not simply recognizing one's diagnosis, but sharing this information with others as a way to normalize and validate one's experiences and to recast one's condition and body in a positive manner. Samson says: "I've lived in a cave and ashamed most of my life. To meet other women has been fantastic for me." Elissa Ford shares a similar experience: "Meeting thirty other women was just so huge because I really didn't think they existed. And then, finding out that they've gone through so many of the same things I've gone through was just amazing. Getting the full picture of my diagnosis, I really came to terms with the whole XY chromosome thing in that meeting."

When asked what should be done for individuals born with intersex diagnoses today, a number of respondents suggested that parents and clinicians should be more honest with children and tell them more about the diagnosis. They also advocated providing more counseling and support. Teresa Diaz notes: "I think the physicians need to be more aware of the resources in the therapeutic community around them. When I talk to my pediatric endocrinologist and the people in their clinic, they say, 'But where could we have sent you? There was no one in this area who could have dealt with you and your family around those issues.' And they don't quite understand that any certified therapist who does work on sexuality and sexual abuse, and works with families on those issues, would be suitable."

Some individuals strongly advocated deferring surgery until adolescence or later, when the child could participate in the decision. Someone expressed a dissenting view in an online forum for parents of children with CAH and adults with the condition: "The first surgery was my parents' decision and the second one was my choice entirely because by that time I knew I looked strange and I only wish I had done it much sooner, I probably would have saved a lot of money on psychotherapists trying to find self-esteem I didn't have. I think the psychological impact of how the child is going to feel isn't being explored; I think that is the bigger issue than consent." For the most part, however, participants shared Daniel Johnson's view: "I say do no surgery and let the child decide at puberty. That would be the only way it would be fair to the individual. Although they may not think it at that time, and you have to assign a sex, you wouldn't do anything surgically because you don't know the psychological. See, there's two aspects—anatomical and psychological—and you're not going to know at such an early age, so you would have to wait until they're old enough. You would need to explain to this person that they're different and ask how they feel as they're growing up, but you'd have to wait for the surgery." Ford now questions whether surgery is necessary at all: "Even if it was guaranteed that I'd have no loss of sensation, I would not have surgery because I am pretty happy with the idea that there is a lot of variety in life. It does scare me sometimes that I am the epitome of variety, but at the same time, I think it's kind of cool." As Alice Dreger has noted, unusual anatomies do not simply force the question of what or who is normal, but much tougher questions such as why we value and prefer the normal (Dreger 2000).

Drawing on their personal experience as well as on social movements concerned with health issues, some adults have reframed intersexuality from an individual issue to a social and political one—one that is deeply tied to cultural ideas about gender and the proper relationships among the body, gender, and sexuality. In doing so, they have opened intersexuality and its medical treatment to contestation and negotiation. As holds true for other social movements around health issues, activists have mobilized to change how intersexuality is understood and treated by raising awareness about their experiences, including the sometimes harmful effects of treatment on their emotional health and social interactions.

In the following chapter I examine these negotiations, showing the different forms of legitimacy involved in interactions between medical viewpoints and lay perspectives. Moving away from the individual level of diagnosis and treatment discussed in previous chapters, I focus on how groups of adults define and understand their intersex conditions and negotiate what constitutes appropriate treatment with groups of physicians. Whereas the medical viewpoint encompasses clinical descriptions and laboratory data, lay perspectives are infused with and affected by the perceived degree of interference with an individual's sense of well-being and with interpersonal relationships, as well as with social roles, and with and by treatment issues. This intersection is highly charged with conflict, which affects all parties, resulting in changes in medical conceptions of illness and in the lay understanding of medicine and science.

EIGHT

The Intersex Body in the World: *Activism and Social and Medical Change*

For forty years after it was introduced, Money's treatment paradigm largely went uncontested by clinicians and recipients of treatment. By the early 1990s, the first generation of those treated according to the protocol was in adulthood. They began to seek information about their condition, examine what had happened to them, and, in some cases, find others who shared similar experiences. While some individuals sought simply to connect with others, others were angry over what they perceived as treatment practices that had not simply failed to improve their quality of life but had caused irrevocable physical and emotional harm. Not content to quietly adjust, these individuals embraced a wider goal: to change how medicine and society respond to intersexuality. Initially rebuffed by doctors, they literally took to the street to change how intersexuality is understood and treated in order and to save future generations from suffering the shame, pain, and anguish they experienced.

These developments emerged in a context of broader social changes in attitudes toward gender and sexuality, as well as in the authority of science and medicine. In this environment, feminist academics and eventually parents of children with intersex diagnoses, sympathetic clinicians, ethicists, and legal scholars began offering alternative views on the meaning and construction of intersexuality, and on appropriate medical responses to it. These views provoked turbulent controversies in the popular media, propelling intersexuality from the shadows to popular consciousness. What was once known only to physicians, researchers, and those affected suddenly became showcased in national newspapers and on network television programs.

Whereas some clinicians had initially characterized advocates as "zealots" (John Gearhart qtd. in Angier 1996) and "green-wellied loonies" (Toomey 2001: 39), the mounting challenges to contemporary treatment practices by lay persons—and, later, clinicians—meant that practitioners were forced to respond to the criticisms. Clinicians and researchers convened symposia and working groups to discuss controversial aspects of treatment, but internecine battles often hampered their efforts, and they failed to recommend any substantive changes to treatment practices.

This changed in 2005 when fifty international experts spanning multiple medical specialties gathered in Chicago to revisit treatment guidelines for infants born with intersex diagnoses. The meeting, jointly sponsored by the Lawson Wilkins Pediatric Endocrine Society and the European Society for Paediatric Endocrinology, marked the first time in over fifty years that physicians so thoroughly reviewed medical treatment for intersexuality. The "Consensus Statement on Management of Intersex Disorders," as it came to be called, marked the first time that medical professionals and advocacy groups worked together to reconsider medical care for those born with intersex diagnoses. Recognizing a role for patient input, meeting organizers invited two patient advocates to participate: Cheryl Chase, the executive director of the Intersex Society of North America, and Barbara Thomas, a member of XY-Frauen, a German support group. Only ten years earlier Chase, unable to begin a dialogue with clinicians over treatment, had taken to picketing medical meetings to have her voice heard.

The consensus statement that emerged from the conference recommends several significant changes to medical care that demonstrate a marked shift in thinking about treatment for intersex diagnoses (Houk et al. 2006; Hughes et al. 2006; Lee et al. 2006). These include revising the medical nomenclature from stigmatizing terms such as *hermaphrodite* to more clinically descriptive terms (e.g., *congenital adrenal hyperplasia*); integrating psychosocial support and professional mental health care for persons with intersex diagnoses and their families at all stages of development; advocating honest and complete disclosure with patients and their families; considering the potential for fertility for all infants; curtailing genital exams and medical photography; and advising a more cautious approach to early genital surgery, including no vaginoplasty in infants with short or absent vaginas (and no vaginal dilation before puberty) and clitoral surgery only in "severe" cases.[1]

In addition, participants introduced several important conceptual changes regarding gender, sexuality, and genital surgery. The document states, for example, that homosexuality should not be construed as an indication of incorrect sex assignment and that a person's wish to change gender should be supported. Undermining a common rationale for early genital surgery, the document notes that no evidence suggests that surgery alleviates parental distress and further underscores that parents should not be given unrealistic expectations about genital surgery. Finally, the participants agreed that outcome studies should explore a broad range of factors relating to quality of life such as social adjustment and participation, as well as the ability to maintain interpersonal and intimate relationships. For someone unfamiliar with the previous debates, these recommendations might appear commonsensical, but they proved groundbreaking for the field.

What made these changes possible? In this chapter I explore the emergence of the support and activist networks that have developed for individuals with intersex conditions and the ways in which activists have sought to change dominant medical conceptions of and treatment for intersexuality. I also assess the effects their efforts have had on clinicians by exploring their responses to internal and external challenges to the dominant medical paradigm.

The Background: Social Movements and Scholarship

In the decades during which Money's treatment protocol dominated, a number of broad social changes opened the way for a discussion about the treatment of intersexuality. Diverse movements such as the feminist and women's health movements, gay and transgender movements, and patient health movements influenced the development of challenges to the traditional treatment paradigm for intersexuality. These movements formed part of the larger sociopolitical and cultural context that made debates over treatment for intersexuality possible, probable, and meaningful: intersex advocacy built on and benefited from the organizations, leaders, tactics, and previous efforts of other movements. On another level, scholarly work in feminist theory especially about gender and the body, feminist critiques of science and medicine, and gay history and queer theory, which emanated from and responded to these social movements, provided an analytical and theoretical basis for the critique of the traditional treatment paradigm for intersexuality.

The women's health movement, which emerged out of feminism's second wave in the late 1960s, placed female sexuality and sexual self-determination at the center of women's health concerns (Ehrenreich and English

1973; Morgen 2002). Providing a gendered analysis of the health system and a critique of medical practices and laws governing female reproduction and sexuality, feminists sought to reshape understandings of female sexuality to emphasize female sexual pleasure.[2] Challenging doctors' authority in matters of female sexuality, feminists provided a history of women's health, alternative views of women's bodies, and new knowledge of women's sexuality that emerged out of their own bodily experiences and ran counter to medical accounts. They also generated important critiques of medical practice, questioning medical orthodoxies, calling for increased patient autonomy, condemning paternalism in medicine, and challenging the view that medico-scientific knowledge constituted the only form of legitimate knowledge about the body. Among the contributions of this movement were woman-centered health services and health information written by and tailored to women. These developments and the critiques that undergirded them later proved important for intersex activism.

While women's health advocacy was burgeoning, feminist academics (many of whom were active in the grassroots women's health movement) began mounting powerful challenges to conventional explanations for gender differences that framed them as natural or biologically caused (e.g., Rubin 1975, 1992; Kessler and McKenna 1978; Fee 1979; Freedman 1982; Fausto-Sterling 1985). In 1975, for example, the anthropologist Gayle Rubin introduced the concept of a "sex/gender system," which she defined as the "set of arrangements by which a society transforms biological sexuality into products of human activity" (159). This paradigmatic shift resulted in powerful arguments against essentialist theories that had used reproduction to yoke gender with sexuality (Vance 1991: 876). Rubin later (1992) built on this work in an influential article in which she proposed an analytic distinction between sexuality and gender.

Women's health activism and feminist scholarship did not emerge in a vacuum, but instead grew out of a social environment in which a wide array of movements (e.g., civil rights, antiwar, sexual liberation, and consumer health) was reshaping the social landscape. In the same period during which feminists were attempting to change mainstream ideas about women's bodies and sexuality, the gay rights and transgender movements sought to broaden the understanding and acceptance of non-normative gender roles and sexual desire.

The beginnings of the gay and transgender rights movements in the United States can be traced to the late nineteenth century, but it was not until the 1950s that they began gaining momentum and became more militant in nature. Around this time, gay and people who we would now call transgender activists organized to challenge escalating police harassment and raids on establishments they frequented. In 1966, drag queens and gay hustlers rioted against police harassment at Compton's Cafeteria, a popular twenty-four-hour diner in San Francisco's poor Tenderloin neighborhood (Silverman and Stryker 2005). The riot marked the first known instance of overt resistance against the oppression and discrimination they suffered for their gender expression. Three years later, a police raid at the Stonewall Inn, a bar in New York City's Greenwich Village, provoked a series of riots by lesbians, gay men, drag queens, street hustlers, and others, many of whom had been politicized by and involved in other movements of the 1960s. (Organizing for gay and lesbian rights had begun years earlier with the homophile movement of the 1950s, but Stonewall is widely viewed as the origin of the contemporary gay rights movement.) The Stonewall riots mobilized a broad, forceful grassroots movement across the country to advocate for legal and social reforms such as stopping police harassment, ending the criminalization of homosexuality, and protecting the civil rights of lesbians, gays, bisexuals, and transgender individuals.[3] Gay activists won a critical victory in 1973 when under intense pressure from groups such as the Gay Liberation Front and critics in psychiatry, the board of directors of the American Psychiatric Association agreed to remove homosexuality from the official list of psychiatric disorders in its highly influential handbook, the *Diagnostic and Statistical Manual of Mental Disorders* (*DSM*).[4]

On the heels of these developments, scholarly work in history and the social sciences began demonstrating the ways in which medicine and science are social enterprises that reflect the ideas, assumptions, values, and beliefs of time and place. Studies of the emergence of biomedicine challenged its status as an unassailable form of truth and paved the way for the examination of the formative practices through which illness and other matters of medical "reality" are formulated and of the manner by which these truth claims are constructed as uncontestable (Foucault 1973). Research highlighted how the rational bases of medical knowledge disguise its operations of power: physicians have the unique authority to determine who is sick, the etiology of illness, and the proper form of treatment. Subsequent scholars noted that scientific findings, appealing to reason and nature, have often confirmed common prejudices about race and gender, enshrining existing hierarchies as proper and inevitable (Gould 1981; Schiebinger 1989).

Working across several disciplines, some scholars turned a critical eye to the ways in which researchers conceptualized, or at times ignored, gender in their scientific theories and studies, leading them variously to interrogate the content of scientific knowledge for gender bias and to reject conventional explanations for gender differences that frame them as natural or biologically caused. They showed that each step in the scientific process—from the choice of research questions to the measures and concepts and the collection and interpretation of data—is laden with cultural presuppositions about gender. For example, many pointed out that scientists started with the assertion that male and female differ both physically and behaviorally, effectively naturalizing these perceived differences by attributing them to biology. The critical reexamination of specific theories has covered fields ranging from animal ethology to human neuroendocrinology (e.g., Haraway 1991; Oudshoorn 1994) and has allowed researchers to reassess "what is taken as natural and normal in connection with the human body" (Lindenbaum and Lock 1993: xi).

The rise in health movements over the past few decades can be read as a response to the growth of medical authority over the twentieth century—as well as the increasing reliance on scientific research in the formation of social policy and regulation—and as the attempt to dismantle this authority (Morello-Frosch et al. 2006). The demedicalization of homosexuality, to give but one example, pointed to a key change taking place in the United States that would prove important for intersex advocacy: the waning of the authority traditionally held by clinicians and the medical establishment (Starr 1982; Rothman 1991). This change was brought about by health-related movements and the rise of input into medical decision making by "outsiders" such as judges, lawyers, legislators, and academics, which took place in the 1960s and 1970s (Rothman 1991). Whereas prior to these shifts clinicians largely made decisions independently, on a case-by-case basis and with little input from others within or outside medicine, these movements brought medical decision making and practices into public awareness and popular discourse. Among other things these movements sought to challenge medical authority; reduce physicians' discretion and enhance patients' autonomy through collective decision making and a collaboration between clinicians and patients; and reframe and contest disease categories and illness experience.

In their typology of health social movements, Phil Brown and his colleagues include a category they call "embodied health movements" (Brown et al. 2004: 52). As in the women's health and disability rights movements, the lived-in physical body is central to the embodied health movements' force, goals, and aims. Embodied health movements are shaped and driven by actors' embodied knowledge and experience of their illnesses. They also depend on a "politicized collective illness identity," which emerges out of a disease process but is also linked to a social critique of the factors that shape, cause, or create that disease, thus transforming a personal problem into a social one (61). Some of these movements seek to gain recognition in medical and scientific communities and institutions (e.g., multiple chemical sensitivities and chronic fatigue syndrome) and to influence medical research and treatment (as in the case of breast cancer or autism). Others have focused on reducing the stigma associated with an illness or disability or on including affected individuals in all levels of service, treatment, and research (succinctly captured by the slogan "Nothing about us without us"), while yet others reject contemporary medico-scientific accounts of what is a disease and explanations for disease etiologies.

Increasingly the self-organization of intersex adults has used a mixture of traditional (e.g., face-to-face encounters) and modern forms of communication (e.g., e-mail and online bulletin boards) to connect, but the use of the Internet has emerged as a major driver of intersex activism. For those suffering from fear or shame about their condition, the Web's anonymity and the ability to reach people across the country and even the world provide a way to connect and share experiences with others. Moreover, it enables the sharing of hard-to-obtain information about conditions and treatments given the constraints of flows of information. Connecting with others and obtaining detailed and accurate information has proven critical to intersex adults' politicization.

Partly as a consequence of health activism, patients today have greater autonomy in health-care decisions than even a few years ago. It is now generally expected that patients (or their surrogates) will participate with clinicians in decisions about medical care and treatment. The move to patient-centered care flows from a desire to make health care more responsive to patients' needs and preferences. It also requires that patients be given sufficient information about their condition and treatment options to exercise a degree of control over treatment decisions that affect them.

Scholarly work and health advocacy movements, which challenged the idea that doctors are the dominant or exclusive authorities in matters of health, and movements by gay, lesbian, and transgender individuals challenging normative ideas about gender and sexuality made it possible and perhaps even legitimate for intersex adults to challenge clinical opinions about their bodies and treatment. The push for access to their medical records, for example, or their criticisms about having critical medical information withheld from them stem directly from criticisms of paternalism in medicine and from shifts in ideas about patient autonomy. At the same time, intersex activists' point that the medical decisions made for and about them have been based on questionable assumptions about the relationship between gender and sexuality owes much to gay, lesbian, and transgender activists' efforts to broaden the acceptance of non-normative sexual desire and gender expression.

Intersex Support Groups

Although the organizations that now provide support to individuals with various intersex diagnoses all to some degree owe their origins to the social and health activism described above, many do not exist for the purpose of bringing about social change: they aim to help individuals live with their diagnosis. As I showed in chapter 7, some individuals with intersex conditions credit the discovery of a support group for people with their condition as a lifesaver, perhaps literally so for individuals suffering from depression. As with other groups for individuals with a variety of medical conditions and disabilities, support groups for those with intersex diagnoses are typically designed to encourage contact with others with the same condition, offer mutual support, and promote the dissemination of information about the causes and treatment of the condition. For Louise Rutherford a support group proved significantly more beneficial than therapy: "No one else has taken the time to really listen to the whole story and to identify in some way. The only people that identify with it are people who've been there."

What I refer to here as support groups tend to be organized by specific conditions. Such groups include KS&A (originally Klinefelter Syndrome and Associates, and more recently Knowledge, Support, and Action), which has described itself as an organization "to educate, encourage research, and foster treatment and cures for symptoms of sex chromosome variations"; the Magic Foundation, an umbrella group for growth failure disorders, which has divisions dedicated to specific diagnoses, including CAH and Turner syndrome; and the CARES Foundation, which supports individuals with CAH and their families, among others. This self-organization of groups by diagnosis has the benefit of allowing people to form relationships with those with similar biological predicaments and treatment histories, but it also tends to reflect and reinforce the medicalization of intersex conditions and a particular contingent relationship of individuals to their biology, wherein complicated biological information serves as the foundation for identity. (To some degree this is unsurprising, as parents of children with some forms of CAH, for example, are likely to be the most concerned about dealing with the potentially life-threatening crises that can arise from the diagnosis.) The general description of CAH on the Web site of the CARES Foundation, for example, makes no mention of genital anomalies (although links on the same web page lead to discussions of genital surgery and gender-typical behavior). As chapter 7 has shown, some intersex advocates feel that parents in such groups focus on medical complications because they are reluctant to discuss sex, sexuality, and the potential complications of genital surgery.

Although these groups are undoubtedly very helpful to parents and affected individuals, they can also contribute to intersex conditions being predominately viewed and understood through the lens of biomedicine. Medical knowledge, no longer exclusive to the physician as the interpreter of obscure information for others, is also actively translated by these groups for individuals and their families. Not simply consumers of expert knowledge, some of these groups also participate in the generation and dissemination of medical knowledge in part through their direct partnership with and financial support of clinical research. These organizations are typically registered charities or nonprofit entities that, in addition to providing support for individuals, raise funds for research and conduct other philanthropic activities. Other diagnosis-specific groups also exist, often in the form of online forums like CAH Support and xyTurners. Such forums may be run by an individual or a small group of motivated individuals with little or no formal organizational structure or funding.

With few exceptions, the support groups and online forums for specific intersex conditions tend not to be activist in nature; their main purpose is to help individuals and their families understand and adjust to living with their diagnoses, rather than to change the ways in which those diagnoses are understood by society or treated by clinicians. They vary in the types of activities and services they offer. Some, for example, hold face-to-face meetings, whereas others exist only online. Lisa Todd noted the difference between the support groups for women with AIS and CAH: "The women with androgen insensitivity have a well-defined support group that meets yearly, and everybody can come: family, parents, sister, lovers. It's a way for them to connect, face-to-face. CAH doesn't have that. What we've got is the Magic Foundation, and that's for kids, so it's rarer that you hear someone with CAH meeting another person."

Groups may vary widely in their philosophies of gender identity and behavior, on the advisability of genital surgery on children, and so on; their orientation is to some degree determined by the views of the most active or vocal members and may change over time as new members join. Even when its intentions are not overtly political, a Web site may spark controversy because the task of putting complex information into a form accessible to lay readers on the Internet—for example, a discussion of whether the hormonal imbalances of CAH dispose individuals toward homosexuality—inevitably involves judgment and interpretation. Moreover, as chapter 7 has shown, discussions may be polarized by generation: some groups are intended primarily for parents of children with intersex conditions, whereas others may be intended for the individuals with the diagnosis. (The CARES Foundation has separate discussion boards for adolescents and parents.) And although the participants are united by their interest in the specific diagnosis, their attitudes may be diametrically opposed, causing conflict. Elizabeth Potts said of one CAH group: "It gets very angry on that board when they discuss intersex or more sensitive topics, like gender identity or sexuality. Certain people don't want their kids to be considered intersex and others that don't want to have CAH be known as the 'gay gene.' Somebody has said that they have heard that there's a lot of gay women that have CAH and that it's from the imprinting of the testosterone on the brain at birth. They don't want their children to be known for that. There was a huge fight on a board one time when some guy said that gayness is a choice; you're not born with it. It got really nasty."

Some individuals have found support groups of this kind inadequate for addressing questions and problems related to their diagnoses. Seeking to challenge views they see as too accepting of conventional medical wisdom and as intolerant of difference (physical or behavioral), they have formed organizations with a more activist agenda. These groups, although welcomed by many individuals with intersex diagnoses as well as by gender studies scholars and some clinicians, have also engendered controversy because of their opposition to surgical and medical intervention and their sometimes strident demands for greater patient involvement in treatment decisions.

Intersex Activism

Intersex activists and advocates may be seen as direct descendants of the earlier social movements concerning health. Their movement differs, however, from other health movements such as AIDS activism, breast cancer activism, or the diagnosis-specific groups named above, which often seek to improve access to care, broaden research agendas, examine causes of disease, and develop better treatment options. Although some activists have recently begun to advocate for these same changes, intersex activism and politicization were initially grounded in its conscious opposition to the dominant biomedical model of intersexuality and were more fundamentally rooted in antimedicalization movements. With its focus on questioning the necessity of medical treatment for gender-atypical bodies, intersex activism echoes the efforts of the women's health movement to demedicalize birth, gay activists' efforts to demedicalize homosexuality, and the campaign of the Deaf movement against cochlear implants.

But intersex activism also shares common ground with movements aimed at garnering civil rights, such as the disability rights movement, whose members both advocate within conventional channels and undertake acts of civil disobedience to promote equality, self-definition, autonomy, and full access to society and to eliminate societal fear, stigma, and oppression. Intersex activism also follows this movement's efforts to reframe views of and attitudes toward differences in individual bodies and capabilities.

What underpins the collective action of intersex activists, then, is not a particular disease process, because individual biologies often vary significantly, but shared familial, social, and medical treatment experiences and subjective responses to these experiences—what medical anthropologists have broadly called the "illness experience" (Kleinman 1988). The biomedical view of intersexuality is overwhelmingly focused on disease and limited to abnormalities in the structure and function of biological systems and organs. For physicians diagnosing and treating clinical pathology is of primary importance. This narrow view often obscures the way in which diseases are also illnesses, that is, lived experiences shaped not simply by clinical pathology but also by the social reality in which a person lives. Illness refers to something different and more than clinical pathology to include the way in which individuals, family members, and society perceive, live with, and respond to symptoms and disability (Kleinman 1988). Thus while the experiences of those with CAH, for example, can result in an identity or self-concept largely shaped by physical signs and symptoms, that identity is also shaped through illness experience, much of which may also be common to those with other intersex conditions. This lived treatment experience, often not known or experientially available to clinicians or parents, is frequently used to challenge medical authority. Intersex activists seek to modify the experience of their conditions by challenging medicine and science on the understanding and treatment of intersexuality, shaping medical research and treatment—in some cases through collaboration with sympathetic clinicians and researchers—and arguing that medicine has failed to offer accounts of intersexuality that are consistent with individuals' experiences, and in some cases has even offered understandings that the patients are unwilling to accept.

The extent to which the illness experience maps onto disease varies with each condition, as does the meaning imbued in that condition. An illness's meaning is intimately tied to different systems of understanding, with their different rules and logic, and thus to time and place. These systems of logic will include some information while excluding other information, and assign importance to some factors while downplaying others. It is precisely the system of logic for understanding intersexuality that activists are trying to change. Thus, for example, whereas a doctor or parent may view genital surgery as a necessary burden for improving life, individuals themselves may give this surgery much more emotional valence, which could include disfigurement and physical debility as a result of complications. Illness experiences are also imbued with meaning specific to the individual; genital variation could have a different meaning for someone who has grown up in a family in which the parents exhibit bodily or behavioral difference than for someone whose family shows no overt difference.

Although in many respects activists' analyses of treatment for intersexuality draw on theories about the social construction of gender—and intersexuality has been used as a prime example of this phenomenon in women's studies and queer studies classes—the two fields' interests are not congruent. Most feminist interest in intersexuality stems from its value as a heuristic device, but the goal of many people with intersex conditions and activists is not to deconstruct or eliminate gender, or to advocate for a third sex or no sex, but rather to change treatment practices and improve the well being of others with these conditions. While this goal necessarily involves a broadened understanding of what it means to be male or female, the cornerstone of this argument does not center on gender: intersexuality, these activists argue, is primarily a problem of stigma and trauma, not of gender.

Nevertheless, gender underpins many of the discourses around intersexuality and the experiences of those with intersex conditions, for a number of reasons. First, the majority of doctors involved in these cases are still male, whereas many patients are female. Second, mothers tend to be the primary managers of the treatment of children with intersex conditions. Third, most intersex activists and academics conducting research on intersexuality are women. Not surprisingly, gender also functions at other levels. As I have shown, the standard treatment paradigm perpetuates traditional, idealized, and stereotypical notions of masculinity, femininity, and sexuality: it portrays some individuals as inappropriately masculine and others as inappropriately feminine. These constructions shape intersexuality not only as a disease category but also as a subjective illness experience. By validating their own modes of embodied knowledge, intersex activists strive to overturn some of these assumptions embedded in the ostensibly scientific dominant medical model.

Intersex activists have called into question not only the broad theoretical basis of medical decision making but also the specific practices and assumptions of the traditional protocol. They have argued that although most clinicians and parents act out of compassion, and most are convinced that they are providing the best care for the children, current clinical knowledge, with its ignorance of long-term outcomes, provides an insufficient justification for treatment. It has simply been taken for granted that surgical and other interventions help in the long term and that individuals with intersex diagnoses view their bodies in the same terms as medical professionals do. The protocols in fact make assumptions and bear implications that are potentially harmful to those with intersex conditions.

The traditional protocol advocates aggressive intervention in the bodies of very small children—individuals incapable of expressing their own choices. Clinicians assume, fundamentally, that when a baby with an intersex condition is born and identified, something must be done to correct the incongruities among the physical sex characteristics. Some clinicians construe activist demands as for "doing nothing," which is not an option once a diagnosis is made. Activists have argued that they hardly recommend "doing nothing"; instead, they call for necessary medical intervention and lifelong psychosocial and other services to help the child and family.

Activists have also rightly argued that John Money's paradigm reinforces, indeed reifies, a heteronormative model of sexuality for the treatment of intersexuality: that is, the notion that genitals, bodies, behavior, sexual desire and behavior are linked in a straightforward and obvious way; that humans naturally fall into two distinct and discrete categories, male and female; that sexuality is naturally expressed only between two people of different genders; and that observed behavioral differences between male and females are given in nature. Social institutions and policies, as well as medical practices, support the belief that physical sex, gender identity, and gender roles in an individual should all align to either masculine or feminine norms, and that heterosexuality (expressed through penile-vaginal intercourse) is the only natural sexual orientation. The surgery recommended by the protocol attempts to shape an individual to conform to a standard conception of the male or female body, with an emphasis on correct morphology. On the assumption that "normal" women wish to, and should, bear children, female reproductive capacity has been privileged and protected: in cases of ambiguous genitalia, it has been more common to reassign a fertile male infant as female than the other way around.[5]

The dominant clinical paradigm is constructed through a generally accepted set of beliefs that the birth of an infant with an intersex diagnosis is an emergency and should be handled as quickly as possible, that intersexuality is exclusively a medical issue, and that good treatment ensures helping the child conform to male or female ideals. It also construes an intersex birth as an individual anomaly with an individualized solution, which may entail gender assignment, genital surgery, hormone supplements at puberty, and other interventions. The belief has been that the "correct" decisions would result in a "good" outcome. But as noted earlier, this narrow, individualized approach fails to address many other factors that make up the illness experience.

The focus on individualized approaches obscures and sometimes entirely conceals the ways in which intersexuality is politically and socially constructed. Thus activists and their allies have focused on offering a counterpoint to this medical understanding and on resituating intersex in its broader social and cultural context. Some seek to provide new cultural models for intersexuality that are more positive than pathologizing and to resist the medical characterization not only of intersexuality but also of what constitute appropriate notions of male and female genitals, bodies, and sexual behaviors. Thus, while medicine has cast intersex as pathological, it has also thereby created the possibility of an intersex illness experience, and even an intersex identity and subjectivity. While some activists strive to improve clinical encounters and treatment experience, others work to assert the validity of embodied knowledge and of their experiences as persons with intersex diagnoses, and to locate the social motivations for treatment linked to cultural norms of sex, gender, and sexuality. Some have asked provocative and productive questions such as what genitals are for and whether atypical genitals should be seen as a medical issue at all.

THE INTERSEX SOCIETY OF NORTH AMERICA (*ISNA*)

In the 1980s, the social psychologist Suzanne Kessler began to ask questions about intersex management protocols that went beyond the bounds of traditional medical concern and practice. In an astute and groundbreaking article published in the feminist journal *Signs*, Kessler reported the results of her interviews with six clinical experts in intersexuality (Kessler 1990). She noted that although clinicians considered biological factors in their treatment decisions, the principles underlying their decisions were aimed at enforcing a vision of gender difference that was culturally contingent, and she thus called into question the supposed "naturalness" of our two-gender system (1990: 3). Some physicians, believing in the immediate necessity of gender assignment, assigned gender before all medical tests had been completed; ultimately, their decisions relied on cultural assumptions about gender difference and the normative relationships among sex, gender, and sexuality. Drawing an analytical distinction between gender and sexuality, for example, Kessler pointed to the presumption of heterosexuality in the process of gender assignment: for an infant to be male assigned, its penis had to be capable of vaginal penetration; a female assignment required surgery to make a vagina capable of "receiving" a penis.

Three years after Kessler's article appeared, the feminist scholar and biologist Anne Fausto-Sterling published the article "The Five Sexes: Why Male and Female Are Not Enough," in which she proposed replacing the current two-sex system of male and female with a five-sex model (Fausto-Sterling 1993). In her scheme, the categories of male and female would be supplemented by three additional categories based on the classificatory scheme for hermaphroditism: herms, for true hermaphrodites; merms, for male pseudohermaphrodites; and ferms, for female pseudohermaphrodites. Although the article was tongue-in-cheek, it incorporated serious criticism of binary views of gender.[6]

The articles on intersexuality by Kessler and Fausto-Sterling had little influence on clinicians treating newborns with intersex conditions. In part this was simply because they did not constitute the primary audience for these articles, and the articles were published in journals that rarely cross a clinician's desk. But it was also because even if doctors felt that there was a problem with their treatment practices—and it seems likely that most did not—these theoretical critiques could not be easily translated into a new technique, protocol, or code of practice. The articles did, however, reach one individual central to changing views of the understanding and treatment of intersexuality: Cheryl Chase, the founder of the Intersex Society of North America (ISNA), the first patient advocacy group for intersex individuals.[7]

At birth doctors assigned Chase as a boy. At age one and a half, however, doctors told Chase's parents that their son's penis was an inadequate size for a male. They reassigned Chase as female, removed what was now understood as her clitoris (what had been too small for a male assignment was nevertheless too large for a female assignment), and told her parents to change her name to a more feminine one. The specialists also advised Chase's family to remove all traces of her life as a boy, move to another town, and keep what had happened secret from everyone, including Chase. They did.

In her teens, Chase, having realized that she was unable to reach orgasm, began to seek out medical libraries to research what had happened to her. At this time she also visited a gynecologist with the hope of obtaining her medical records.[8] At age twenty-one Chase obtained a three-page summary of her medical history; she finally learned of her diagnosis as a "true hermaphrodite" and of the history of her gender assignment and genital surgery. Shocked and terrified by what she had learned and fearful of ostracism by others, she did not tell anyone about her newly discovered history for the following decade.

Fifteen years after obtaining her medical records, Chase had what she calls an "emotional meltdown" and sought psychological help. She also contacted specialists in intersexuality to see if there was a way to restore her clitoris; she was told there was not. She expected that physicians would be interested to hear about her situation, but she found they were not. When Chase told a well-regarded surgeon about her complications due to surgery, including her inability to reach orgasm, the surgeon remarked that she should not complain; hers was a successful outcome. Frustrated and depressed by clinicians' indifference, Chase turned to more sympathetic figures: Kessler and Fausto-Sterling. In 1993, Chase contacted Kessler about her earlier article. Chase then wrote Kessler a long letter, complete with a bibliography and a packet of articles. The letter included the seeds of what would be ISNA's platform: "Intersex specialists claim that surgery is intended to save the child from the pain and stigma of growing up as a hermaphrodite. . . . A recent personal correspondence from a well-published innovator of surgical techniques for intersex correction . . . [says] I was well-served by having 'your feminine appearance improved' even at the expense of an entire lifetime of normal sexual gratification. Several life stories I have thus far collected indicate that medicine does not achieve that goal. . . . Medicine thus inflicts exactly the harm it claims the surgery is intended to prevent. It hurts. It shuts people up." Kessler was intrigued by what Chase had written—she had assumed that genital surgery was largely successful—and they kept up a correspondence and became friends. This was around the same time that Fausto-Sterling's article "The Five Sexes" was published. Chase wrote a letter to the editor praising it (Chase 1993). In the letter, Chase announced the formation of a support group, which she initially called

Intersex Is Not Criminal and eventually named ISNA, and provided contact information. (A few years later Chase also found a willing ally in the historian of science Alice Dreger, who helped her make her initial connections with clinicians; she and Dreger worked closely for almost ten years.)[9]

Early on, Chase met with members of support groups for individuals with diagnoses such as Turner syndrome, Klinefelter syndrome, and AIS, but she found she had little in common with the members of these groups. She felt their aims were to help individuals adjust to their situation and to work with clinicians to improve clinical treatment and extend research. Some argued for increased medical attention; others resisted discussing gender identity or sexual orientation. None were politicized in the way she wanted ISNA to be. Chase later wrote: "I was less willing to think of intersexuality as a pathology or disability, more interested in challenging its medicalization entirely, and more interested still in politicizing a pan-intersexual identity across the divisions of particular etiologies" (Chase 1998: 199).

Initially, Chase's goal was simply to enable people with intersex conditions to connect with one another, many for the first time in their lives, by forming a network for peer support. By telling acquaintances of her medical history, she met a surprising number of individuals with intersex diagnoses. She found others by placing ads in local gay papers in San Francisco where she was living. She contacted journalists who she felt would report responsibly on intersex issues, and she spoke to everyone who called ISNA. Some days she spent hours on the phone listening to the stories of those with similar experiences.

Although many accounts of ISNA's beginnings overwhelmingly focus on Chase, it is important to note that she worked very closely with a number of activists who were instrumental to both ISNA's development and intersex advocacy more generally. These people included Morgan Holmes, Hida Viloria, Angela Moreno Lippert, David Vandertie, Sherri Groveman Morris, David Cameron Strachan, Max Beck, Tamara Beck, Kiira Triea, and Lynnell Stephani Long—all of whom not only helped to shape ISNA's early platform, but also spoke on behalf of the group and in many instances supported it financially. Though their role in the intersex movement has not received the same attention as Chase, in part because some of these individuals moved away from the group over the years due to differences of opinion—a story that has yet to be told—their efforts and support were critical to ISNA's development.

In the winter of 1994, ISNA published the first issue of its now defunct newsletter *Hermaphrodites with Attitude*, which it dubbed "the world's only newsletter by and for intersexuals." Intersex individuals filled the first issue with poetry, personal essays, and even drawings. It also displayed a sense of humor: the front page showed a "before" and "after" drawing of Rudolph the Red-Nosed Reindeer. The "before" picture shows his "disfiguring hypertrophy of nose" and the "after" shows Rudolph sans nose with the commentary, "Post-surgical aspect. An excellent cosmetic result was achieved," mimicking the language of scientific accounts of genital surgery to reduce hypertrophied clitorises. (Over time Chase collected any sort of lead related to intersexuality—journalists, intersex individuals, clinicians, researchers, organizations—and sent them the newsletter whether they asked for it or not.)

In addition to providing peer support, ISNA began challenging standard medical protocol. Its long-term goal was (and has remained) to change the medical treatment of intersexuality to reduce the "emotional and physical carnage resulting from medical interventions" (Chase 1998: 198). From the beginning, ISNA members claimed that the treatment paradigm was focused almost exclusively on producing children with gender-normative bodies and heteronormative sexual desires, and their early critiques zeroed in on the problems raised by genital surgery. Drawing on personal experience, with accounts published in *Hermaphrodites with Attitude*, its members objected to the timing and necessity of genital surgeries, arguing that genital surgery emphasized a child's genital appearance at the expense of sexual pleasure. Surgery, they asserted, harmed erotic response and resulted in lifelong pain, discomfort, and emotional scarring. They advocated against all medically unnecessary surgery, and stated that no one should be allowed to decide that the cosmetic appearance of the genitals should take precedence over sexual pleasure before the child could agree to this trade-off.

The medical justification for early genital surgery, ISNA has observed, is the risk of psychological harm, but there is no proof that children with atypical genitals are harmed psychosocially any more than those who have had surgery. Indeed, some would argue that those who have undergone surgery suffer more than those who have not had surgery. The other motivations for early genital surgery—to allay parental anxiety, keep the world safe for dichotomous gender, and ensure heterosexuality—have been valued over the integrity of the infant's body. Acknowledging that some adult women do opt for surgery, ISNA and its supporters have said that surgery should be available when individuals are old enough to consent.

On another level, members also argued that the standard treatment protocol, with its focus on quick decisions and on withholding information from the parents and those treated, led to individuals' not knowing the truth about their treatment or bodies, and thus to a deep sense of shame. Not being told the truth, they argued, often led to children's distrust of their physicians and parents and caused deep rifts in families. They called for counseling and support for individuals and families and suggested introducing children to a support group early on; they also encouraged parents to talk openly about the child's diagnosis and treatment at age-appropriate levels.

The group's objections and criticisms—often delivered in the form of experiential narratives readily accessible to the lay public—attracted attention from clinicians as well as the broader population. Drawing on tactics used by gay and AIDS activist groups—such as protests outside medical meetings and conferences and corresponding with clinicians privately and through letters to the editor in medical journals—ISNA challenged medical authority and demanded change. The group also worked with more mainstream gay and lesbian organizations such as the Gay and Lesbian Alliance against Defamation, the National Gay and Lesbian Task Force, and the International Gay and Lesbian Human Rights Commission, which provided introductions to others and in some cases included intersex concerns in their agendas. In addition, ISNA found support from gay and lesbian caucuses within professional medical associations, as well as from the Gay and Lesbian Medical Association.

Not all of ISNA's early undertakings proved successful. Chase tried to reach out to clinicians through the newsletter she sent directly to them, but these encounters remained frustrating. A 1995 edition of the newsletter included a pamphlet for medical professionals, explaining the organization and its position on intersex management. In the next issue, David Sandberg, a psychologist then in the psychoendocrinology program at Buffalo Children's Hospital, shared his reactions to ISNA's literature, printed under the optimistic heading "First Dialog with Medical Community!"

The dialogue proved difficult; Sandberg chipped away at ISNA's recommendations, writing that it was not evident whether genital surgery was helpful or harmful because there was little long-term follow-up data. While true, the overall effect of his words was to diminish the concerns raised by ISNA members. He also argued that ISNA likely represented a vocal minority harmed by surgery and that ISNA's goal should be to call for more research because "there is very little data to support either Money's or ISNA's approach. We are unfortunately flying by the seat of our pants, clinically" (Intersex Society of North America 1995–96: 8). (Sandberg has since become a strong advocate for improving medical care for persons with intersex diagnoses.) Despite Sandberg's discouraging response, the newsletter showed that ISNA's circle of supporters had grown; Kessler, Fausto-Sterling, and the psychologist Howard Devore (who earned his PhD under Money) all responded to Sandberg's piece. Three other clinicians seemed sympathetic to the experience of ISNA members, but they expressed the same skepticism as Sandberg: were ISNA members the vocal minority?

The group experienced other failures during this period as well. In 1996, when a congressional bill to ban female genital mutilation was proposed in the United States after several highly publicized cases of women seeking asylum in this country to escape the practice, ISNA attempted to have it amended to include a ban on intersex genital surgeries (which it deftly called "intersex genital mutilation"). However, in its final form the ban excepted operations "necessary to the person's health and . . . performed by a person licensed as a medical practitioner in that state," which effectively meant surgeries on intersex infants. That same year, Chase learned of a symposium on surgical advances for children with ambiguous genitalia to be held at Mount Sinai School of Medicine in New York City. Chase wrote the organizers asking for permission to speak at the conference. They declined. Not ready to give up, and on the recommendation of Fausto-Sterling, Chase contacted William Byne, a psychiatrist at Mount Sinai who had recently written an important article evaluating biological theories of homosexuality (Byne 1994), for help. Chase relayed that ISNA was contemplating picketing the conference, but she also wanted to know whether he would be willing to talk to the conference organizers on her behalf. Byne spent months going back and forth with the organizers, but they would not change the schedule to accommodate ISNA members. This experience, Chase said, made her realize that doctors "were absolutely not going to listen to us, and helped me to make the decision I was going to picket them." She later wrote: "ISNA has found it almost entirely fruitless to attempt direct, non-confrontational interactions with the medical specialists who themselves determine policy on the treatment of intersex infants and who actually carry out the surgeries" (Chase 1998: 202).

Reeling from the experience at Mount Sinai and the failure to get a legislative ban on intersex genital surgeries, ISNA organized a protest of the American Academy of Pediatrics annual meeting in 1996. Chase knew protesting would turn physicians against ISNA, but they were not listening to ISNA anyway, she reasoned. The protest generated a surprising amount of media coverage; newspapers as diverse as the *Urology Times* and the *Boston*

Phoenix covered it (Barry 1996; Scheck 1997; "Is Early Vaginal Reconstruction Wrong" 1997).[10]

It was the launch of ISNA's Web site in 1996, however, that established an intersex activist presence and helped to get ISNA's message out to anyone who was curious. (Due to Chase's facility with technology, she was able to manipulate search-engine rankings so that ISNA's Web site would appear first when *intersex* was used as the search term. Wikipedia only recently surpassed it.) The Web presence not only facilitated Chase's connection with others with intersex diagnoses but also that with journalists writing stories on the topic: ISNA was the first online information they found.

The following year a confluence of events would push ISNA and Chase into the media spotlight. In March 1997, Milton Diamond and Keith Sigmundson published their update to the John/Joan case, generating the kind of publicity ISNA needed to put intersexuality on the public map. Journalists began looking for intersex individuals to interview, and through the Internet they easily located ISNA, enabling Chase to refer them to others receptive to ISNA's views. As a result, national media outlets such as *Newsweek*, the *New York Times*, *Rolling Stone*, *Mademoiselle*, NPR's *All Things Considered* and *Fresh Air*, NBC's *Dateline*, and ABC's *Primetime Live* told stories sympathetic to activist concerns (Angier 1996, 1997; Colapinto 1997; Cowley 1997; "I Gotta Be Me" 1997; Gross 1997; "Boy or Girl?" 1997; White 1997; Moreno and Goodwin 1998). Chase and other ISNA members made for articulate patients who gave compelling stories about the plight of intersex individuals. Still, ISNA had few medical allies; most were reluctant to work with what they perceived as an activist group hostile to clinicians.

To achieve its goal of changing the medical management of intersex infants, ISNA has developed a diverse and complex group of organizational, institutional, and individual allies. Initially, Chase's negative encounters as both a patient and an advocate discouraged her from trying to work with clinicians. Over the years, however, her group has acknowledged the value of alliances with clinicians and made efforts to form them; since 1999 ISNA has worked almost exclusively to reform medical education and care and more recently to ensure the implementation of the guidelines suggested in the 2006 "Consensus Statement on Management of Intersex Disorders."

The pediatric urologist Justine Schober was one of ISNA's earliest allies within the medical profession. In a 1997 appearance on *Dateline*, Schober challenged the American Academy of Pediatrics policy for the treatment of intersexuality. In her contribution to a textbook on pediatric surgery, she was one of the first specialists to criticize early genital surgery (Schober 1998). In

1999, she argued that no studies had adequately addressed the quality of life of adult patients with intersex conditions (Schober 1999).

Realizing that her early exchanges with clinicians were unsuccessful, Chase attempted to learn how to speak to different medical specialists—an effort largely facilitated by Dreger who, with her doctorate and position in academia, could open doors Chase could not. With Dreger's help, Chase built up more clinical contacts, who helped her make referrals for families, offered clinical endorsements of ISNA's message, wrote reviews of books sympathetic to activist concerns (many of them authored by Dreger), spoke to reporters covering intersexuality, and referred them to other sympathetic clinicians. Over a period of several years, Dreger and Chase spent hundreds of hours brainstorming tactics, hashing out arguments, writing articles, grants, and handouts, fundraising, recruiting allies, and doing what Dreger calls "road work," traveling to give talks at diverse venues such as universities, gay and lesbian centers, and hospitals. Dreger would do much to promote intersex rights: she published voluminously on the topic (e.g., Dreger 1998a, 1998b, 1998c, 1999) and served in several official capacities at ISNA. (She left ISNA in late 2005.)

Improved relationships with clinicians slowly strengthened ISNA's credibility in the world of medicine: after years of strained contact, Chase was beginning to be asked to give grand rounds with specialists and medical students. One sign of this rapprochement came when Chase was invited to present the patient perspective at the 2000 meeting of the Lawson Wilkins Pediatric Endocrine Society. This marked the first time that the society's annual symposium was devoted to intersexuality. It was also the first time that organizers of a major medical conference solicited the patient-advocate's perspective on the medical treatment of those with intersex diagnoses. Chase's invitation largely resulted from the actions of a sympathetic and well-respected clinician in the group. Close to 80 percent of the world's pediatric endocrinologists were present; Chase's participation proved the more striking because she had organized a protest against this very meeting in 1996.

Around the same time, the North American Task Force on Intersexuality (NATFI) asked Chase to participate. The newly formed NATFI had grown out of the 1999 meeting of the American Urological Association. Endorsed by eight medical organizations, NATFI brought together medical specialists and patient advocates to conduct retrospective reviews of the long-term psychosexual status of patients; establish guidelines for the management of children born with atypical sex anatomy; conduct outcome studies; assess the ethics of current treatment protocol; and develop new treatment guidelines. The group as it was originally conceived folded several years later in part due to difficulties in designing and carrying out outcome studies. Nevertheless, NATFI's creation as well as the increasing focus on intersexuality at medical meetings can be seen as the outcome of years of ISNA's activism.

By 2000 the field of intersexuality found itself in deep crisis, and increasing numbers of clinical articles were acknowledging many of the issues raised by ISNA and other groups, noting that surgery might not be as beneficial as originally thought and urging caution in surgical decision making (Creighton 2001; Creighton and Minto 2001a, 2001b; Creighton, Minto, and Steele 2001; Daaboul and Frader 2001). Clinicians began to conduct more follow-up studies of genital surgeries whose results suggested that vaginal surgery should be delayed, that repeat clitoral surgery proved damaging to sexual function, and that satisfaction with cosmetic appearance was variable (Creighton, Minto, and Steele 2001a, 2001b; Schober 2001).

By 2003 ISNA had announced the creation of a medical advisory board whose members included experts in pediatric endocrinology, urology, general pediatrics, nursing, psychiatry, psychology, and social work. With Dreger acting as project manager, and under the guise of the Disorders of Sex Development Consortium—consisting of clinical specialists, intersex adults, and their families—ISNA published clinical guidelines and a parents' handbook in 2006 (see www.dsdguidelines.org). The clinical guidelines stress "patient-centered care" for persons with disorders of sex development (DSD) and set out treatment guidelines following from this principle. The document also provides scripts for talking with parents. The parent's handbook covers background information on intersex diagnoses, suggestions on how to talk openly with one's child and others about the child's condition, answers to common questions, and resources on where to gather more information. Shortly after the publication of these documents, Chase was invited to participate in the development of the 2006 "Consensus Statement on Management of Intersex Disorders."

These collaborations with clinicians and Chase's participation in the drafting of treatment guidelines, both of which necessarily involved compromises between clinical specialists and patient advocates, have pleased some ISNA supporters and angered others. For clinicians who are receptive to ISNA's message, the attempt to work more closely with physicians is seen as an important advance. However, those questioning, for example, the medicalization of intersexuality and gender nonconformity more generally, see ISNA's collaboration with clinicians as a betrayal and an abandonment of what they regarded as the larger goal of depathologizing gender nonconformity.

Interestingly, one of the more contentious issues among activists has been the proposed change in medical nomenclature. While many do not object to this idea in principle, it is the suggestion that the term *intersex* be abandoned in favor of *disorders of sex development* (DSD) that has led to heated debates. All indications signal that the shift to DSD has deeply angered some individuals, but has pleased a great deal of parents and clinicians. ISNA, clinicians, parents, and others see DSD as a way to avoid potentially pejorative and confusing terms such as *intersex*, *pseudohermaphroditism*, *hermaphroditism*, *sex reversal*, and other gender-based diagnostic labels. As ISNA sees it, *intersex* labels the person, not the condition; DSD is meant to refer to the underlying physiology that causes atypical sex anatomy. ISNA has argued that the use of DSD is not an attempt to paint intersex as a medical issue solely, but that, as they state, "we are addressing people working in a medical context. We have found that the word DSD is much less charged than 'intersex,' and that it makes our message of patient-centered care much more accessible to parents and doctors. Our aim is to meet them where they are" (www.isna.org/node/1066). The term *DSD* thus attempts to reframe intersex as genetic or as an endocrine disorder no different from many other physiological disorders. This approach marks an attempt to "bracket intersexuality" by ignoring or reinterpreting conventional signs of intersexuality and by not giving atypical genitals credibility as anything but a sign of an underlying medical condition (Kessler 1998). But as desirable as this goal may be, we are not there yet. Although some may be able to rethink the meaning of genital variability and the profound entanglements between genitals and gender, most clinicians and parents are either unprepared or unable to do this. Moreover, since the movement's beginnings, intersex activists have challenged the idea that intersex bodies are abnormal, diseased, or pathological. It is no surprise, then, that some have vehemently argued against using the term *disorder*, feeling that it reifies intersex conditions and bodies as in need of fixing. They generally find it equally stigmatizing and pathologizing, as succinctly captured by the statement, "I'm a human not a disorder," and believe that this semantic shift works against much of what the intersex movement has advocated over the years. The debate, which one person has referred to as a rift between assimilationists and nonassimilationists, appears to be an inevitable result of ISNA becoming more mainstream and has proven incredibly heated because it is deeply, achingly personal, centering on who gets to define, determine, and label the truth and authenticity of one's life.[11]

COVERING ALL THE BASES: NEW GROUPS

As ISNA moved away from peer support and toward working with clinicians on reforming medical care, other groups formed to provide peer support, pursue overt direct action, and promote a broader appreciation of gender diversity and variation. The three most prominent advocacy groups have been Intersex Initiative (ipdx), founded in 1999 by Emi Koyama, a longtime activist and a former intern at ISNA; Bodies Like Ours (BLO), formed in 2002 by Betsy Driver and Janet Green; and Organisation Intersex International (OII), founded in 2003 by Curtis Hinkle. Each group has a Web site with information about intersexuality. Members speak at conferences, universities, and other organizations to raise awareness about intersexuality and to change medical practice, although in recent years these groups have been more oriented toward grassroots activism than ISNA. The groups share similar views of intersexuality and of changing treatment practices. The groups BLO and ipdx have collaborated, shared information and knowledge, and referred media and individual inquiries to one another.

In a conscious effort to provide the peer support no longer offered by ISNA, BLO developed a Web site (www.bodieslikeours.org) and had members respond personally to letters, e-mails, and phone calls from adults and parents seeking assistance. (The Web site has been offline since the fall of 2006, but the forums remain open.) The group has provided advice and help to individuals wanting to obtain their medical records, has matched up parents for peer support, and has supplied advice and wording for organizations wanting to include intersex in their mission statements. The group has no desire to work more closely with clinicians. As Driver said, "I am much more apt to go and chain myself to the doors at a medical conference than get in there and speak to them."

Koyama began ipdx in 1999 with the goal of raising awareness about intersexuality. She and members of her speakers' bureau also travel to speak at conferences and universities. Like BLO, ipdx is activist oriented and has organized protests, for example at the American Association of Pediatrics in San Francisco in 2004. Koyama has also developed pedagogical materials on intersexuality: feminists' predominant interest in intersexuality as an example of the social construction of gender has come, in her view, at the expense of exploring its meaning as a lived experience (Koyama and Weasel 2002). Whereas ISNA has often relied on speakers already knowledgeable about intersexuality, ipdx trains novices. Koyama wants ipdx to remain small, funded through speaking honoraria and the online sales of materials.

More than any other group, OII views itself in opposition to ISNA. It argues that ISNA has been too accommodating about the medicalization of intersexuality. For OII members, intersex is not so much a condition as an identity: intersex is not something one has, but who one is. Its platform resembles ISNA's earlier stance: OII "resists all efforts to make intersex invisible, including genital mutilation, medicalisation and normalisation without consent and offers another face to intersex lives and experience by highlighting the richness and diversity of intersex identities and cultures" (www.intersexualite.org). This last point regarding gender diversity marks the greatest distance between OII and ISNA today. Whereas ISNA focuses on direct involvement with doctors to change medical practices, OII's efforts are focused on improving the societal acceptance of gender diversity, which its members hope will result in less medical intervention. As my brief summary makes clear, there is no single voice or viewpoint among individuals with intersex conditions. Nevertheless, ISNA has been viewed as the primary representative of intersex individuals in large part because it represented the first, and for many years the only, intersex advocacy group.

Although some individuals with intersex diagnoses (and their parents) have welcomed the establishment of intersex activist groups, others echo the criticism often leveled at them by physicians: that they do not speak for a representative population of those with intersex conditions, but only for the most troubled and angry minority. One woman with PAIS feels that ISNA's opposition to genital surgery is unhelpful: "I have yet to meet a single person who has contacted ISNA and received support. Two have said, 'Stay away from ISNA.'" Another person refers to them as the "IS-NAZIS. ISNA tends to discourage adults who are, by choice, having genital surgery. They also accept an intersex identity." My reading is that ISNA's take on genital surgery promotes choice. Since adults can choose surgery whereas infants cannot, ISNA has opposed genital surgery on children but takes no issue with adults choosing the procedure for themselves. Interestingly, it is ISNA's historical acceptance of an intersex identity, and the perceived rejection of this stance through its promotion of the term *disorders of sex development*, that has driven some longtime supporters away from the group.

The question of whether intersex refers to a condition one *has* or to a person one *is* has proven contentious. In its early tactics ISNA promoted the wider acceptance of gender diversity and embraced those with an intersex identity. In doing so, however, ISNA did not appeal to most parents or to individuals who do not identify as intersex or consider their diagnosis an intersex condition and therefore avoid association with so-called intersex groups. Hannah Samson, who has CAH, comments: "Intersex may apply to some other people, but I don't feel like I fall in that group. But I'm also finding that they [ISNA] are against early genital surgery, and I was like, 'Oh, my God. This is what's happened to me.' I like that they are working to stop that." Precisely for this reason ISNA began distancing itself from the term *intersex:* to reach a broader group of clinicians, parents, and individuals with intersex diagnoses themselves who dislike the term but are otherwise sympathetic to ISNA's views. In making the group more appealing to those with less politicized views, however, ISNA has lost support from individuals who continue to align themselves with the group's earlier, more radical stances.

These same issues surface with respect to sexuality. Some individuals have avoided any contact with ISNA because they do not like its association with the gay community: "I can understand that homosexuality is involved in intersex, but the whole homosexual rights thing is something that I don't want to get involved in," claims Helen Minot. Lesbian and gay activism has been an important factor in efforts to change the treatment for intersexuality, in part for its challenge to gender-nonconforming sexual desire and behavior, but also for providing a foundation on which to understand the demands to changes in medial approaches to intersexuality. This is because the management of intersex conditions—and indeed what calls our attention to them as a "problem" in the first place—is centered on a narrow view of gender conformity in which anatomies, behaviors, and identities must be unequivocally consistent with gender assignment. The narrow focus on atypical genital anatomy, whose correction is seen as essential to a successful gender assignment because it will help to produce gender-typical boys and girls, has echoed the aims of lesbian, gay, bisexual, and transgender (LGBT) groups to depathologize all types of gender nonconformity. Fears of gender atypicality (of which homosexuality is only one part) thread through the treatment of intersexuality, and they are precisely why some individuals and groups react so strongly to ISNA's and others' linkage between intersexuality and homosexuality. It is also why ISNA has moved away from overt connections with LGBT concerns in an effort to reach a broader audience, perhaps especially parents. (It is precisely these types of barriers that prompted ISNA to close its doors in June 2008 to support the development of a new organization called Accord Alliance, which would not be hamstrung by ISNA's perceived biases, its previous activism, the widespread misrepresentations of its positions, and its association with the term *intersex*—a medical term ISNA adopted for lack of a better one, preferring it, if only slightly, to *hermaphrodite*.)

The Role of the Media

Since the mid 1990s intersex activism has been a frequent topic on television and in the national print media. Prurient interest in gender-atypical bodies and in the sexualization of intersexuality has generated far more media coverage than that accorded other conditions with the same prevalence. The media play an important role in disseminating information (and misinformation) and fostering debate, and activists have become skilled at working with journalists and other media figures to generate publicity. The media response has generally been quite sympathetic to activists' concerns, and advocates credit the media with publicizing their efforts: "In '96, when I started talking to people about intersexuality, maybe three out of fifty people had heard of it, whereas now . . . it's very rare that people don't have some relationship, at least to the word [*intersex*]. That change has been because of ISNA pushing it into the media and making doctors talk about it" (Teresa Diaz).

In many instances, however, media coverage tends to perpetuate medical and clinical paradigms. The media share many clinicians' confusion, conflating the demand for a moratorium on genital surgery with a demand that children with intersex conditions not be assigned to or reared in a specific gender. Media coverage of David Reimer's story seemed especially prone to perpetuating the medicalized view of intersexuality and of a biological basis for gender more generally. Although Reimer did not have an intersex diagnosis, his story is often tied in with intersex activists' concerns, especially surrounding early genital surgery. (To call attention to their concerns, activists have also drawn connections between Reimer's experiences and their own.) In retelling Reimer's story, the media has perpetuated a very specific medicalized view of intersex, which is that gender—broadly conceived to include identity, behavior, and sexuality—is biologically determined and that current treatment protocols are bad primarily because they insufficiently attend to this fact.

Clinicians have had a mixed response to the media coverage of intersex issues. Although they frequently claim not to read social science or other nonclinical work on intersexuality (such as the writings of Kessler or Dreger), many are aware of popular coverage of the issue and even notify colleagues of magazine articles and television programs. In particular, they know about or have read John Colapinto's book about Reimer, *As Nature Made*

Him. One clinician says that his decision to change his treatment practices was based on both his clinical experience and coverage in the popular media. He specifically cited the 1997 article about Reimer published in *Rolling Stone*: "I copied it and gave it to everybody in the division. Because of a similar case we had, which preceded Colapinto's article, we had already done a lot of talking about the issue. Everyone in the department very quickly developed the same concerns I had. I wound up concluding that this idea that you could assign gender, and as long as the parents weren't ambiguous about it, it would result in the patient accepting whatever gender, was nonsense. . . . I decided not only is this nonsense, but it's based on science that could almost be construed as scientific fraud" (Dr. B). Others, however, take a more skeptical view. One leading surgeon is emphatic about his dislike for media coverage of intersex issues: "I don't like it at all. These are very private, very tragic birth defects that affect young parents, and I don't think it's anybody's business but that family. It's not anything to be on television, in the newspapers, because it's a tragedy. It's exploitative, and there's too much focus on a freakish nature, and an overfocus on ambiguous genitalia" (Dr. P). Although he correctly points to the media's tendency of sexualizing intersexuality, his idea that the condition should be hidden is precisely the type of thinking that has led to much of the silence many adults and parents have complained about. As Alice Dreger notes, it will be familiarity with intersexuality that will take away its supposed strangeness (1998a: 201). Rather than not cover intersexuality in the media, a more productive tack would be to cover it in a way that seeks to reduce stigma by refraining from sensationalizing the issue. This might be a tall order for some media outlets, but it has been done successfully by some journalists.

Clinicians are more likely to say, as did a prominent pediatric urologist, that media coverage "makes it more difficult to be authoritative," a concept on which another clinician expanded: "More and more parents come with media or Internet information. The problem is that sometimes it's more confusing than anything else. That's difficult, because the more information they get, the less controlled that information is" (Dr. M).

Concern about the reliability of the information available from the media, the Internet, and other sources is not unique to intersexuality and signals a broader erosion of clinical authority over the past several decades. In the case of intersexuality, however, the media and the Internet present knowledge not available from other sources—such as clinicians themselves, who likely would have preferred to resolve these issues internally—which has been critical for fostering dialogue among clinicians and between physicians and parents.

Changing Clinical Practice

What effects have intersex activist groups and media coverage had on the clinical practice dealing with intersexuality? Changes in clinical medicine happen through internal and external means. Examples of internal changes include new research that challenges clinicians' old methods or understandings, prevailing theories, personal beliefs, and clinical experience. External pressures for change might include critiques by, or dialogue with, patients, social scientists, ethicists, and others seeking to influence medical understanding, knowledge, and practice. The boundaries between internal and external means are permeable—outsiders may become insiders—but they are also interactive: external challenges can lead to internal theoretical reformulations.

In the case of intersexuality, internal and external actors have called the traditional treatment paradigm into question. The primary external influence has come from intersex adults' efforts to reformulate both the medical understanding of intersex conditions and their treatment, and in doing so, to challenge the exclusive authority of medical professionals to make judgments. Intersex adults' interventions have rendered intersexuality not simply a medical but also a social and political controversy: the evaluation of scientific data is important, but it is not the only criterion for decision making about treatment.

Activism—from the first confrontations organized by intersex demonstrators at a medical convention to the inclusion of ISNA in medical conferences and meetings to develop guidelines for treatment—refigures ideas about how scientific questions should be resolved. Patient advocates and activists have compelled clinicians to discuss the status of intersexuality with the individuals affected. When medical views and treatment practices change as a result of these efforts, it demonstrates that medical practice rests not only on supposedly scientific truths but also on the subjective views of intersex adults. Early "dissident" clinicians (Joffe, Weitz, and Stacey 2004) willing to reshape their views and practices in response to these perspectives showed the ability to be flexible in their thinking and to acknowledge and value nonmedical and nonscientific viewpoints. Moreover, as Carole Joffe and her associates have noted of the abortion rights movement, in some cases the roles of activists and clinicians have converged: as some activists have become more professional, some clinicians have become more politicized. This is especially true of ISNA, which abandoned more confrontational approaches almost a decade ago, paving the way for new groups to pursue these efforts.

Very few clinicians I interviewed attribute changes in intersex management directly to activism. Not surprisingly, they are much more comfortable with attributing changes in their practice to medical and scientific influences such as clinical studies, colleagues' conference presentations, or clinical experience (their own or that of colleagues). This does not mean that advocates have not had an impact on clinical practice. Indeed, what clinicians often see as a gradual evolution of their own beliefs and practices, and as a product of reason, activists see as a result of their activism.

Although until recently few clinicians openly supported ISNA, some have been receptive to ISNA's recommendations, particularly to its calls for fewer genital exams, limits on medical photography, and telling parents and patients the truth about a diagnosis. The recent "Consensus Statement on Management of Intersex Disorders" reflects this shift, recommending an increased focus on psychosocial support, open communication with parents and patients, and a decreased focus on atypical genitals by advocating a more cautious approach to early genital surgery.

Despite these changes, ISNA's credibility and reach continually come up against various arguments by clinicians. Many rightly state that the group does not represent all persons with intersex conditions. While ISNA acknowledges this, it disagrees that it only represents the "disgruntled minority," as many physicians have charged.[12] Two clinicians I interviewed commented: "ISNA is the disaffected few, but there's all these happy women out there who've had babies. They're the quiet majority," claims Dr. S. Dr. O held a similar view: "These people are very angry. The patients who are happy don't want to be spoken to; the parents don't want it. They say, 'That's in the past, we don't want to discuss it. My kid is well adjusted. Everything is fine.'" There is, however, scant evidence for this perspective. As one observer quipped, "Let them do their own studies interviewing all the happy intersex people out there, recruiting them through a specially themed happy campaign to indicate that they are looking for satisfied people who want to fade into the woodwork, not carp and complain about monster doctors and their unhappy lives."

It is likely correct that intersex activists and advocates represent only a small proportion of the total number of individuals with intersex conditions who have received treatment. The same holds true in other health advocacy movements—such as the women's health movement, AIDS activism, and the environmental breast cancer movement—for which time and a willingness to go public are required. However, it is mistaken to assume that all those not involved with ISNA or working to change medical treatment are necessarily satisfied with their treatment experiences. There are many reasons why individuals might not choose to speak out: some may have minimal information about their treatment or may not even know their diagnosis; they may feel shame about their condition (or their child's condition) and be reluctant to share their experiences publicly; or they may not know that intersex advocacy groups exist. Also, ISNA or OII will not appeal to those who object to the term *intersex* or its connotations, which is especially true of parents of children with intersex diagnoses, who are integral to making treatment decisions and to whom advocates would thus like to appeal. Finally, some may reject ISNA in particular because they feel it is too radical—a view tied more to its early protests than to its stated aim to reduce shame and stigma, a goal with which few parents or clinicians would disagree. Of course it is also possible that some patients are happy, but given the above discussion it would be imprudent to generally assume that.

Several clinicians have criticized ISNA's position on genital surgery as misleading and counterproductive: "The major disservice has been in making the public believe that there is malice and mutilation going on. I can't think of one person who does this work who isn't thoughtful about what they're doing" (Dr. K). This statement exemplifies a critical and frequent assumption: that good intentions are synonymous with good outcomes. This may make for the most difficult argument for intersex advocates to counter, because clinicians and parents sincerely believe that their interventions will improve a child's quality of life and thus find it hard to believe that they could cause harm, or that individuals might come to regret the decisions made on their behalf. It is easier and more comforting to assume that those adults unhappy with their care received inferior treatment, had poor decisions made on their behalf, or were simply unlucky—all explanations that locate the problems with care in individuals rather than with the treatment paradigm and the assumptions undergirding it.

Other surgeons argue that ISNA's opposition to surgery can unduly influence parents, resulting in decisions not in their children's best interests: "We've had children for whom I think it's very appropriate to do reconstructive surgery, but the parents will look at the [ISNA] Web site and say, 'Are you sure this is okay?' It creates a great deal of difficulty for families" (Dr. L). Dr. H concurs: "The activist argument has been that doctors should have a hands-off approach in surgical correction, because it's not a life-threatening condition, and not health versus disease, it is an ill-conceived cosmetic procedure, which for that person becomes maiming. That argument doesn't acknowledge that physicians can sit down with parents, go through all the information, and come up with a plan that includes surgery, makes sense, and everyone's happy with." This physician's notion of "everyone," however, explicitly excludes the child, and this omission lies at the core of activist complaints. Paradoxically, the intervention that is irreversible and arguably has the most potential for harm—genital surgery—is the one that has taken the longest for clinicians to reconsider, if they do so at all.

Another clinician argued that ISNA "recognizes that there are very few instances where there's mistakes." By "mistakes" he means specifically cases in which physicians assigned a gender in infancy that the individual did not identify with later in life. Intersex advocates, however, take a different view of mistakes. Mistakes are not those decisions that concern gender but those that perpetuate shame and stigma and those that are irreversible. These advocates argue that all gender assignments are provisional, even for individuals without intersex conditions. Most individuals will grow up to identify with the gender to which they were assigned—but not everyone will. Intersex advocates argue against early genital surgery in part because of the possibility that an individual may want to change gender later in life. Either way, their view is that genital exams, surgery, dishonesty, and silence prove far more damaging than an "incorrect" gender assignment.

Clinicians acknowledge that some of activists' criticisms of the medical management of intersex patients need addressing. This may in part derive from the fact that these criticisms are in keeping with larger changes in medical practice and culture such as shared decision making, respect for patient autonomy, and a full disclosure of treatment options, alternatives, and risks. Dr. A noted that "I've talked to quite a few patients since I've been more active in this area, and I've heard the consistency of their feelings about problem areas in their medical care in terms of communication, knowledge sharing, the way they get treated when they go in for care."

Some say that although they do not like ISNA, the activist position has caused them to rethink their practice somewhat. Although "activist groups raise very legitimate concerns," Dr. L added that "they tend to be extreme, and what they think doesn't apply to every patient with that condition. I put what they say in perspective. It has raised questions about things. Say [I've got] a 46,XX child we're deciding to raise as a girl, do they need clitoral surgery? It makes me think a little more hard about that."

Despite these acknowledgments many clinicians still view the gender-atypical anatomy that characterizes intersex conditions as requiring treatment, and this treatment most often consists of ensuring gender conformity somatically, behaviorally, and sexually. The inability to question these aspects that undergird treatment have less to do with an unwillingness to reimagine the relationships between anatomy and identity than with an inability to do so. For many, genitals are naturally dimorphic and an essential sign of gender (if not *the* essential sign), and they assume that gender is both causal and linear, that is, that one's gender and sexuality necessarily derive from one's biology or "sex." Although feminists and others have often highlighted the limitations of this thinking—and in doing so have focused less on the more practical problems with treatment such as repeated genital exams—activists and others have had the most success changing those aspects of treatment that still manage to leave intact deeply ingrained ideas about dichotomous gender.

QUESTIONING CREDIBILITY

One of the major issues in current debates concerns what counts as evidence. Clinicians and researchers often argue that ISNA and other activist groups offer only anecdotal evidence, which has no validity when compared with scientific studies, and that they will not change their practice in the absence of compelling "data." But given the paucity of any form of data on outcomes for intersex patients, they often rely on anecdotal or clinical experience to support their own current treatment practices. The problem is not that these physicians are uninterested in facts. The problem is how they go about marshaling those facts.

The scientific method is the process by which clinical researchers investigate particular phenomena and acquire, correct, or integrate knowledge about those phenomena. Although the body of techniques and procedures varies from one field to another, clinical researchers and scientists use specific principles of reasoning to formulate hypotheses, which they then test by gathering observable, empirical, and measurable evidence collected through observation and experimentation. Among other aspects shared by various fields of inquiry is the conviction that the process must be objective to reduce bias in the data collection or the interpretation of the results. Although this method is highly valuable, when the acquisition of knowledge is narrowly construed to include only data gathered using the scientific method (and with insufficient attention to the limitations of how this method has been applied), however, it can constrain clinicians' thinking, rather than expand it.

Because physicians seek to base their clinical decisions on data gathered using scientific methods and reasoning, it becomes nearly impossible to falsify clinicians' working hypotheses about treatment efficacy or outcomes when these data are lacking, as in the case of intersexuality: data gathered using other methods are viewed as insufficient and therefore are often dismissed. Nevertheless physicians often gather shards of evidence derived from their own clinical experience or from that of colleagues to support their clinical decisions and practices. Although they tend to view these data points as less subjective, they could equally be called anecdotal. Moreover, clinicians attach different values and validity to information (or evidence) based on how it is delivered. That is, clinicians are likely to view information delivered by a patient in the clinic as a type of evidence. Yet if a journalist delivered that same information in the form of a quote by the same patient in an article, clinicians would not view it as evidence.

Some clinicians and researchers do accept the anecdotal and subjective accounts from intersex individuals as valid data even when they are not part of a research study. One pediatric endocrinologist felt that "Cheryl [Chase], frankly, has as much credibility in this area as anybody. I would sooner believe her than most urologists" (Dr. B). Other physicians, however, believe that such evidence is invalid, because it was not gathered using standard scientific techniques. Commenting on Chase's presentation at the Lawson Wilkins conference, one clinician observed: "She went way overboard. . . . She presented her talk as a scientific talk, 'Here's the facts,' overheads and what not. There was [*sic*] no data. If we claim to be scientists and that we're objective, how do we separate emotion and opinions from fact? We need facts to make educated decisions, so I want to know the facts. She had no facts to share, just opinions. What is the information, the data that you're basing these conclusions on? Otherwise, she's getting up and telling a story. It was inappropriate" (Dr. D). This clinician's comment raises the larger question of what counts as scientific fact. What are the outcome measures used? How were they decided on, and how were the data obtained? Clinicians are not necessarily objecting to the subjectivity of anecdotal evidence (i.e., first-person testimonial): many say that they would like to see data gathered from patient self-reports. Rather, their central concern seems to be that the data were not gathered using traditional scientific methods (e.g., sampling frames) and that thus the results may be biased toward poor outcomes.

Chase says that she has spoken to roughly five hundred people with varying diagnoses. Yet clinicians reject the information she has gathered because she did not collect it systematically or tabulate it statistically. However, as one clinician acknowledges, all medicine is based to some degree on anecdotal data: "The reasons for dismissing Cheryl's knowledge about outcomes aren't very well founded. When we take care of diabetics, we do things that have proven to be effective, or we don't do things that have been proven to be ineffective in very carefully carried out scientific studies. In most cases, those scientific studies were precipitated by a limited amount of anecdotal information coming from patients. Why is this area different?" (Dr. B).

One reason this area differs is that treatment rests on naturalized notions of gender difference and on cultural ideas about what traits are essentially masculine or feminine. The decision not to perform genital surgery, for example, not only goes against fifty years of standard care—something difficult but certainly not impossible to overcome—but potentially means leaving a child to live as male or female with somatic traits that defy cultural understandings of male or female. This latter issue proves much harder for many clinicians and parents to overcome, and it is one they feel the children will never overcome. The thinking goes something like, "She will not really be a girl if we leave her with a large clitoris," giving genitals a uniquely deterministic role in the development of gender role and behavior, and one that does not bear itself out empirically.

Although some physicians still argue that early genital surgery will likely prove more helpful than harmful, increasing numbers of physicians are willing to acknowledge that this remains an empirical question. Faced with such uncertainty, surgical decisions now often hinge on assessments of the probability that a child will fair well with surgery. Of course, the risk of harm persists with any surgery. Even if researchers could overcome the many barriers to conducting outcome studies—securing funding and sampling participants are but two—theories of probability or statistical inference cannot deliver the information that both parents and clinicians seek: what the outcome will be for any particular child.[13] As a result, physicians and parents rely on subjective probability, that probability people generate to express uncertainty about the occurrence of various events or outcomes. How people generate subjective probabilities, the cognitive heuristics they use, and the biases to which they are subject: this has been an area of inquiry for cognitive psychologists since the 1970s (see, e.g., Kahneman and Tversky 1972; Kahneman, Slovic, and Tversky 1982).[14] In general, this research shows that people are terrible statisticians. We overvalue low probability events and overreact to small samples. It is not that individuals are more influenced by single-case data than by statistical data; rather, we find single case data far more compelling than we should. Compounding this problem is that people often overestimate how much they actually know and infer a great deal from minimal or incomplete bits of information. These findings, evident in my own data, may explain why single cases such as John/Joan, published case studies, or clinical cases (whether one's own or those of a colleague) have proven unusually influential in shaping treatment practices.

IMPEDIMENTS TO CHANGE: UNCERTAINTY AND THE CULTURE OF MEDICINE

Notably, although clinicians may be quick to reject anecdotal data reported by patients and laypeople, they are prepared to base their practice on very small numbers of case studies reported by physicians, especially if the physicians are well known. Dr. N commented: "The John/Joan update has been incredibly important in changing practice, even though it's only one case. Hearing it failed made me and many others think, 'Maybe gender [identity] is not so malleable.' . . . I don't think anyone will ever unwittingly assign a normal 46,XY child to a girl since that case failed, and they'll think much harder about doing it to 46,XY intersex infants as well."

This case likely became so influential also because social and scientific thinking about how gender identity develops has shifted over the intervening decades from nurture to nature. This case fits contemporary ideas that gender role and behavior derive primarily, if not exclusively, from biological sources (especially from prenatal hormonal effects on the brain). Of the cases reported by the well-known pediatric urologist and psychiatrist William Reiner, Dr. P observed: "Bill's work has made a huge impact on me. Here he is saying all these kids assigned as girls are growing up wanting to be boys. You cannot argue with that. I think any clinician would reconsider that given what his research has shown."

This ready acceptance of limited and incomplete data reflects the importance of the social hierarchy in clinical medicine. The long training period and specialized nature of many fields that deal with intersexuality mean that most residents will apprentice with someone senior and well known in their field. Those in a junior position are expected to follow the treatment protocol already established by their mentors, most of whom have trained and practiced during the dominance of Money's model. Moreover, the prevailing practice at an elite institution can prove enormously influential. One clinician offers an example: "There was an article from Harvard indicating that for little girls with urinary tract infection, the absolute treatment of choice was cystoscopy [a method of examining the bladder] and urethral dilatation. For a decade, little girls with UTIs got cystoscopy and urethral dilatation because of that one paper. . . . When someone from these [institutions] says something and is dogmatic about it, it is very influential" (Dr. N).[15]

As they gain seniority, individual clinicians exert power within an institution and within their field to promote their style of practice, see new patients, and influence the decisions of other clinicians. A newly trained surgeon describes the influence of a senior surgeon at his institution: "Everyone knows [a senior surgeon] is very strong willed and that he establishes the way things should be done. Some other clinicians have been concerned that this might not be the best way to do it. Yet nobody's been able to do anything about it" (Dr. J). He adds that when a parent requested a consultation with another doctor because of reservations about this physician's use of vaginal dilation, he was "very, very upset."

Some individuals working to change intersex management protocol have argued that the only way to counter the influence of traditional, dogmatic practitioners is to wait until they retire. Dreger is convinced that this is the only way change will happen: "People are going to have to die. . . . What I'm saying is that the older generation of clinicians is going to have to retire. I'm speaking as a historian of science. This is why things change. There's the whole issue of patriarchy that has to go its merry way. One clinician [who trained under a student of Lawson Wilkins], when we met with him the first time, said, 'I'm about to commit professional suicide because I'm about to question my mentor.' You just don't question your mentor in medicine" (author's interview, East Lansing, June 1, 2000).

Not everyone shares this view. One clinician believes that younger clinicians may in fact be more likely to hew to the traditional treatment paradigm. He noted that when he gave a talk at a large teaching hospital in Boston, "the two most open-minded people in the room were older. One is in her late fifties; the other's over eighty. And the two people my age were much more rigid. I think that what happens may be the opposite, which is: when you're young, you're pretty sure you're right. As you get older, you find out that you're not always right" (Dr. B).

Irrespective of whether a clinician is willing to reevaluate clinical treatment or resists revaluation, all clinicians feel that they are providing the patient with the best possible care and believe that their treatment will allow for the best possible life. In consequence, some caution against criticizing earlier practices in hindsight, adopting a protective attitude toward their predecessors: "You can't be so quick to condemn the way things were done, because the doctors had been really working with the best interest of children in their mind. It's not worthwhile to expose the surgeons as doing the wrong things for all these years, and destroying their lives" (Dr. J).

This stance is perhaps related to physicians' concerns for their own reputations. Several physicians I spoke to offered that it is difficult for some clinicians whose careers have been built on a certain management style or surgical technique to acknowledge they might have made treatment decisions that harmed their patients: "Some of the big names don't come out as advocating change. . . . If you go, 'I did wrong by these kids, previously, even if it was with the best of intent,' it's still an uncomfortable feeling. The people who've published are resistant because to do that means that they ought to go back and publicly retract what they said before" (Dr. A). Guilt or pain about having potentially caused harm may also keep some clinicians quiet about criticizing previous practice: "For the surgeon who actually did the cutting, how can you say, 'Really blew this one.' How much does that hurt? I've run across of a lot of intersex folks having trouble getting their records, and I think that has to do with being embarrassed by what's in the record, what you did. . . . You can want to hear what you ought to do differently and still feel awkward when you get confronted with the actual patient. It's always easier to deal in the abstract" (Dr. A).

Despite the pressures favoring the traditional treatment approach, a number of the clinicians I interviewed believe in a future shift in practices. Dr. A, a pediatric endocrinologist, told me that "a lot of people feel less comfortable going public with how their thinking has changed. Many tell me after grand rounds or after meetings that they agree with me, but they have trouble saying it publicly." As a consequence of this reticence, Dr. B believes that change will occur gradually, without public fanfare: "What'll happen is that people will change their practice, and not publicly advocate any change. You make your decisions, on the spot, in a room with a patient. The patient's right to privacy protects you from having to discuss that with other people if you don't want to. Practices will change before the public portrayal of these changes, because that's harder to do."

The "Consensus Statement on Management of Intersex Disorders" would suggest otherwise; in it, clinicians publicly advocate changes to treatment practices. As with all treatment guidelines, however, there are no enforcement or oversight mechanisms, and thus no way to know whether the changes are actually implemented and to what extent they improve the health care experiences and quality of life for affected persons and their families. Moreover, there are no obvious incentives or adverse consequences in place to foster or drive implementation. At present, the new standard of care exists as little more than an ideal on paper. For this reason ISNA has begun to focus on finding ways to ensure that physicians use these guidelines. Nevertheless, the new guidelines reflect a shift in thinking about the treatment for intersexuality, albeit not a radical one. More established physicians may begin to base their treatment practices on them, and newly practicing clinicians could look to them for guidance.

FORCES FOR CHANGE: THE DEMOGRAPHICS OF CLINICAL MEDICINE

The passing on of medical and scientific knowledge through formal training is also supplemented by informal instruction and conversations with colleagues. One clinician I interviewed described how a single conversation with a colleague made him question and even change his own treatment practices: "A competitor of mine, we had a conversation over drinks, and he said, 'I'm going to keep the phalluses on these boys [with cloacal exstrophy]. I don't know whether it's right or not, but I'm going to do that.' That really influenced me. . . . I give him credit for changing my views around cloacal exstrophies" (Dr. N).

The presence of one or two clinicians willing to argue for new approaches may thus radically influence an entire medical team. Interviewing all four of the specialists dealing with intersex patients at one prominent research university gave me a glimpse of the power of peer influence. One of the pediatric endocrinologists arrived at the institution several years ago, deeply skeptical about the necessity of early genital surgery for infants with atypical genitals. She had trained with very senior personnel with whom she disagreed about the necessity of clitoral reduction. In her new position, she was able to recommend against clitoral surgery for her patients and gradually shift the views of some of her colleagues. Now she and a colleague are working to change the opinion of a third pediatric endocrinologist with whom they work, as well as the views of the pediatric urologist, who recently changed his position and in one case advised against clitoral surgery: "There's been a change in how he's approaching this, and that's with our influence. Especially [the pediatric endocrinologist's]. She just talked him into the ground, would not let up, very strong views about not operating. And he didn't. Three years ago, with a similar patient, he did, so that is really a change" (Dr. B).

Another clinician opposed to early genital surgery is trying to influence recently trained clinicians through grand rounds, which he has done in several communities: "I explain new ways of thinking about management. I talk about 'do no harm,' that we don't know about [genital] sensation after surgery or future gender identity. The students and clinicians have been very receptive" (Dr. A).

How well new treatment paradigms are received depends on multiple factors, including the medical specialty of the individuals involved. This same clinician, Dr. A, has found that pediatricians are very receptive to alternative views, perhaps because they are often not directly involved in the gender assignment and surgical decision making: "I've found they are much more willing to be open and learn, and I think this is because they are not directly involved in these debates. The treatment doesn't rest on the big names in their specialties or its long history." Another very prominent pediatric endocrinologist has found surgeons particularly resistant to alternative views: "The whole thing is that this needs to be taken out of the hands of surgeons. They create a problem for a few reasons. They're aggressive and want to be in charge. So if the patient gets to them first, if they're called first, then it's their patient. And I think professional pride and hubris make them say, 'I can fix this, I can do this.' But maybe it doesn't need to be done, or maybe they aren't the ones to do it" (Dr. S). Some clinicians have attributed changes in intersex protocol to the increasing number of women now in pediatric endocrinology (pediatric urology still has very few women):

> More women in endocrinology had a lot to do with the recent changes. . . . About seven years ago there was a symposium on what to do with a microphallus, which was basically make them female. This was before John/Joan. All of the presenters were male, above fifty-five years of age. They were some of the most senior, eminent people in our field. A young woman in the question period got up and said, "You make it sound easy to make someone a woman, but maybe your patients won't tell you as freely as they'll tell me what they are going through with the, quote, new equipment that you gave them. With the dryness, and the dilatations, and their discomforts and this is not, 'We've made it female and now it's fixed.' It's horrendous for many of them." . . . I think it had a profound effect on people in the audience. . . . There's a greater willingness to ask these questions and be more forthright because of the women. (Dr. G)

That an increased presence of women in these medical specialties is at least provoking controversy is borne out by the remarks of one of the most senior pediatric urology surgeons in the country, who felt that a female pediatric urologist colleague did not have sufficient experience or expertise to challenge respected (male) colleagues in the field: "She is a very nice young lady, but she does not have the background or wherewithal of others in this field. I have enormous respect for Bill Reiner and Mike Mitchell. They're very careful investigators, and their data is [*sic*] very compelling. I'd trust their data, not hers. When they say something, I listen. I like [her] very much, but I'm not going to listen to what [she] has to say because she doesn't . . . for her, it's reading stuff. For us, it's doing stuff. You've got to have years of surgical experience doing this!" (Dr. K). These remarks convey the impression that this surgeon is a junior physician with no surgical expertise and reveal something about the status of women in the surgical fields. In fact, the woman in question has been practicing in the United States for about twenty years and has performed well over one hundred surgeries in some way related to intersexuality. The surgeon also discounts the fact that the two individuals he names as credible in fact have diametrically opposed opinions: one has found that males with some form of exstrophy have done well raised as females, while the other says they have not.[16]

FORCES FOR CHANGE: CLINICAL EXPERIENCE

The factors that may lead clinicians to change their opinions and methods include not only training, the medical literature, and peer influence but also their own clinical experiences and knowledge. Many researchers and practitioners have noted that medicine is both an art and a science and have underscored the importance of clinical anecdotes—single cases or experiences—in shaping treatment decisions (e.g., Rothman 1991). When clinicians face clinical uncertainty or social and ethical dilemmas, it is common for them to make treatment decisions based on a previous clinical encounter. Ironically, these are some of the same physicians who disparage activist groups for advocating changes in treatment by presenting anecdotal data. Many clinicians recall single cases that proved enough to change their views.

> One of the patients that I learned around was a child born in the late 1970s who got referred back to me two years ago. I was a resident when that patient was born and assigned female. When she became thirteen years old, she started telling everybody that she was a boy. A definitive diagnosis was never made, but I think she's a partial AIS. When she was sixteen, she contacted a colleague and said, "The doctor's been forcing me to take estrogen, I've developed breasts, but I'm going to change to a boy. Do you know anybody who can help me?" She was referred to me and I oversaw "her" change to "him." That was my introduction to all the controversial aspects of this area, which I had never paid attention to. What does it take, then, for somebody to be confronted? In my case, it was that one case, because I remembered everything that transpired when the kid was born. (Dr. B)

Dr. N tells a similar story: "In my training, we had a boy with ambiguous genitalia who was raised female. At ten he said, 'I'm not a girl, I'm a boy, I want to play soccer, I don't want to be in the female locker rooms, I feel uncomfortable.' The whole game changed. They had removed his testicles, but they hadn't done a very good clitorectomy on him. He had genital reconstruction, they moved to a different town, and it was sort of a Joan/John story. These little anecdotal experiences were powerful. They suddenly made you feel nobody had all the answers. As a result, I never assigned a kid like this as female."

Not all such experiences change doctors' minds: sometimes one case is considered sufficient confirmation of a clinician's long-held opinions. One pediatric endocrinologist I interviewed discussed a case that supported his belief in the wisdom of the traditional protocol and in early surgery: "She's heterosexual, female gender identity, and she says it horrifies her to remember as much of the surgeries as she's had. She's had four. She said, 'If they could have fixed it all when I couldn't remember, I'd have welcomed it, even though I might have had a less sensitive clitoris than I have now. I could never have gone through my childhood as uneventfully if I had been. . . . It would have been criminal to leave me with the anatomy that I was born with because I couldn't have hidden it. If I had attempted to hide it, I would have been incredibly fearful' " (Dr. G). The doctor's recollection of the patient's pain of having to remember her surgeries is powerful, but equally powerful (and jarring) is the number of surgeries she underwent (a clitorectomy and a repeat vaginoplasty) and their outcomes. The doctor points out that she wishes she had had surgery before she could remember, but given the frequency of repeat surgeries for vaginoplasty, this is an unrealistic hope. Finally, it appears that this patient would willingly have forgone sexual sensation, but as the personal narratives of some intersex adults reveal, not all patients would concur. A treatment philosophy extrapolated from this one case could prove harmful to many other patients. Nevertheless, as noted earlier, clinicians can often find support for their views in some cases while downplaying, dismissing, or failing to take note of evidence that might challenge these views.

A pediatric surgeon I interviewed argued forcibly that he did not need clinical studies to assess outcomes: "I don't need studies. I have my own clinical practice where I can see long-term outcomes when people come back and talk to me about their outcomes. A girl just came across my desk today I'd done a bowel vaginoplasty on. It was an intersex baby and she does very well as a—she's a gender-reassigned individual, [and she] does beautifully" (Dr. K). But the criteria by which we judge such things are highly variable. Should we rely on the physician's subjective assessment alone, or should we include the patient's subjective experience? Does he mean that she has had no surgical complications? That she is happy? That she has a female gender identity? That she is dating? That she is pleased with her surgical outcome? What are the criteria by which we judge that someone has "done beautifully"? Moreover, one can think of multiple reasons why a patient would not have discussed a poor outcome with her doctor such as discomfort with the topic or a desire not to create strife or jeopardize her medical care, or to end an uncomfortable or embarrassing visit as quickly as possible. One way to overcome some of these issues is through assessments done by outside evaluators not invested in the results, not in charge of clinical care, and able to approach sensitive topics and elicit information by asking carefully formulated questions. This means asking questions that are not narrowly framed around the surgery itself but that concern the effect of surgery on an individual's comfort and well-being. I am not suggesting that surgeons cannot evaluate their own patients, but that these assessments alone remain insufficient.

Another clinician I interviewed explained that certain problems with the management of intersex patients became evident in his clinical practice, but because these results did not fit what he knew about intersex treatment, he assumed they were anomalous and looked for other explanations for the complications. This provides a good example of how we can be misled by the "representative" heuristic, when we are overly influenced by what is typically true, or the "affective" heuristic, shaped by what we wish were true (Kahneman, Slovic, and Tversky 1982; Croskerry 2006). Thus clinicians can filter disturbing or even incomprehensible information to see something entirely different from what the patient is experiencing or presenting. These conclusions, in turn, can shape the doctors' treatment of all subsequent patients. A pediatric endocrinologist describes her mistaken interpretation of what her patient, assigned female, was telling her about her gender identity: "One day she comes to me and says she's a boy. . . . She had excellent muscles, and when she told me she was a boy, I was feeling like it was such a failure of everything that I went into the examining room and I lifted my skirt and showed him my muscles. I said, "I'm a woman, and I have muscles. It's okay for girls to have muscles." But the point is, I should have listened differently. I should have paid more attention. . . . She was getting at something that I didn't want to pay enough attention to" (Dr. E). When I asked this clinician why she had not heard what her patient was saying, she said, "I wanted her to be happy with what had happened to her, and it was hard for me to deal with the fact that everything might have been really screwed up. You talk theories and what we should do as a medical profession, but this was my patient sitting there. She had already had enough bad breaks in life. I wanted to do everything I could to get her to like herself" (Dr. E).

The Clinical Response

Perhaps the greatest impediment to clinical change, and one cited repeatedly in current debates, is the lack of clinical research (evaluation or follow-up studies) to see if interventions work. One clinician I interviewed believes that the lack of data is unusual given research in other areas of biomedicine: "If one thing is obvious now, it's that we need the kinds of follow-up studies that we have in lots of other areas. We can gather incredible amounts of data on short stature and how growth hormone works, but it's amazing that we've been able to gather almost nothing. I think that it's because the people who have access to the patients and could provide the data don't want to" (Dr. B). Given the investment that people have in these children doing well, I have often sensed a deep resistance toward hearing that treatments may not have delivered the hoped-for benefits. Disagreements over study design, including clinicians' unwillingness to let others conduct outcome studies of their patients, are part of what led to NATFI's demise.

In addition to calling for more follow-up data, clinicians have also sought to develop consensus statements and protocols for the treatment of infants with intersex diagnoses (e.g., LWPES/ESPE CAH Working Group 2002; Lee et al. 2006). However, such undertakings have been accompanied by so many disagreements, difficulties, and controversies, ranging from logistical to ethical issues, that it is difficult at times to see to what degree treatment guidelines (or studies, for that matter) are likely to benefit individuals born with these conditions. Moreover, following guidelines is not mandatory. Even if clinicians were eager to adopt the guidelines, there is no mechanism to facilitate their implementation. For example, although the guidelines (and many clinicians) advocate psychosocial care for individuals with intersex diagnoses and their families, clinicians readily acknowledge significant barriers to getting these services covered by insurance carriers, thus making this support highly unlikely for most families. Without the cooperation and support of a wide variety of stakeholders, health payers in this example, the guidelines will remain an unrealized ideal for care.

The push to conduct outcome studies and develop standards for care should be understood in light of the movement toward outcomes-based medical treatment, which began in the 1990s. One aspect of this movement is evidence-based medicine (EBM), which aims to establish clinical medicine as a more scientific endeavor by more uniformly applying the standards of evidence gained from the scientific method to medical practice. The hope is that a greater reliance on scientific assessments of the risks, benefits, and efficacy of treatments—and a lesser one on clinical experience, anecdotal evidence, intuition, judgment, or consultation with colleagues—will improve patient care. Those who advocate EBM recognize that certain aspects of medical care are highly individual, such as quality of life; however, the aim is to clarify those aspects of medical treatment and practice that in principle can be studied using scientific methods to ensure the best prediction of outcomes in medical treatment, even if debates about desirable outcomes continue.

Acknowledging medical uncertainty, gaps in medical knowledge, and wide variations in treatment, outcomes research and EBM strive to standardize clinical judgment and care by deriving treatment guidelines and recommendations for patient care from the systematic evaluation of scientific evidence: as one article notes, EBM seeks to employ "the conscientious, explicit, and judicious use of current best evidence in making decisions about the care of individual patients" (Sackett et al. 1996: 71).[17] The call for EBM stems from a belief that clinicians may fail to keep up to date with advances in medical knowledge because of a lack of time (which is true for the clinicians I interviewed) and because of difficulty in accessing and assessing information: EBM aims to close the gap between what is known and its application in the clinical setting.

Standardization, while attractive, makes for a contested solution to manage uncertain knowledge (Timmermans and Angell 2001). Clinical guidelines are far from infallible: they are hampered by several limitations. Although the principal goal of guidelines is to improve patient care, it remains unclear whether they can achieve this goal. For one, patients, doctors, and health payers are likely to define quality of care differently. Guidelines may provide authoritative recommendations aimed at reassuring practitioners about the appropriateness of their treatment practices, but whether patients experience this as an improvement in their quality of care or health outcome remains largely unknown.

The evidence on which the guidelines are based presents another set of concerns. Although guidelines are often based on a critical appraisal of scientific evidence to clarify which interventions are beneficial, current evidence about the effectiveness of medical treatment is often incomplete. This conundrum—what if there is no evidence on which to base a clinical decision?—led to a humorous article in a medical journal suggesting several alternatives. These include: *Eminence-based medicine*, in which greater seniority requires less evidence; *Vehemence-based medicine*, which entails the substitution of volume for evidence; and *Providence-based medicine*, in which outcomes are essentially left to God (Isaacs and Fitzgerald 1999).

Whereas guidelines can alert clinicians to gaps in knowledge, in some cases the desired evidence may be unattainable because no funding is available or because some forms of testing are ethically unacceptable. (It would be unethical, for example, to conduct a randomized clinical trial to evaluate genital surgery in children.) As a result, guidelines may not be backed up by evidence or else may rest on evidence that has methodological inconsistencies or flaws contributing to bias or limiting their generalizability. Hence an important limitation of guidelines is that the recommendations may be wrong, or at least wrong for some patients.

Guidelines may also be viewed as the last word, with the belief that all issues and concerns have been sufficiently addressed when in fact they really constitute a first attempt to derive treatment practices from limited evidence. As one researcher said to me about the "Consensus Statement on Management of Intersex Disorders," "Once this comes out, everyone will think this is the last word on the topic."

On another level, the guidelines and recommendations are heavily influenced by the experiences, knowledge, opinions, beliefs, and composition of the group and the process by which the group develops them. If those developing the guidelines do not have the time, resources, or skills to gather and analyze all evidence, the guidelines will reflect these oversights and may even contain errors. Practitioners may also advocate treatments that they believe are good for patients, but that are in fact based on misconceptions and personal recollections that do not represent population norms (Woolf et al. 1999). This may hold especially true when data are absent or conflicting. If a clinician or a patient advocate is not invited, chooses not to participate, or has an unpopular opinion, consensus-based guidelines will not reflect his or her viewpoint. Moreover, because consensus is required, what is agreed to will often represent the least divisive option. Finally, the needs of the patient are not the only priority in making recommendations or clinical treatment decisions. Practices may be recommended for many other reasons including controlling costs, serving social or cultural needs, and protecting the interests of doctors, risk managers, and politicians.

While EBM encompasses many seemingly beneficial moves, social scientists analyzing it have argued that it can be seen as a form of external control over the clinical encounter, a way to control costs, and an approach that promotes a cookbook rigidity in clinical care discounting the needs of the individual patient (see Mykhalovskiy and Weir 2004). Others caution that the push for hard evidence means a greater pressure to regard patients as experimental subjects essential to the production of medical data rather than as that data's potential beneficiaries (Epstein 1996). Others have examined the broader effects of the need for evidence at the level of markets (Mykhalovskiy and Weir 2004).

Evidence-based medicine can be read as a discourse arising in response to specific contemporary challenges to established medical authority, an authority not neutral but culturally and politically situated (Denny 1999: 247). While some argue that EBM attempts to limit individual clinical authority, it actually reinforces medical authority in general at a time when health movements have presented contemporary challenges to this authority. In some respects, then, EBM constitutes an attempt to provide a technological fix, solution, or justification for what might otherwise be resolved in the social realm.

Many assume that the evaluative standards of EBM are transparent, neutral, objective, and universal, yet this is hardly ever the case. Perhaps not surprisingly, guidelines overlook the way in which knowledge is always socially shaped and can be reinvented, reconfigured, and reimagined in different contexts. Guidelines for treating intersexuality promote a narrow view of the conditions, one in which gender-atypical bodies are trapped in the discourses and rhetoric of medicine. Absent a wider sociocultural take on intersexuality, the guidelines rest on the assumption that physicians alone can and should judge how we understand what intersexuality is and what we, as a society, propose should be done for those born with gender-atypical anatomies. Guidelines are cases of "doctors talking to doctors about ways of improving medical practice" (Denny 1999: 252), and they provide a closed and self-confident narrative of what constitute important issues in the treatment of intersexuality.

The advocates' challenge has been to pierce this narrative. Running up against medicine's unequaled power to define what it is to be normatively human, advocates' task has proven difficult. In many ways their seat at the table has had a high price, namely, the ability to ask these broader social questions. To work with clinicians, advocates have had to take intersexuality on medicine's terms. So although patient advocates were invited to participate in reformulating the guidelines, their participation still left questions about who belongs "inside" to evaluate medico-scientific problems, to what extent they are allowed to participate, and the types of issues they are able to raise. One attending patient advocate had the impression that her and other advocates' very inclusion was contentious and that their participation was tolerated rather than meaningfully solicited.[18] Knowing that other groups had requested permission to participate and were denied only reinforced their "special status" (Leight 2006), making them painfully aware of the proximity of exclusion.

In the case of intersex, where there have been tremendous challenges to clinical authority, outcome studies and consensus statements seek to reinforce, restore, realign, and stabilize clinical authority over this issue. Who can argue with treatments based on "evidence"? Studies purporting to explore whether genital surgery improves psychosocial adjustment or harms sexual sensation, for example, effectively make these questions and their subsequent resolution the purview of strictly scientific inquiry. Outcome studies and consensus statements value one type of knowledge over other types of knowledge and ways of knowing. But they also reinforce clinical knowledge and authority as legitimate and even self-evident because science already constitutes a dominant discourse and a source of clinical authority in this realm.

In this context certain questions remain outside the scope of investigation, such as whether atypical genitals should be seen as a medical issue at all, or what social forces make certain treatments seem not only beneficial but necessary. Evidence-based medicine can be construed as a productive power using scientific knowledge to buttress particular social views, and thus treatment practices, while at the same time disqualifying other forms of medical knowledge as unscientific.

It seems plausible to state that the near universal demand for outcomes research and a concomitant effort to draft consensus statements, while deriving from activists' efforts to reshape clinical practice, in effect reinforce clinical authority in the domain of intersexuality. While intersex adults may participate in the outcomes research as subjects, and advocates have sometimes been consulted by medical authorities, their role has remained limited: they often do not participate in study design, data collection, analysis, or the choice of measurements. Moreover, there are serious doubts about whether outcome studies are feasible and what the value of their findings might be. The structural and financial barriers to conducting outcome studies, as well as the methodological limitations, prove significant. These obstacles make it unlikely that comprehensive studies will be completed any time soon. It would be virtually impossible, for example, to conduct conclusive studies of sexual sensation following surgery.

The organizational meetings for NATFI epitomized the logistical, ethical, and political barriers to any effort to share information on or standardize the medical management of intersexuality. The group's primary aims were retrospective and prospective reviews of treatment, a broad interdisciplinary consensus on treatment and the issuance of guidelines, and the establishment of a patient registry. Founded in 1999, NATFI by 2001 found itself essentially paralyzed by fierce professional wrangling over how such studies might be carried out. Interdisciplinary disagreement made for one of the persistent difficulties. The group was begun by urologists (who constituted two-thirds of those participating); pediatric endocrinologists had been slow to get on board and some had declined to participate at all citing concerns about patient confidentiality and the loss of investigator autonomy (Aaronson 2001).

The most crippling dispute arose over centralized studies in which clinicians would name patients to be evaluated by a team. Many senior clinicians objected to surrendering the medical records of their patients to this group; others thus accused them of being reluctant to relinquish control over their patients and of fearing professional scrutiny. Further questions cropped up about the feasibility of locating adult patients and financing the studies. As of early 2001, NATFI had no patients to evaluate for research purposes.

One set of issues centered on finding patients to participate. There was concern that even the name of the task force might deter participation by patients who did not consider themselves intersex. (This has proven a valid and persistent concern that in part has led to the new nomenclature of "disorders of sex development.") Some task force members pointed out that adult patients were difficult to locate because they often moved away or severed contact with their physicians after adolescence (clinicians also frequently change institutions). Furthermore, those with poor outcomes or those traumatized by memories of early medical interventions may well decline any involvement with medical researchers.

A related concern was that selecting patients by reviewing their medical records and then soliciting their participation might reveal to them details of their diagnosis and medical history possibly not disclosed to them earlier: for instance, some patients whose gender assignment in infancy went against their karyotype might never have been told this. Some doctors feared that the disclosure of this information might prove psychologically damaging to the patients. Some clinicians regard this type of disclosure not as a research procedure but as a form of medical intervention because of its potential effect on patients: as Dr. J commented to me in an interview,

"Ethically you're opening a càn of worms becaúse you have no idea what it's going to mean to these patients." However, a patient arguably cannot give informed consent to participate in a study without knowing everything the administrators know about his or her medical history. Various ways around this problem involving different degrees of disclosure were explored, but no consensus emerged.

A second set of issues centered on physicians' concerns about possible litigation and about damaging their relationships with the patient or parents. This reluctance has impeded follow-up studies of patients outside NATFI too. Arguments about protecting patient confidentiality can be valid, but they can also serve as masks for a larger concern, a fear of exposure of poor outcomes: "Some of our most prominent pediatric endocrinologists clitorectomized a lot of girls. They made female assignments out of many somewhat microphallic males. They've got a lot of patients out there who were treated in a way that they wouldn't do now. There's a reluctance on the part of some of those people to do follow-up studies. They say they're interested in the data, but they won't be forthcoming with the patients" (Dr. G). Dr. H asserts:

> If I'm seventy-five years old and over my career I had female-assigned thirty to forty boys with small penises, many of these men are in between twenty and forty years of age. Their parents told them they were born without ovaries and were infertile, and they're now living as women. You haven't seen them, because the endocrinologist stopped seeing them when they were twenty-one, and now you know the unhappiness of some people, how readily would you say, "I'm going to give you these thirty names of people that we did this to, for you to do the follow-up on." Think about it. There's been litigation. There's been publicity. Now there's self-doubt. (Dr. H)

Despite the reluctance of many clinicians to refer their own patients for follow-up studies, most readily agree that such studies would provide important data for refining treatment. They are curious about how children interact with their peers in school and out of it, and how adults fare with regard to fertility, sexual satisfaction, long-term relationships, and relationships with their doctors.

> We don't have a lot of data yet, but as more people are coming through, it shows things like locker rooms can be survived, and the people can be psychologically healthy with a small penis. When you look at it, if you're

going to argue, "We're just listening to a few activists who are upset about their care," then we need to hear from people who are satisfied with their care. We need to hear from patients who didn't have interventive care, but had supportive care, to know which of the three paradigms works the best. There's no data to say that surgery improves the reaction of the family to their child any more than not doing surgery. There's very little data about how people with kids who are intersexed react, and what the families need in the way of support. (Dr. A)

In the absence of such data, the best treatment for infants born with intersex conditions remains open to debate. One response has been to develop consensus statements and protocols, but these raise other complications and issues. First, consensus statements tend to be vague and inconclusive because they are born of the need to avoid contentious debate. And even when protocols are clear and widely accepted, as Money's was, clinicians may depart from them for a variety of reasons. Moreover, clinicians do not automatically accept the results of new studies and change their practice accordingly; research is selected, adapted, and refigured with respect to individual circumstances, including those of the practitioner and the family.

Intersex activists are fundamentally challenging how change takes place in science and medicine. To what degree they will succeed in part depends on the willingness of clinicians to hear their arguments and evidence. While some clinicians dismiss activists' data and recommendations as anecdotal and flawed, others have called for further studies based on these arguments. In the end, clinicians may conduct follow-up and prospective studies to confirm or deny activist arguments. Whatever the outcome of such projects, activists will have played a fundamental role in calling attention to the need to reexamine treatment practices for intersexuality in the United States and elsewhere.

Conclusion

Bodies are not only biological phenomena but also complex social creations onto which meanings have been variously composed and imposed according to time and place. The history of thinking about the body demonstrates that its functions, specific organs, and other attributes can be highly politicized and controversial. Since the early twentieth century, intersexuality has been framed exclusively as a medical issue, but intersexuality makes for a cultural and historical phenomenon as much as for a biological one. The penchant to view gender development and difference as natural and biological, and the concomitant resistance to understanding gender-atypical bodies in terms other than biomedical, has framed intersex bodies as ones for which treatment is necessary and necessarily more beneficial than harmful.

For much of the early twentieth century treatment was ad hoc and often arbitrary as clinicians held conflicting views about the true markers of maleness and femaleness. Beginning in the 1950s, John Money sought to improve treatment for those born with gender-atypical anatomies. By complicating our understandings about gender development, he hoped to guard against facile and idiosyncratic gender assignment and treatment for individuals with intersex diagnoses. Although Money and his colleagues were attuned to gender variation—for example, outlining seven variables of sex (chromosomes, gonads, hormones, external genitals, internal reproductive structures, assigned sex and rearing, and gender role)—they noted that in most people all of these seven different aspects agree with one another in being all female or else all male, creating a gendered and heterosexual ideal for the treatment of individuals with intersex diagnoses. Moreover, despite Money and his colleagues' evidence that biology alone did not determine gender role, the researchers' sequential and linear arrangement of these variables—from the biological to the social—implied that the latter derived from the former to some degree and that gender development was above all a scientific topic.

Then as now understandings of and the treatment for intersexuality are steeped in ideas about what bodily parts and configurations are required to be male or female. Diagnosis and treatment both frame and are framed by a dichotomy, an absolute separation, between those traits of women and those of men. Falling between the interstices of idealized male and female anatomies, the bodies of infants with gender-atypical anatomies have become embattled sites where clinicians and parents rely on folk rules about gender when interpreting biological information to make treatment decisions. Their ruminations about "good" outcomes reveal the underlying logic of how different bodily parts and functions and ways of being "normally" go together, and what different cultural logics can be used to determine the supposedly next best outcome.

Contemporary cultural beliefs about gender assume an agreement between a series of somatic characteristics (chromosomes, gonads, genitals, and secondary sex characteristics) and more phenomenological processes such as gender identity, gender role, and sexuality. In this matrix, gender identity and behavior differentiate unproblematically and supposedly naturally from biological sex. Intersex infants violate these gender rules, forcing clinicians and parents to make sense of seemingly unnatural gender combinations—for example, an infant with XY chromosomes who has female-typical genitalia.

Money's paradigm did much to reify ideas about idealized and naturalized masculinity and femininity. As implemented, the traditional treatment paradigm has been concerned with reshaping the material body so that its markers of gender become congruous. The diagnosis and treatment of intersexuality are characterized by what Suzanne Kessler has called "lookism": treatment focused primarily on the body's exterior and meant to give patients a visually coherent gender (Kessler 1998: 109). But what clinicians and parents see on the body's surface is profoundly shaped by cultural ideologies and values. Indeed, various aspects of the traditional treatment paradigm such as surgically reshaping gender-atypical genitals and administering hormonal treatment to suppress contradictory secondary sex characteristics are effects of the need both to maintain and to produce dichotomous gender. Parents' and clinicians' willingness to implement these treatments, too often with inadequate evidence as to their efficacy and mounting evidence of their harm, speaks to the importance of maintaining binary gender.

The medical construction of and approach to intersexuality are sustained by and sustain a particular gender narrative. This narrative begins with dichotomous gender and assumes that much gender difference derives from biology. In this framework, atypical anatomies and genitals emerge not as a sign of human variation, or as a marker of diversity, but as a social threat to the stories we tell ourselves about gender difference. Given that a good deal of treatment has little to do with the health of the child and much to do with his or her appearance, we are left with some fundamental questions: What is accomplished by making children's' somatic traits fit our idealized notions of what constitutes males and females? Can genital surgery make a patient more male or female, more masculine or feminine? Why is it so important to make gender variance invisible? To what extent should chromosomes or genitals be relevant determinates of who we are?

The treatment for intersexuality has been overwhelmingly focused on gender; clinicians have assumed that it is enough to carefully examine the body to reveal the "true" or "best" sex, make genitals that approximate the assigned gender, and rear the child in that gender to ensure a healthy individual. In short, the key to treatment has been centered on providing a coherent and consistent physical and psychological gender. But this rigidly normative understanding of gender, in which only the perfect agreement of physiological, psychological, and social factors can produce a well-adjusted individual, has proven terribly shortsighted. Clinicians have failed to appreciate how many aspects of treatment—repeated genital exams, medical photography, the withholding of critical information about one's own body, the stigma associated with the perceived oddity of having a gender-atypical body—have shaped the lives of those treated. This is not to say that the treatment of intersex is not about gender—it is, and deeply so. Yet it is also about so much more than gender.

Until recently, treatment was viewed as a medical necessity, but what intersex adults and advocates have forcefully demanded is that we look beyond treatment to examine the actual stakes. What, for example, is so threatening about leaving a child with an atypically large clitoris? And what is gained for the parents, the child, and society by making her clitoris smaller? Although genital surgery and other modes of treatment seem commonsensical, this is only because they fit our contemporary ideas about what organs and parts are required to make us male or female. What intersexuality is and how it should be treated (if it all) are questions always subject to the constraints of time and place; that is, understandings of these conditions and their treatment are culturally and historically contingent and thus subject to change. This holds no less true for the current debates than it did for those in the nineteenth century and the early twentieth. The new standard of care, then, is only the most recent iteration of approaches to intersexuality. It is not necessarily the best approach, nor will it be the last one.

NOTES

Introduction

1 The consensus statement (or a summary of it) was published in three venues simultaneously: Houk et al. 2006; Hughes et al. 2006; Lee et al. 2006.

2 The proposed revised nomenclature is:

PREVIOUS	PROPOSED
Intersex	Disorders of Sex Development (DSD)
Male pseudohermaphroditism —Undervirilization of an XY male —Undermasculinization of an XY male	46,XY DSD
Female pseudohermaphroditism —Overvirilization of an XX female —Masculinization of an XX female	46,XX DSD
True hermaphroditism	Ovotesticular DSD
XX male or XX sex reversal	46,XX testicular DSD
XY sex reversal	46,XY complete gonadal dysgenesis

3 A perusal of several online message boards in the fall of 2007 suggests that these surgeries continue to be performed on girls with "severely virilized" genitalia often when they are between four and nine months old. What counts as "severely virilized," however, is open to interpretation (see chapter 5).

4 The chapter on disorders of sex differentiation in the tenth edition of the *Williams Textbook of Endocrinology*, a widely used and authoritative text, lists roughly forty intersex diagnoses (Grumbach, Conte, and Hughes. 2003: 880). Clinicians I interviewed frequently said there were sixty to seventy. Until recently intersex nosology was primarily based on a system developed in the nineteenth century that categorized individuals according to the presence or absence of certain gonadal tissues (see chapter 1). In this framework, individuals with both ovarian and

testicular tissue are labeled true hermaphrodites; people with testes, genital ducts, and/or external genitalia that are either present as more female typical or else as a "completely undifferentiated male" are considered male pseudohermaphrodites. Those with ovaries and more masculine-appearing external genitalia are called female pseudohermaphrodites.

5 This is assuming that the individual was able to survive with his or her condition. Most individuals with salt-wasting CAH, for example, would have died within several weeks of birth.

6 Very few intersex diagnoses are directly life threatening. However, some conditions warrant immediate medical intervention to save the life of the child, such as salt-wasting CAH, which is characterized by severe salt and hormonal imbalances and accounts for roughly two-thirds to three-quarters of all cases of classical CAH. In other diagnoses that fall under the rubric of disorders of sex development, immediate medical attention may be needed to allow for the elimination of urine or feces or to close an open abdomen or bladder (e.g., cloacal exstrophy). These are often not thought of as intersex diagnoses, but in some instances individuals with these conditions have been treated according to the traditional treatment paradigm for intersexuality.

7 I also conducted interviews with the scholars Suzanne Kessler and Alice Dreger, the journalist John Colapinto, the urologist and child psychiatrist William Reiner, and the intersex advocates Cheryl Chase, Betsy Driver, and Emi Koyama. I interviewed Kessler and Dreger because of their scholarly contributions to cultural understandings of intersexuality and to the ethical issues raised by the traditional medical protocol, both of which have proven important in intersex debates. I also wanted to explore their working relationships with intersex activists. I interviewed Colapinto because he authored the book *As Nature Made Him*, in which he detailed a case widely discussed in debates over the medical management of intersexuality. Reiner was interviewed due to his prominent role in debates over gender assignment and gender-identity development beginning in the late 1990s. Chase, the founder of ISNA, has been instrumental in challenging intersex treatment in medical and popular venues. Driver cofounded the intersex advocacy group Bodies Like Ours (BLO) and Koyama founded Intersex Initiative (ipdx). I discuss ISNA, BLO, and ipdx in chapter 8. John Money and his foremost critic, Milton Diamond (discussed in chapter 3), both declined to be interviewed for this project.

8 Some clinicians feel that DSD refers to a greater number of diagnoses than those traditionally thought of as intersex (e.g., cloacal exstrophy, some types of hypospadias). I am using the term intersex in this broader, more inclusive sense.

9 Important epistemological, representational, and methodological differences exist between clinical (or quantitative) research and qualitative research. Unlike quantitative studies that focus on the measurement and the analysis of causal relationships between variables, qualitative research explores processes and values in an effort to explain how social experience is created and given meaning (Denzin and Lincoln 2005). Whereas quantitative approaches often use data to

test existing models, hypotheses, or theories, qualitative techniques are used to develop detailed understandings of phenomena from patterns found directly in the data. The inductive nature of qualitative approaches makes them especially suitable when the phenomena under study are emerging or not fully understood. Qualitative methods do not purport to have the generalizability that larger and more random samples provide. However, what is lost in terms of sample size and generalizability is gained in the depth, expanse, and novelty of the findings. Ethnographic and interview data, for example, can be used to contextualize participants' understandings, make comparisons, and track variations in meaning across settings. The extent to which clinical and scientific researchers understand and accept qualitative research remains an open question. My own fieldwork suggests that it is not uncommon for researchers more accustomed to quantitative approaches to view qualitative research as a marginal methodology. For one example of the tensions between these two approaches in the case of intersexuality, see Meyer-Bahlburg 2005b.

10 Because I can never know the denominator from which I gathered this sample, I cannot generate a response rate. I interviewed adults and parents until I reached my quota. I did not turn people away or select participants because I did not know how many people would respond. After I reached my quota, I continued to receive inquiries about the project (approximately ten adults and ten parents). I did not interview these individuals, though I spoke to all of them. Based on these conversations, the individuals did not appear to markedly differ from those participants I interviewed. Of course I cannot make any conclusions about those who did not contact me for interviews.

11 The following are estimates of incidence for some conditions considered intersex (Grumbach, Conte, and Hughes 2003):

Turner syndrome (XO)	1 in 2000
Klinefelter (XXY)	1 in 800–1000
Classic CAH (after screening)	1 in 15,000
Averages from screening studies worldwide for specific groups:	
Yupik Eskimos	1 in 282
Washington State residents	1 in 17,942
CAIS	1 in 20,000
PAIS	unknown
True hermaphroditism	uncommon; reported in 400 individuals

These statistics do not include other diagnoses, such as cloacal exstrophy, that have traditionally been treated according to the treatment protocol for intersex conditions, even though clinicians would not identify them as such per se.

12 The disagreements between Sax and these authors run deeper than statistical accuracy. Sax further charges that some researchers are pushing the notion that "true intersex conditions" are not pathological and that their "insistence that all combinations of sexual anatomy be regarded as normal is reminiscent of Szasz's view of mental illness. . . . Like Fausto-Sterling, Szasz was suspicious of the

distinction between normal and pathological. Fausto-Sterling follows the example set by Szasz in her belief that classifications of normal and abnormal sexual anatomy are mere social conventions, prejudices which can and should be set aside by an enlightened intelligentsia" (Sax 2002: 177). After tightly drawing the boundaries of intersex to include no more than 0.02 percent of humans, Sax, a physician and psychologist, concludes that since "more than 99.98% of humans are either male or female," these data "support the conclusion that human sexuality is a dichotomy, not a continuum" (2002: 177). Note his quick and troublesome leap from biology to behavior: he extends ideas about sexual dimorphism to sexual behavior by assuming that individuals with gender-typical anatomies will naturally express opposite-sex attraction. His view that biology is responsible for a host of observed differences between girls and boys, which stands in stark contrast to Fausto-Sterling's, is expressed more fully in his 2005 book, *Why Gender Matters*. Not surprisingly, Sax's view has been promoted by the conservative Christian group the Traditional Values Coalition in a series called Homosexual Urban Legends, in which it seeks to debunk the claim that hermaphroditism is a separate sex (see traditionalvalues.org/urban/ten.php; accessed December 9, 2005).

13 The diagnosis CAH, including NCAH, is an autosomal recessive genetic condition affecting males and females. For a child to be born with CAH, both parents must carry a gene for the condition. Parents are usually asymptomatic carriers of the recessive gene, although one or both parents may also be affected. To diagnose CAH and NCAH and detect carriers of the gene mutations DNA testing is now available.

The adrenal cortex makes three main steroids that are necessary for health and that are involved in CAH: cortisol, aldosterone, and androgens. Cortisol is needed to cope with emotional and physical stress and helps to control blood sugar levels. Aldosterone is necessary for maintaining normal fluid volume and potassium balance, which stabilizes the heart. Androgens, such as testosterone, are produced in varying quantities by the adrenal cortex in both males and females and contribute to sebaceous gland and pubic hair development. While the majority of the androgen production (testosterone) in males comes from the testes, the adrenal glands contribute a large portion of the androgens involved in female sexual development. Excess production of these androgens, as is seen in CAH, can result in genital virilization.

Roughly three-quarters of newborns with classical CAH have what is called salt-wasting CAH. If not diagnosed and treated promptly after birth, babies with salt-wasting CAH become very ill soon after birth and may die. The nonclassical form of CAH is not life threatening, but it can affect puberty and growth in children and can cause infertility in males and females, as well as other symptoms affecting the quality of life.

14 This statistic is based on an analysis of a Danish patient registry that included only hospitalized cases, so the true incidence of AIS may be higher (Bangsboll et al. 1992). The condition CAIS appears more commonly than does PAIS, although

exact figures are unavailable. For more information on AIS, readers should consult the AIS Support Group (AISSG) Web site (www.aissg.org, accessed April 18, 2008).

15 With AIS, genetic testing is possible if the mother is known to be a carrier. The condition has been diagnosed in the first trimester by chorionic villus sampling (sampling tissue from the fetal side of the placenta). By the fifteenth week or so, AIS often can be detected by ultrasound and amniocentesis. Klinefelter syndrome involves a variation of the twenty-third pair of chromosomes that determine biological sex. The possible chromosomal variations are numerous, but they always involve the presence of one or more extra X chromosomes (e.g., 47,XXY). Today the condition is frequently diagnosed by amniocentesis.

16 Prenatal imaging is a highly operator-dependent technology: it requires a great deal of expertise for a sonographer to interpret these low-resolution images accurately. Moreover, the visualization of any fetal part depends on a host of factors, such as fetal position and the thickness of the abdominal wall. The sonographer must be reassured that she or he actually sees the penis and scrotum or the labia ("the three lines"). Most of the time, one can read the genitals at about twenty weeks (and often at sixteen or even twelve weeks with some types of ultrasound equipment). By the time the genitals develop to the point at which they can be distinguished by fetal imaging, however, the pregnancy will be rather far along. As a consequence, the question of terminating the pregnancy proves a very fraught one for the family, if it is possible at all.

17 Although prenatal genetic testing is theoretically widely available, in fact, tests for only a few extremely common genetic conditions are available off the shelf. Genetic testing for intersex conditions is infrequent for several reasons. First, it is expensive; second, its use is shaped and constrained by economic, cultural, moral, and ethical factors (Rapp 1999). Moreover, prenatal sex differentiation and development are highly complex processes that involve many genes, most of which are autosomal, not sex-chromosomal: that is, they are on chromosomes other than the sex-determining ones (i.e., the X and Y), and only a few have been mapped and are well understood. Because most intersex diagnoses are relatively rare, the number of commercial laboratories offering testing for such conditions remains limited.

18 If CAH is diagnosed relatively late in the pregnancy, some doctors put the mother on high-dose dexamethasone, which is unlikely to have the desired effect because the genitals have already formed.

1. *Taxonomies of Intersexuality to the 1950s*

1 Far from being natural or inherent, concepts of the psychologically or physiologically normal or abnormal have been crafted since the mid-1800s, when the British scientist Francis Galton put forth his eugenic principles and the term *normal*, which had previously meant "perpendicular," began to be applied widely to the human body and psyche.

2 Londa Schiebinger observes that sixteenth-century anatomists were interested in similarities rather than differences between males and females: Vesalius did not differentiate the nonreproductive parts of the male and female body. In his view, all other organs were interchangeable between the sexes (Schiebinger 1989: 184).

3 Throughout this section, my account of nineteenth-century understandings of hermaphroditism is drawn from Dreger 1998a.

4 Mak's revelation that European clinicians at the turn of the century began to consider the "sex-gender consciousness" or a subjective sense of oneself as male or female stands in contrast to arguments by Dreger (1998a) and Hausman (1995) that medical opinion about objective sex dominated until the 1920s.

5 The term *hormone* was coined in 1905. Estrin was isolated in 1923, progesterone in 1929, and testosterone in 1935 (Oudshoorn 1994).

6 It still was unclear why male embryos subverted the development of the female twin, but females did not affect their male twin. To explain this, "Lillie pointed to evidence for the earlier appearance of steroidogenic cells in males, and concluded that differentiation of males occurs early enough to influence female development whereas development of females occurs too late to affect development of the male co-twin. He also made the important suggestion that the extent of transformation of the female twin depended on the stage at which extensive anastamoses exposed her to circulating male hormones" (Capel and Coveney 2004: 854–55).

7 William Didusch, a medical illustrator trained at Johns Hopkins, made the drawings for this book while observing Young performing surgeries in the operating theater.

8 Around 1960, scientists determined that the Y chromosome was the determinant rather than the marker of maleness (Kevles 1986: 241–45). Thus chromosomes governed sex determination, and hormones dictated sex differentiation.

2. *Complicating Sex, Routinizing Intervention*

1 In his autobiography, *A First Person History of Pediatric Psychoendocrinology*, Money wrote, "The list of ten cases I had studied in detail needed to be expanded so as to provide sufficient data for publication. This accounts in part for the fact that my first data-based reports from Johns Hopkins were not in print until 1955. Responsibilities for their design and writing was mine, but the published authorship was shared" (Money 2002: 35). Not long after the publication of these essays, the Hampsons left Hopkins for Washington State University and soon thereafter stopped conducting research on gender development.

2 A hint as to Money and his colleagues' understanding of gender role as both a subjective and a social expression can be gleaned from their suggestion of how to assess it, which included an appraisal of "general mannerisms, deportment, and demeanor; play preferences and recreational interests; spontaneous topics of talk in unprompted conversation and casual comment; content of dreams, daydreams, and fantasies; replies to oblique inquiries and projective tests; evidence of erotic

practices and, finally, the person's own replies to direct inquiry" (Money, Hampson, and Hampson 1955a: 302n).

3 The range of study subjects available to Money was skewed toward individuals with CAH (then know as hyperadrenocorticism), in particular those with the simple virilizing form, who had the best chance of surviving before the advent of glucocorticoid replacement therapy. This is likely because of the high prevalence of CAH and because Wilkins's pioneering research into the causes of and treatment for CAH made Johns Hopkins a magnet for these cases (see, e.g., Hampson 1955; Money 1955; Money, Hampson, and Hampson 1957). Since Money did not always analyze his results by syndrome, instead referring to the entire sample, subjects with CAH were overrepresented in his analysis, which as a result may have underestimated the incongruence between gender role and sex assignment and rearing in other diagnoses.

4 Almost all of the individuals referenced in these articles were seen by the team at Hopkins during the time that Money and his associates were crafting the protocols. They were primarily adolescents or adults who were gender-assigned and raised before Money's arrival at Hopkins. Consequently, although these studies are undergirded by empirical data, they are not outcome studies of patients treated according to the proposed protocol. Instead, the protocols as first formulated derived from Money's observations and analyses of these individuals.

5 According to Money, the "penis is the only organ in the body that contains the spongy tissue (corpora cavernosa) responsible for erection. Thus there is no possibility of enlarging a small penis with a graft of spongy tissue transplanted from elsewhere in the body. There is no particular corresponding obstacle to feminizing reconstructive surgery" (Money 1994: 45). This statement is partially correct but also potentially misleading. The penis is the only organ in the *male* body in which the corpora cavernosa is located. However, because the penis and the clitoris are homologous structures, the clitoris also has this tissue, which is one of two tissues that make erection possible. The inherent problem with Money's statement is his view of feminizing genitoplasty, which is reductively viewed as reducing a phallus to the size of a typical clitoris, as opposed to, for example, the ability to create a clitoris de novo (or enlarge one), which poses the same obstacles as those for enlarging a penis.

6 On rare occasions, Money suggested stature as a criterion for sex assignment. Discussing cases associated with growth-hormone deficiency, Money noted that the treatment produces "an estimated maximum adult height of only 5 feet, more suitable for a girl than a boy."

7 A micropenis is a normally formed penis, except for its small size. To qualify as a micropenis, the organ must have a median raphe, a foreskin, and a urinary hole (urethral meatus) at the tip of the glans, which is what allows a boy to urinate standing up. This last trait, however, is not as common as one might think. In one recent study of five hundred men, almost half of them (45 percent) had openings whose positions were considered abnormal (Fichtner et al. 1995). Any opening of

the meatus other than at the tip of the penis (e.g., located on the glans, shaft, scrotum, or perineum) is called hypospadias, making the penis a microphallus (because of its abnormal formation) as opposed to a micropenis. Despite these definitional differences, these terms are often used interchangeably in the medical literature and by clinicians. Kessler has pointed out that according to the standard definition of micropenis "a phallus without a urethral tube, *no matter what the size,* is a clitoris" (1990: 41; original emphasis). Urology texts cite numerous penile abnormalities (see, e.g., Walsh et al. 1992), such as phimosis (inability to retract the foreskin); inconspicuous penis (one that *appears* too small); webbed penis, concealed, buried or hidden penis; trapped penis; penile torsion; lateral curvature of the penis; and diphallia (duplication of the penis). These morphological abnormalities are not considered grounds for gender (re)assignment.

8 The reason for this is that the micropenis and the microphallus are symptoms rather than diagnoses per se. A micropenis, for example, is characteristic of a variety of syndromes and etiologies, including growth hormone deficiency, hypo- or hypergonadotropic hypogonadism, and partial androgen insensitivity (see, e.g., Danish et al. 1980; Lee, et al. 1980a; Lee et al. 1980b; Aaronson 1994). Consequently, infants with a micropenis or a microphallus may have any one of several etiologies that, for hormonal or genetic reasons, can result in infertility. Most cases of micropenis are due to fetal testosterone deficiency. This condition can occur in gonadotrophin deficiency, either as an isolated finding or in conjunction with multiple pituitary hormone deficiencies. Other causes include the decreased synthesis of testosterone (or DHT), decreased testosterone sensitivity, and growth-hormone deficiency. A micropenis can occur as an isolated idiopathic anomaly or in association with other anomalies. I thank Heino Meyer-Bahlburg for helping me clarify this point.

9 In the traditional medical protocol, penis size has typically been the primary factor in decisions about gender assignment and subsequent genital surgery for infants with a 46,XY karyotype (other than those with, say, CAIS). In the absence of any significant options for penile enhancement through surgery, and given the belief that feminizing genital surgery is easier and yields better results, clinicians have traditionally (though not always) recommended female gender assignment for males with a too-small phallus regardless of karyotype, hormone exposure, or other factors. Money explained the rationale in one study: "The decision in favor of surgical feminization and rearing the baby as a female [is] based primarily on the prognosis of adult erotic and sexual functioning. An adequate sexual life as a female could be fairly well guaranteed, after surgical reconstruction, whereas that as a male would have been dubious. As a male, the penis would always have been too small in coitus (Money, Potter, and Stoll 1969: 210)." For females with CAH assigned as males, there is little chance that the penis will reach an acceptably large size for penetrative vaginal intercourse. Curiously, this issue did not appear to disturb Money, even though he argued that adequate phallus size was necessary for intromission and a male gender identity.

10 Around the time when Money and his colleagues developed their protocols, the detection and assessment of certain intersex conditions in infancy was made possible by the development of a test for sex chromatins in humans. This was first done through a skin biopsy (Moore, Graham, and Barr 1953) and shortly thereafter through cells obtained from the interior of the cheek (Moore and Barr 1955).

11 There are various reasons why families and individuals may not have received psychological care; some researchers have pointed to a lack of clinical mental-health practitioners with the training or expertise to deal with intersexuality, as well as to the difficulty of obtaining reimbursement from insurance companies for such support (Melvin Grumbach, personal communication, April 21, 2005; Heino Meyer-Bahlburg, personal communication, July 15, 2005).

12 Money and Ehrhardt appear to have been aware of the importance of their work for feminists of the time. In the original edition of *Man and Woman, Boy and Girl*, the index has an entry for "Woman's liberation, quotable material," which, rather comically, refers to most of the book (1972: 311). I thank Bo Laurent for sharing this gem with me.

3. *From Socialization to Hardwire*

1 Diamond's challenge was not without precedent. In a 1959 paper published in the *Canadian Psychiatric Association Journal*, three Toronto physicians charged that the Johns Hopkins team's statistical and research methods were not sufficiently rigorous and that their clinical treatment recommendations were based on "shaky theory" (Cappon, Ezrin, and Lynes 1959: 90). They were particularly concerned about the recommendation of female assignment for males with a small or no penis. To guard against researchers leveling similar criticisms at their own work, they employed methods that they felt would further reduce observer bias (by having different team members assess endocrinological and psychological factors), and they used a control group of individuals without intersex conditions with which to compare their findings. The paper made little impact, perhaps in part because of the established reputation of Johns Hopkins as a center for the treatment of intersexuality.

2 Depending on how it is used, and whether it is made explicit, the term *psychosexual development* or *differentiation* can refer to gender identity, gender role, sexual orientation, or sexual identity. *Gender identity* refers to one's sense of being male or female. *Gender role*, used extensively by developmental psychologists, refers to behaviors and traits that are culturally designated or understood as masculine or feminine and thus considered typical for a male or female social role. *Sexual orientation* is often used to describe the sex of those to whom one is attracted sexually. *Sexual identity* is often viewed as distinct from sexual orientation in an effort to acknowledge distinctions between one's behavior and one's identity.

3 The reader will note a slip here from what might be called gender identity and role to sexuality. While Money was interested in the development of gender role (under

which he subsumed what we currently understand as gender identity, behavior, and sexual orientation), Diamond was interested in explaining sexual behavior through an understanding of what he called psychosexual development. This interest stemmed from his graduate studies at the University of Kansas, where he worked with William C. Young, who specialized in research on sexual behavior and development, and Robert W. Goy and Charles H. Phoenix, who investigated the origins of behavioral sex differences.

4 The history and critical analysis of this paradigm can be found in the following sources: Bleier 1984; Byne 1995; Fausto-Sterling 2000, esp. chap. 8; Young 2000.

5 After repeated close readings of many of Money's papers published prior to this period, I have never come across the phrase *psychosexually neutral at birth* in his writings. In the 1965 paper, as in subsequent ones, Diamond presented several quotes from Money's papers from 1955 to 1963 to support his claim, but none of those quotes include this phrase. In the 1955 papers, Money and his colleagues use the phrase *undifferentiated at birth*, but it is not obvious that they understood this to mean "neutral." My own reading of Money's early papers suggests that Money was already proposing an interactionist model of gender development, although his phenomenological descriptions and etiological theories remained vague at times.

6 Diamond wrote: "It has been shown that hermaphroditic individuals in our society find it possible to assume sexual roles opposite to their genetic sex, morphological sex, etc., and they can function socially as 'normal' members of society, engage in erotic activities, and receive pleasure in the reared roles" (1965: 149). The combination of male and female biological characteristics evident in intersex individuals, he reasoned, might also be manifest in their behavior.

7 It remains exceedingly questionable whether there could ever be appropriate analogous animal models for gender identity and sexual orientation in humans, given the highly phenomenological and subjective nature of these phenomena. In addition, it seems likely that rodents are not a good neuroendocrine model for human primates; furthermore, nonhuman primates are not a good model for gender identity in humans. The reason is simple: there is no analogue for human gender identity in mammals.

8 Aside from Daniel Cappon and his colleagues and Diamond, two English physicians, Christopher J. Dewhurst and Ronald R. Gordon, criticized Money's theory and treatment paradigm in their 1969 book *The Intersexual Disorders*, in which they expressed uncertainty about the idea that sex of rearing would always prevail over biology. One of the only others to challenge Money's theory prior to the 1990s was the pediatrician and child psychiatrist Bernard Zuger, who treated young gay men and was a specialist in what are now called gender identity disorders. Based on his clinical experience, Zuger questioned the prevailing view that gender role was shaped primarily by rearing and environmental influences (a view that rested in part on Money's early papers) and proposed a biological basis for gender role and homosexuality. Zuger reexamined Money's clinical evidence from fifteen years

earlier; referring to recent biological findings such as more detailed chromosomal types, he argued that of the sixty-five cases of "ambiguously sexed people" that Money had examined in his first article, in all but four that classification could be challenged on some grounds, calling into question the extent to which the individuals' sex of rearing and their biology were discordant. He concluded that "the data from hermaphrodites purporting to show that sex of rearing overrides contradictions of chromosomes, gonads, hormones, internal and external genitalia in gender role determination are found unsupportable on methodological and clinical grounds" (Zuger 1970: 461).

In Money's biting rebuttal of Zuger's essay, which appeared in the same issue of the journal, he chided Zuger for finding scientific fault with an article published fifteen years earlier and for failing to provide clinical data of his own to support his argument. His remarks extended to Diamond as well: "What really worries me, even terrifies me, about Dr. Zuger's paper, however, is more than a matter of theory alone . . . it will be used by inexperienced and/or dogmatic physicians and surgeons as a justification to impose an erroneous sex reassignment on a child . . . omitting a psychological evaluation as irrelevant—to the ultimate ruination of the patient's life" (Money 1970: 464). He made a similar criticism in *Man and Woman, Boy and Girl*: "The tyranny of the gonads . . . renders [a physician] incompetent to pay attention to the coital capacity of the phallic organ, or to the importance of gender identity," adding that the writings of his detractors had "become instrumental in wrecking the lives of unknown numbers of hermaphroditic youngsters, by authorizing or denying sex reassignment" (Money and Erhardt 1972: 154).

9 John Colapinto, in his book-length retelling of the story of John/Joan, notes that doctors in Winnipeg and at the Mayo Clinic had raised the possibility of raising the infant as a girl before the parents saw Money on television (2000: 19).

10 HarperCollins took advantage of this angle when promoting the book. In the discussion topics provided at the back of the book, the first question states that "nature versus nurture" is one of the most hotly debated topics today and asks readers why it is important socially and how the outcome of this debate might affect policy. The second question asks if or how the book might have changed the reader's opinion about the role of hormones in the formation of gender identity.

11 The authors mentioned that this woman worked in a blue-collar job "practiced almost exclusively by men," had a "strong history of behavioral masculinity during childhood," as well as of "tomboyism," and a "predominance of sexual attraction to females in fantasy" (Bradley et al. 1998: 4). The authors cautioned that these traits should not be read as signs of an unsuccessful gender assignment and conclude based on this case that perhaps gender role and sexual orientation are more strongly influenced by biological factors than is gender identity.

12 Meyer-Bahlburg has examined the published literature to assess the extent to which female gender assignment was accepted or rejected by "genetic males" with these diagnoses. He concludes that his findings "clearly indicate an in-

creased risk of later patient-initiated gender re-assignment to male after female assignment in infancy or early childhood, but are nevertheless incompatible with the notion of a full determination of core gender identity by prenatal androgens" (Meyer-Bahlburg 2005a: 423).

13 Other clinicians and researchers have published similar accounts of the rejection of female gender assignment and the successful rearing of males with atypical or no penises (Meyer-Bahlburg et al. 1996; Reiner 1996, 1997; Diamond and Sigmundson 1997; Slijper et al. 1998; Diamond 1999; Van Wyk and Calikoglu 1999; Zucker 1999; Phornphutkul, Fausto-Sterling, and Gruppuso 2000; Schober et al. 2002; Zderic et al. 2002). The authors disagree about whether a child with a malformed or small penis or no penis at all should be reared as female.

14 The call for a moratorium on early genital surgery may appear to come out of nowhere, but at the time this article was published, intersex activists had been voicing arguments against early genital surgery for several years. To his credit, Diamond was one of the first professionals to question the wisdom and necessity of early genital surgery.

15 Some have pointed out that even after his genital surgery, the position of his meatus was such that he would have had to push his genitals down to urinate while sitting, thus making standing urination easier.

16 In a 2002 book, *A First Person History of Pediatric Psychoendocrinology*, Money said that he remained silent "to allow the fire to burn itself out rather than add fuel to the flames by engaging in adversarial argument in print or television" (76). He devotes a chapter entitled "David and Goliath" to detailing his involvement in the John/Joan case. In this chapter he questions the accuracy of Colapinto's depiction of events, suggests that Colapinto was primarily interested in the story for financial gain, and charges that in the journalist's account he was "sprayed by the blinding venom of a spitting cobra" (76).

17 Colapinto likely promotes Diamond's viewpoint exclusively because he conducted multiple interviews with Diamond, whereas Money declined to be interviewed. Colapinto has acknowledged a possible bias: "Some critics have said it [the book] suffers from the fact that I don't even *speak* to Money, because I demonize him. And maybe that does happen. It does happen. If you speak to a person, you cannot paint them as a pure villain. There's something about the human contact of meeting someone, you become more sympathetic to their story" (Colapinto, interview, June 14, 2000).

18 The pediatric urologist Justine Schober's study of gender identity among children with cloacal exstrophy produced results irreconcilable with Reiner's; she found that all have a "feminine typical core gender identity" even though "they do appear to have more masculine typical gender role behavior in childhood, tend to prefer [masculine] toys, such as vehicles over dolls, and show a strong interest in athletic activity. Thus, they demonstrate masculine childhood role behavior but feminine gender identity" (Schober et al. 2002: 302). The psychologist Kenneth Zucker also noted that he followed two individuals with cloacal exstrophy raised as females,

then aged twelve, "with no clear evidence for marked gender dysphoria" (Zucker 2002: 11). The pediatric urologist Michael Mitchell writes: "We are still working on an adequate assessment vehicle. In general those 46,XY patients kept in the male role seemed to do well, however were limited in some cases by phallic deficiency problems (this was true for older patients). Those 46,XY converted to females tended to retain interest in aggressive activities and behavior, however, most seemed to be content (some were not aware of their genetic sex). I am quite convinced there is gender specific imprinting. But how important this is to the individual is unclear" (personal communication, July 18, 2001; for the published study, see Mukherjee et al. 2007). Another researcher following fifty-eight genetic XY individuals with multiple diagnoses (including cloacal exstrophy) found that the vast majority of individuals were satisfied with their gender (irrespective of assignment and rearing as male or female), although one-third had periods of uncertainty about their gender identity over their lifetimes (Meyer-Bahlburg 1999).

19 These theories move from explicit biological determinism to more tempered claims of "biological potentiality." For a more detailed review of the limitations of this body of research, see Doell and Longino 1988; Fausto-Sterling 2000; Roberts 2000; and Young 2000.

20 Five-alpha reductase is a condition that affects individuals with a 46,XY karyotype who are unable to convert testosterone to dihydrotestosterone, the hormone that masculinizes the external genitalia in utero. These infants may be misdiagnosed as having PAIS because of the way the genitalia form, which can vary from typically male to predominantly female. The infant has testes, as well as other male-typical internal reproductive structures. If there is no intervention, the activation of an isoenzyme at puberty will cause the individual to develop secondary sex characteristics typical for a male: the phallus will grow, the shoulders will broaden and the hips narrow, muscularity and body hair may increase, facial hair will develop, and the voice will deepen.

21 Because of their complete insensitivity to androgens, individuals with CAIS, who are phenotypically female at birth, are assigned as females and exclusively grow up to identify as women. Adults with PAIS may grow up to identify as either female or male, making this a diagnosis for which gender assignment proves very difficult. I interviewed both male- and female-identified individuals with PAIS, all of whom are now living as the gender opposite to that which they were assigned as infants.

22 Many of these studies are solely concerned with gender identity and pay scant attention to the individuals' quality of life.

23 As noted earlier, a new consensus statement regarding medical care for newborns with intersex diagnoses was published in 2006, but it provides mostly general guidance regarding genital surgery and gender assignment due to the dearth of evidence to support any particular treatment recommendation. Although the statement notes that the evidence supports the recommendation to raise markedly virilized 46,XX infants with CAH as female (based on the finding that the vast

majority of 46,XX CAH patients and all 46,XY CAIS assigned female in infancy identify as females), it is less clear how to gender-assign newborns with other diagnoses. For example, it notes that roughly half of those with 5-alpha reductase deficiency assigned female in infancy and virilizing at puberty (and all those assigned male) live as males, thus suggesting that the potential for fertility and a male gender identity should be discussed when making gender assignment decisions. Yet it falls short of suggesting a particular gender assignment. The consensus also stops short of specific gender-assignment recommendations for persons with PAIS, ovotestes, and incomplete or mixed gonadal dysgenesis, instead listing factors to consider—diagnosis, genital appearance, surgical options, the need for lifelong hormone replacement therapy, the potential for fertility, as well as familial and cultural views.

4. *Boy or Girl?*

1 Because narratives about developmental biology play a critical role in our understandings of how we become male and female—succinctly captured in the title of a recent article, "Sexual Differentiation: From Genes to Gender" (Migeon and Wisniewski 1998)—a brief review of fetal development may be useful. According to contemporary scientific understandings, the first step of human sex differentiation takes place at fertilization, when an egg combines with sperm to produce an embryo with what is usually considered a female (46,XX) or male (46,XY) chromosomal type. In the first weeks of development, fetuses are distinguished only by this karyotype: all have identical internal genitalia—two gonadal ridges and two sets of internal ducts (Müllerian and Wolffian)—that will develop into testicles or ovaries. One part of the Y chromosome, called the sex-determining region (SRY), carries a gene responsible for the gonads' development into testicles. The absence of this gene permits the expression of other genes, which trigger the gonadal ridge to develop into ovaries.

If the gonads develop into testes, the hormones they secrete (Müllerian inhibiting substance [MIS] and androgens) and tissue responsiveness determine how the internal and external genitalia will develop. These hormones will either promote or inhibit the growth of the Müllerian ducts (uterus, fallopian tubes, and upper vagina) and the Wolffian ducts (epididymis, vas deferens, and seminal vesicles). At roughly the ninth week of gestation, the external genitalia—a urogenital sinus, genital tubercle, genital folds, and genital swelling—form. Androgens from the testes masculinize the external genitalia such that the genital tubercle grows to become the penis and the genital swellings fuse to form the scrotum. The absence of androgens permits the external genitalia to remain feminine: the genital tubercle becomes the clitoris, the genital swellings become the labia majora, and the genital folds become the labia minora.

This development is typical, but not definitive: at any point, development may proceed differently. Such a variation is often referred to as a "disruption," "disorder," or "breakdown" in the medical literature, reinforcing the idea that humans are naturally divided by sex. These "disorders of sexual differentiation" can involve "sex chromosomes" (e.g., 47,XXY or 45,X in Klinefelter and Turner syndromes, respectively), genetic and endocrine systems that affect enzyme levels (e.g., congenital adrenal hyperplasia), receptors (e.g., androgen insensitivity syndrome), and the genitals or sex organs (e.g., absence of a uterus or phallus). Such a disruption is frequently, though not always, signaled by the presence of what clinicians call ambiguous genitalia. Atypical genitals seen at birth are often the first clue that a newborn has an intersex diagnosis.

The fact that these disruptions can occur on so many levels raises the question of what our categories of sex are meant to refer to: chromosomes, gonads, hormones, genitals? The ideals and requirements making us male or female begin with cultural assumptions about the activities and practices that males and females are supposed to fulfill and engage in—including the correct positions for urination, proper genital conformity within sex, proper sexual activity (i.e., heterosexuality), and hence the appropriate organs for said sexual activity—all of which are more requirements of gender than of biological sex (Warnke 2001).

2 Paradoxically, medical technologies such as obstetric ultrasonography that often (but not always) enable radiologists to "see" the fetal sex organs at about the eighteenth week of pregnancy, and thus to sex the fetus, can make the experience of the birth of a baby with an intersex condition even more strange and confusing than in the days when parents could not know their baby's sex until birth. Fetal ultrasound is instrumental to the personification of the fetus and, as Barbara Katz Rothman has shown, accelerates the progress of a pregnancy; ultrasonography turns the fetus not simply into a "real" baby but into a fully gendered person, with pink or blue clothing ready at the moment of birth (Rothman 1991). Part of the impetus to know the sex of the baby is linguistic: referring to their child as "it" feels distancing for some parents during pregnancy. The English language has no terms for uncertain gender that are not depersonalized. Moreover, knowing a baby's sex allows parents to choose a name. The revelation at birth of "ambiguous" sex (which is infrequently revealed prior to birth) can thus shatter expectations and plans formed months before.

3 Discrepancies among sex characteristics are relatively easy to spot at birth if the genitals are affected. Some conditions, however, such as CAIS are characterized by incongruencies among bodily sex characteristics but involve no obvious genital atypicality. These conditions may become apparent only later in life when clinicians discover that the child has chromosomes or gonads that are discordant with the genitals or gender assignment.

4 Depending on the severity of hyperandrogenism, a female infant may have genitals that are "mildly" virilized, "ambiguous," or virilized so as to appear to be male. At birth, infants with the greatest degree of genital virilization may be assumed to be boys with undescended (unpalpable) testes and assigned male. In most cases, the diagnosis of CAH will be made later when, for example, signs of

salt wasting develop, or in the absence of the testes descending. In rare instances, these infants have been assigned and reared as male; in other instances some reared female have transitioned to live as male later in life. More than 90 percent of 46,XX CAH assigned female in infancy, however, identify as females as adults (Dessens, Slijper, and Drop 2005).

5 Based on a study of 125 infants aged zero to five months, the researchers found a mean phallic size of 3.9 centimeters (Schonfeld and Beebe 1942). However, these results could not give standards for neonate penis size because a surge in testosterone production in males stimulates penile growth during the first three months of life, and this study did not disaggregate the newborns from the older infants. In 1975, two studies specifically established standards for phallic size in neonate males. In a study of one hundred newborn males, penis size ranged between 2.8 and 4.5 centimeters (Flatau et al. 1975). The other study assessed the phallic size of sixty-three premature and full-term male infants (Feldman and Smith 1975). Using the results of both studies, the median length for a neonate penis is around 3.5 centimeters (Lee et al. 1980a).

6 The normative data available for healthy full-term newborn male penis size come from two widely referenced studies performed in mainly Caucasian babies (Feldman and Smith 1975; Flatau et al. 1975), prompting some clinicians to wonder whether the definition of micropenis should vary according to ethnicity. A recent Canadian study compared penis size among infants identified as Caucasian, Chinese, and East Indian (Cheng and Chanoine 2001). Using the definition of micropenis as roughly 2.5 centimeters in length (which corresponds to 2.5 standard deviations below the mean), the authors obtained values of 2.6, 2.5, and 2.3 centimeters for newborns of Caucasian, East Indian, and Chinese background, respectively. The authors concluded that mean penile length and diameter are slightly but significantly smaller in newborns of Chinese origin compared to newborns of Caucasian and East Indian origins, but concluded that "from a practical point of view, the differences reported in this study have relatively minor implications" (281).

7 The creation of the nomogram comes on the heels of the ascendancy of normalization in the nineteenth century and of the idea that myriad aspects of deviance could be read on and through the body. By the 1940s normalization and socialization were firmly linked and apparatuses of both science and public policy. The physical body revealed one's social identity. During this particular period, a much larger American project to create statistical averages for the "typical" American body was at its peak (Urla and Swedlund 1995). This effort, which aimed to describe the parameters of the normal (i.e., typical) male and female American body, contributed to the shift. Anthropometry, which refers to techniques of measuring the human body using dimensions, proportions, and ratios, dates to the nineteenth century, and its development was linked to ideas about nationalism, citizenship, and race purity since it was used for racial classification and the identification of criminality. It was seen as a way to compare but also to interpret

variability in different parts of the body and to permit systematic comparisons of human bodies. Measurement has frequently been used as a technique in resolving critical border disputes such as the boundaries between races, between the normal and the abnormal, and between male and female (Urla and Swedlund 1995: 287). Many studies have shown that more than an accurate appraisal of the actual body, ostensible attempts to quantify and typoligize human bodies were infused with cultural notions about gender and racial difference (Gould 1981; Russett 1989; Schiebinger 1989, 1993). Moreover, such standards have the effect of pathologizing the bodies of those who fall outside the norms because the average and the ideal are routinely conflated (Urla and Swedlund 1995).

8 These individuals were born in the Netherlands before Money's treatment paradigm was developed. Depending on the degree of the genital virilization of the CAH individuals, Money might have recommended a male assignment. It seems unlikely, however, that he would have suggested a male assignment for the male with a micropenis. At the time of evaluation, all three individuals identified with their sex of rearing. These findings could be read as both supporting and not supporting Money's theory and treatment paradigm.

9 Had their ovaries not been removed, the two individuals with CAH in principle would have been fertile and could have, with appropriate hormonal manipulation, become biological parents through egg donation. However, this scenario is unlikely for several reasons. In the presence of high testosterone—either by testosterone treatment, as is typical for male-raised 46,XX CAH individuals, or by no or insufficient glucocorticoid replacement and, thereby, non-suppression in CAH, as some male-raised 46,XX simple virilizers with CAH elect to do—ovulation does not take place. Moreover, ovaries chronically exposed to increased testosterone are at increased risk of developing polycystic disease, which may also render the individual infertile. This situation may change in the future as reproductive technologies advance.

10 The two quotes by pediatric urologists raise the question of possible retrospective revision regarding clinical practice now that the traditional paradigm has come under fire.

11 It is rare to hear of such cases, but one clinician-researcher told me of a case in which a mother purposefully removed her young son's penis. In this case, the specialist was consulted regarding appropriate care, which included the question of whether to continue raising the child as a boy.

12 This study also found a shift away from the traditional policy of assigning "genetic males with ambiguous genitalia" as female and the continued recommendation for early genital surgery to normalize atypical genitals among both groups.

13 Prader stages are a scale system for describing the degree of genital virilization in girls with CAH. The scale runs from I to V, with I indicating an enlarged clitoris and a slightly reduced vaginal opening (this degree may often go unnoticed) and V indicating complete genital virilization, with a "normally formed penis" with a urethra at the tip and a scrotum (Prader 1958). Heino Meyer-Bahlburg, who

conducts research on gender-identity formation in individuals with intersex diagnoses, notes that if Prader had included male-typical and female-typical genitals on the scale, zero would be normal female and VI would be normal male (personal communication, March 25, 2002).

14 In this study, Reiner compared gender identity to sex of rearing in a clinical sample of eighty-four individuals with various disorders of sex development. This sample included forty-five individuals with "male-typical prenatal androgen effects but an absent or severely inadequate penis" (e.g., cloacal exstrophy or aphallia); twenty-eight individuals with "inadequate prenatal androgens and a Y-chromosome" (e.g., PAIS, hermaphroditism); and eleven individuals with "inappropriate prenatal androgen effects and a 46,XX karyotype" (that is, CAH). The groupings suggest his interest in the association between male gender identity and having a Y chromosome and androgen exposure. Based on the finding that of the eighty-four individuals, of whom sixty-nine were reared female, thirty-two lived as female, while twenty-nine lived as male (the gender identity of the others was less conclusive), he concluded: "Active prenatal androgen effects appeared to dramatically increase the likelihood of recognition of male sexual identity independent of sex-of-rearing. Genetic males with male-typical prenatal androgen effects should be reared male" (Reiner 2005: 549). With studies such as these, the potential for misconstruing association and causation heightens. Intersex diagnoses allow for the study of gender identity in subjects who have atypical biologies and who were raised contrary to chromosomal or hormonal sex. Because the gender change occurs in individuals with male-typical biologies, it is assumed that "male" biology and gender identity are associated. Risks come with attempting to compare both within and across diagnoses, thus glossing individual biological variation and virtually ignoring all factors contributing to gender-identity formation other than biology. Too often findings of gender change are interpreted as evidence that a particular aspect of biology has caused the gender change—a biological input and an identity output—although the possibility that other factors contributed is equally possible.

15 In the general population, testicular tumors in children are rare, and there is negligible risk of cancer before puberty. For this group, the vast majority of the changes that take place in testicular tissue are benign including hamartoma, a benign tumor, and Sertoli cell adenoma, which typically appears late in life and is also benign. More uncommon and serious are testicular seminoma, a type of testicular germ-cell tumor, and gonadoblastoma, both of which are malignant. The risk of malignant testicular tumors in individuals with MGD is considerably higher. Studies have indicated that individuals with a mosaic 45,X/46,XY karyotype have an increased risk of developing a germ-cell neoplasia in the gonads (Gantt et al. 1980; Robboy et al. 1982; Wallace and Levin 1990; Nonomura et al. 1991; Ramani, Yeung, and Habeebu. 1993). Researchers have also detected carcinoma in situ and gonadoblastoma (or invasive cancer) in these individuals prior to puberty or in adolescence (Doll et al. 1978; Muller et al. 1983; Muller et al.

1999). Most of these studies have included relatively few individuals, the majority of whom have had female phenotypes. Nevertheless, early removal of the streak gonad and in some cases bilateral gonadectomy seem justified in cases of MGD, irrespective of gender assignment.

If and when to remove testes is also a concern for infants born with partial or complete AIS. As with MGD, the risk for individuals with PAIS is higher than for both the general population and for those with CAIS (the earliest reported malignancy in CAIS occurred at age fourteen). For those with PAIS, early gonadectomy seems justified due to the increased risk of malignancy (as high as 50 percent for PAIS, versus 15–35 percent for MGD). For a concise summary of malignancy risk, see Lee et al. 2006. In CAIS cases, some clinicians recommend against the removal of the gonads in childhood, when the likelihood of tumor development is quite low; instead, they suggest leaving the testicles intact until puberty and growth are complete, believing that the natural estrogen produced by the testes is better than estrogen therapy to promote breast development at puberty, for example. After this, the testes may be removed if desired because of the cancer risk. Although the risk of a malignant tumor developing increases with age, it remains low throughout early adulthood (roughly 2 percent).

5. *Fixing Sex*

Portions of this chapter originally appeared in Karkazis 2006.

1 Feminist discussions of female genitalia differ significantly from clinical discussions in this respect. Feminist health and sexuality books, as well as Web sites and videos, commonly acknowledge a range of variation in female genitalia, noting in detail the striking differences in anatomy and appearance among women (see, e.g., Chalker 2000; Federation of Feminist Women's Health Centers 1981). In contrast, clinical literature almost never mentions such variety, leaving one to surmise that this level of detail is unknown or unimportant. An important exception to the conservative metrics of medical textbooks is found in the fascinating *Human Sex Anatomy: A Topographical Hand Atlas* by the gynecologist Robert Dickinson (Dickinson 1949). This work shows extraordinary variety in the appearance, shape, and size of women's external genitalia. Dickinson also compiled clitoral measurements made by anatomists from the late nineteenth century and the early twentieth, which ranged from roughly 5.5 to 8 millimeters in length. One sample of 100 women found the average nonerect clitoral length to be 5 millimeters. Some women in this sample, however, had clitorises 15 millimeters in length. Dickinson's text is striking for its largely descriptive nature: in contrast to the studies he cited, Dickinson catalogued clitoral size with no proscription as to what was normal.

2 See, e.g., Creighton and Minto 2001b; Creighton, Minto, and Steele 2001a; Creighton, Minto, and Steele 2001b; Creighton, Minto, and Woodhouse 2001; Minto et al. 2003a; Minto et al. 2003b; Crouch et al. 2004; Lloyd et al. 2005.

3 Before 1980, surgeons typically performed procedures on children three years or

older because the larger phallus size makes the procedure easier. More recently, the trend has been towards performing surgery as soon as possible because surgeons feel that emotional and psychological results will be better if surgery is performed before the child has an awareness of the congenital deficiency and the experience of surgery. To compensate for the difficulty of operating on a smaller phallus, surgeons may administer testosterone or human chorionic gonadotropin (hCG) to enlarge the penis prior to surgery.

4 There are numerous reports of an increase in the rate of hypospadias in the United States over the last thirty or so years, which in part may be explained by an increase in surveillance and the reporting of minor grades of hypospadias.

5 One Swedish study from 1997 puts the costs of a single procedure at $14,000 (Svensson et al. 1997).

6 Surgery has long been the recommended treatment for hypospadias. During the first millennium, this meant an amputation of the penis to the urinary opening (Smith 1997). At the end of the nineteenth century, however, at least one medical text suggested no surgical treatment was required for its mildest forms (Erichsen, Beck, and Johnson 1895).

7 For studies of the psychosocial aspects of having a hypospadic penis or undergoing surgery for hypospadias, see Berg, Svensson, and Astrom 1981; Berg, Berg, and Svensson 1982; Berg and Berg 1983a, 1983b; Mureau et al. 1995a; Mureau et al. 1995b; Mureau et al. 1995c; Mureau et al. 1996; Mureau et al. 1997; Walker 1998; May 2003.

8 Freud's claims that there were two locations from which women derived sexual pleasure and orgasm and that vaginal orgasm was the norm for adult women were anomalous when he made them. For centuries before his controversial postulation, it was widely and commonly believed that clitoral orgasm was the only possible orgasm. Masters's and Johnson's revelation that female orgasm is almost entirely clitoral would have been no surprise to many in the seventeenth century. After Freud, however, the significance of the role of the clitoris in orgasm waned in scientific circles to the point that claims like those of Masters and Johnson appeared novel in the second half of the twentieth century.

9 The authors write that Hampson studied six women with "clitoral hypertrophy" and that none had orgasm. After clitorectomy, five of these six women reported "normal sexual gratification" (Gross, Randolph, and Crigler 1966: 307). Regarding the work of Gulliver, they write: "It is the custom of a number of African tribes to excise the clitoris and other parts of the external genitals at pubertal ceremonies. Yet normal sexual function is observed in these females" (307). No information is provided as to how Gulliver reached this conclusion or on the type of data on which he based his conclusion.

10 Alfred Kinsey's *Sexual Behavior in the Human Female* (1953) challenged long-standing assumptions about female sexuality. Backing up their claims with numerical data, Kinsey and his associates asserted that the vaginal orgasm was biologically impossible, thereby contributing to the notion of the clitoris as the site of female

erotic sensation. Beginning in 1961, William Masters, a gynecologist, and Virginia Johnson, a psychologist, followed in Kinsey's pioneering footsteps: by providing detailed and scientific studies of female sexual anatomy, their research directly challenged prevailing views of female sexuality as exclusively bound up with reproductive practices. In their 1966 book *Human Sexual Response* they devoted an entire chapter to the role of the clitoris as a receptor and transformer of sexual stimuli, effectively shifting the focus of female orgasm from the vagina to the clitoris (Masters and Johnson 1966). They reframed sexuality as a healthy human trait and the experience of sexual pleasure as a socially acceptable goal.

11 Other studies have assessed adult clitoral size finding mean lengths of 15.4 mm (with a range of 11.7 to 20.3 mm) (Verkauf, von Thron, and O'Brien 1992) and 19.1 mm (Lloyd et al. 2005). Verkauf and his colleagues found no correlation between clitoral size and age, height, weight, or use of oral contraceptives, but that women who had given birth had "significantly larger measurements." In 1998, Helen O'Connell released a study in which she lamented the incomplete and inaccurate information on the anatomy of the clitoris in most anatomy texts (O'Connell et al. 1998). Using dissection, she observed that the body of the excised clitoris is 1 to 2 centimeters wide and 2 to 4 centimeters long. O'Connell conducted this study of clitoral anatomy in collaboration with John Hutson, a leading intersex surgeon at Royal Children's Hospital in Melbourne.

12 In one survey of twentieth-century medical texts for their accuracy of illustrations of the clitoris, the majority of depictions of the clitoris were shown to be simplistic, and several texts from the 1950s and 1960s omitted the clitoris entirely (Moore and Clarke 1995). Descriptions of the labia are even scarcer, aside from the perhaps obvious observation that they should not be fused. A striking exception was Dickinson (1949), who catalogued the variety of female genitalia. Drawing on a sample of 100 women, he found that the typical clitoris was 3 to 4 millimeters in width and 4 to 5 millimeters in length. He also found that the majority of women, 75 percent, had a clitoris between 2.5 and 6.5 millimeters. Five percent of women had a clitoral glans between 0 to 2.5 millimeters, and 20 percent had a glans between 6.5 and 15 millimeters. He cites previous researchers who obtained average clitoral lengths of 5.6, 6.7, and 8.0 millimeters. He attributes the size variation to whether individuals with intersex conditions were included in the sample data, which would account for a larger average size.

13 A recent study from England compared clitoral size, labial length and width, vaginal length, and other genital features in fifty premenopausal women undergoing gynecological procedures not involving the external genitalia. Not surprisingly, the authors noted a wide range of values for each measurement (Lloyd et al. 2005).

14 A number of procedures for extending the vagina do not involve surgery. These are often called pressure-dilation techniques, in which the woman herself uses special dilation instruments to apply pressure to expand the vagina. The simplest method, devised in 1938 by Robert Frank, involves the woman's applying inter-

mittent pressure on the vaginal "dimple" with a series of dilators for twenty minutes three times a day (Frank 1938). However, this technique may require as long as seven months to achieve the desired result, and it must be done in an awkward squatting position. Seeking to overcome these drawbacks, James Ingram developed a method using the woman's body weight to achieve pressure, which allowed patients to remain clothed and sit upright (perhaps performing other tasks) while dilating. He devised a special bicycle seat used in conjunction with graduated Lucite dilators for two hours a day (Ingram 1981).

15 In infants with a 46,XX karyotype, the vagina develops during fetal development from two different primordial structures, the Müllerian ducts and the urogenital sinus. Although there is clinical disagreement about their relative contribution, the Müllerian ducts roughly form into the upper vagina, whereas the lower vagina develops from the urogenital sinus (Grumbach and Conte 1998: 1326). In 46,XY infants (with AIS, for example) the testes do not produce enough testosterone to masculinize the external genitalia, but they may produce enough MIF (Müllerian inhibiting factor) to curb the development of part of the vagina. Consequently, some women with AIS may have no upper vagina, but their lower vagina may be fully developed, a few centimeters long, or what clinicians call a "dimple."

16 The Sheares method uses different tissue (a perineal skin flap) that is placed into a space tunneled between the rectum and the bladder and uses different placement and incisions (Sheares 1960). Other types of vaginoplasty (vulvovaginoplasty) use the labia or tissue expansion to create a vagina (e.g., Williams 1964; Flack, Barraza, and Stevens 1993; Chudacoff et al. 1996). These are not widely used, however.

17 Some hold out hope that tissue engineering will enable surgeons to treat boys with no penis or traumatic loss of the penis. Anthony Atala, a pediatric urologist, is exploring a technique that would use the patient's own cells to grow the penis, which would then be surgically attached to the individual. So far this work has been carried out in the lab using rabbits. The researchers took smooth muscle and blood vessel cells from rabbits' penises and seeded them onto a collagen matrix—a scaffold-like structure upon which the tissue regenerates. Researchers removed the penises from the rabbits and replaced them with those grown in the lab. According to Atala, the new penises have blood vessels and nerves that enable them to become erect and mate (Chen, Yoo, and Atala 2006).

18 While similar to the Frank procedure, this method involves surgery as well. The original method allows for the creation of a neovagina through uninterrupted dilation using a traction device surgically attached to the abdomen. Modifications to this method have used laparotomy for the surgical portion of the procedure (e.g., Fedele et al. 1994; Veronikis, McClure, and Nichols 1997).

19 This information was taken from the Department of Pediatric Urology at Cornell, www.cornellurology.com/pediatrics/genitoplasty.shtml#treatment (accessed November 15, 2005).

20 If knowledge about female genitalia and sexuality remain "black boxes" for many

clinicians, some of the blame can be placed on a society in which funding for sexuality research has been severely curtailed, sexuality education programs provide false, misleading, or distorted information, and frank discussions of sexuality are discouraged. Blame can also be placed on medical education and clinical training, which devotes minimal teaching time to the topic of sexuality (generally three to ten hours, most of which cover the causes of sexual dysfunction and their treatment and issues of sexuality in illness or disability), even though it is critical for many areas of medicine (Solursh et al. 2003). As a result, clinicians often lack the knowledge, skills, and attitudes necessary for communicating effectively with patients and parents about issues of sexuality.

21 For a detailed review of problematic issues in outcome studies of clitoral and vaginal surgery, see Fausto-Sterling 2000: chap. 3.

22 See Randolph and Hung 1970; Kumar et al. 1974; Fonkalsrud, Kaplan, and Lippe 1977; Randolph, Hung, and Rathlev 1981; Allen, Hardy, and Churchhill 1982; Mininberg 1982; Rajfer, Ehrlich, and Goodwin 1982; Oesterling, Gearhart, and Jeffs 1987; Sharp et al. 1987; Pinter and Kosztolanyi 1990; Newman, Randolph, and Parson 1992; Van der Kamp et al. 1992; Bellinger 1993; de Jong and Boemers 1995; Gearhart, Burnett, and Owen 1995; Costa et al. 1997; Joseph 1997; Alizai et al. 1999; Minto et al. 2003b; Crouch et al. 2004.

23 The authors describe unsatisfactory results as both clitoral atrophy, where the clitoris has shrunk too much, and a prominent glans, where the clitoris essentially remains too large, suggesting a very small window of acceptable clitoral size.

24 See Hendren and Crawford 1969; Sotiropoulos et al. 1976; Garcia and Jones 1977; Farber and Mitchell 1978; Nihoul-Fékété 1981; Allen, Hardy, and Churchill 1982; Hecker 1982; Rock et al. 1983; Wiser and Bates 1984; Azziz et al. 1986; Lobe et al. 1987; Mulaikal, Migeon, and Rock 1987; Buss and Lee 1989; Johnson, Batchelor, and Lilford 1991; Bailez et al. 1992; Borruto 1992; Martinez-Mora et al. 1992; Newman, Randolph, and Anderson 1992; Wesley and Coran 1992; Hitchcock and Malone 1994; de Jong and Boemers 1995; Hendren and Atala 1995; Hojsgaard and Villadsen 1995; Alessandrescu et al. 1996; Fliegner 1996; Costa et al. 1997; Di Benedetto et al. 1997; Premawardhana et al. 1997; Hensle and Reiley 1998; Rink and Adams 1998; Alizai et al. 1999; Sakurai et al. 2000; Wisniewski et al. 2000; Creighton, Minto, and Steele 2001b; Farkas, Chertin, and Hadas-Halpren 2001; Graziano et al. 2002; Minto et al. 2003a; Creighton 2004.

25 Some of these studies assess outcomes of vaginoplasty in women with diagnoses where the vagina has failed to develop. Because of the methodological variations, valuable comparisons across these studies are impossible. Moreover, these studies raise other methodological concerns: for example, many of these studies do not assess long-term outcomes into adulthood and do not use subjective patient assessment.

26 In this study, it is unclear if "no sexual activity" means no sexual activity whatsoever or simply no coitus. If, as seems likely, it refers to the latter, it may be that these women participate in other sexual behaviors that they find satisfactory.

27 He writes: "Most women born with ambiguous genitalia are masculine in behavior. Masculine gender behavior in childhood excludes such women from the social world of their sexmates. In adolescence and adulthood, this behavior can continue and lead to problems in making sexual contacts" (Slijper 2003).

28 In his 1979 study of the socialization of surgical residents, *Forgive and Remember*, Charles Bosk examined the values used to assist in residents' growth from novices into competent, independent surgeons (Bosk 1979). He focused especially on the ways in which surgeons understand and address error, finding that they differentiate between technical, judgmental, and normative, or moral, errors. The technical errors, such as stitching poorly, and judgmental errors, such as operating when one should not or not operating when one should, relate to the trainees' experience and theoretical knowledge, and are the least blameworthy. Often the outcome, rather than an absolute standard, determines whether the judgment was correct. The normative errors, by contrast, violate codes of conduct established by the senior surgeons, while quasi-normative errors involve deviating from the ways attending senior physicians prefer things to be done, whether backed by evidence or not. Errors of technique could be "forgiven" because it was assumed they would diminish with experience, but normative errors reflected on the character of the residents and were inexcusable. Surgeons' definition of errors may not include actions or decisions that others would clearly consider mistakes (Millman 1976).

6. *Wanting and Deciding What Is Best*

1 Society, and women themselves, often hold mothers accountable for a pregnancy's outcome and for giving birth to children who in some way differ from societal norms (Rapp 1999). The notion that women are responsible for these atypical births stretches back centuries in medicine, theology, and philosophy and often defies contemporary biomedical knowledge about how these conditions arise.

2 One diagnosis requiring critical care is congenital adrenal hyperplasia (CAH). In the salt-wasting form of CAH, which accounts for about 75 percent of CAH cases, infants lack the salt-retaining hormone aldosterone, which is necessary for salt reabsorption. Without sufficient aldosterone, too much salt and water are lost in the urine, leading to dehydration and salt deficiency. High levels of potassium can cause serious problems with heart rhythm and even cardiac arrest. Newborns with salt-wasting CAH develop symptoms shortly after birth, including vomiting, dehydration, electrolyte changes, and cardiac arrhythmias. Untreated, this condition can lead to death within one to six weeks after birth. Female infants with CAH typically receive immediate medical treatment because they exhibit atypical genitals. However, newborn males, and females who because of the degree of genital virilization are assumed to be male, show no other outward signs of the disorder and are often sent home without treatment. Frequently, these babies are brought

back to the hospital for care when symptoms present. As of late 2007, all fifty states in the United States as well as the District of Columbia include testing for CAH in newborn screening programs. In 2008, the Newborn Screening Saves Lives Act, which provides funding for states to expand and improve their newborn screening programs, was signed into law.

3 In one study carried out in the period after the birth of non-intersex newborns, researchers found that the majority of the parents' comments concerned the sex of their newborn (Wollett, White, and Lyon 1982). In another study, "parent collaborators" were instructed to call their friends after the birth of their (non-intersex) newborns. They found that 80 percent of the initial questions concerned the baby's sex; the most frequently asked question was, "Is it a boy or a girl?" (Intons-Peterson and Reddel 1984).

4 Of course, in medicine there are numerous examples where the treatment goal is to adjust the child to meet social norms. Some examples include the administration of growth hormone or limb-lengthening surgery for short stature, the surgical separation of conjoined twins, and the surgical implantation of cochlear implants for children who are hearing impaired.

5 This quote highlights something that a fair number of women with intersex diagnoses allude to: namely, that their treatment experience parallels intrafamilial sexual abuse in terms of how women describe their "routine" bodily vulnerability, extensive sexualization, and the role they feel their parents, especially their mothers, play in this. A recent study of psychological distress in individuals with DSD compared their rates of self-harm and suicidal tendencies with a community-based sample of women that included subgroups of women with a history of physical or sexual abuse. Its results suggest that adults in the DSD sample had rates of self-harming behavior and suicidal tendencies comparable to those in women with a history of physical or sexual abuse (Schützmann et al. 2007).

6 It reads: "Posts related to the more controversial topics of: the use of the term 'intersex,' the pros and cons of cosmetic surgery and CAH in relation to homosexuality, need to be posted on the new Controversies Board. All messages posted here on those topics will be removed." Discussions regarding concerns about gender identity and gender-atypical play are also confined to the Controversies Board.

7. *Growing Up under the Medical Gaze*

1 Laing notes: The individual in the ordinary circumstances of living may feel more unreal than real; in a literal sense, more dead than alive; precariously differentiated from the rest of the world, so that his identity and autonomy are always in question. . . . He may feel more insubstantial than substantial, and unable to assume that the stuff he is made of is genuine, good, valuable. And he may feel his self as partially divorced from his body" (1960: 42).

8. The *Intersex Body in the World*

1 This was defined as Prader stages III-V.

2 Some of the groundwork for a new understanding of the female body had been laid in the 1950s and 1960s with widely reported studies by pioneering researchers of human sexuality. Among these works are Alfred Kinsey's *Sexual Behavior in the Human Female* (1953) and William Masters's and Virginia Johnson's *Human Sexual Response* (1966). For more on these two works, see chapter 5, note 10.

3 Gay, lesbian, and transgender movements share related goals such as ending violence and discrimination against people who express gender non-normative identities, gender roles, or sexualities. Although in recent years major gay, lesbian, and bisexual groups have changed their mission statements to include transgender people, mainstream gay and lesbian groups have not always welcomed the latter as members nor have they adopted issues important to transgender people as part of their agendas.

4 By 1980, however, psychiatrists had introduced several diagnoses that were troubling in new ways. One diagnosis, "ego-dystonic homosexuality," was defined by both a persistent desire to increase heterosexual arousal and a persistent distress from a sustained pattern of unwanted homosexual arousal. Not surprisingly, mental health professionals, among others, criticized this new diagnosis, which they argued was designed to appease psychiatrists who still considered homosexuality pathological. In 1986 the diagnosis was removed from the *DSM* because it still suggested homosexuality as a disorder. The American Psychiatric Association (APA) also argued that due to the widespread prejudice against homosexuality in the United States, virtually all homosexuals go though an initial period of ego-dystonic homosexuality. For an account of the events leading up to the 1973 and 1986 decisions, see Bayer 1981. The other diagnoses introduced in 1980 were "gender identity disorder of childhood" (GIDC), which understood gender-variant behavior in children (especially effeminate behavior in young boys) as pathological and requiring treatment, and "transsexualism." By 1994, these diagnoses were collapsed under the category "gender identity disorder" (GID), with further specifications for variants in children (largely echoing the older category of GIDC), adolescents, and adults (basically the earlier diagnosis "transsexualism") (Bryant 2006). Bryant notes that collapsing these categories and renaming transsexualism GID arguably complicated efforts to remove the diagnosis from the *DSM* (Bryant 2006: 23). The diagnosis GID has different implications for those so categorized. Some have argued for the removal of the GIDC diagnosis, arguing that it unfairly targets individuals who may be likely to grow up to be gay or transsexual. The blanket elimination of GID from the *DSM*, however, would negatively impact transsexual adults, for whom a diagnosis of GID is usually necessary to get hormones and surgeries, or to get reimbursed for transition-related care. For critical analyses of the GID diagnosis, see Valentine 2000: chap. 4; Bryant 2006, 2007; Feder 2007: chap. 3.

5 In this connection, Hausman has noted that while overall Money's theory was antiessentialist for drawing distinctions between biology, psychology, and gender role, Money "established a new essentialism that fixed gender role and orientation with an exclusively heterosexual framework" (Hausman 1995: 98). The early articles certainly provide evidence of this assumption. For example, in Money's now classic definition of gender role—"all those things that a person says or does to disclose himself or herself as having the status of a boy or man, girl or woman. . . . It includes, but is not restricted to, sexuality in the sense of eroticism"—sexuality is subsumed under gender. In this formulation, a male gender role assumes (and even requires) desire for females and vice versa. Hausman has concluded that Money's treatment paradigm "had as one explicit object the maintenance of heterosexuality among intersexual subjects" (1995: 98). (Interestingly, although this may have been his view when the protocol was implemented, Money later became a champion for the acceptance of non-normative sexual practices of all sorts. Healthy patients and good outcomes were heterosexual not as much by design as by convention.)

6 Several scholars have noted that this still makes for a biologically based system of understanding and categorizing sex, and thus gender. In response to these criticisms, Fausto-Sterling revised this system in her later book *Sexing the Body* (2000).

7 For Chase's published account of the birth and development of ISNA, see Chase 1998.

8 In an interview Chase said that this female gynecologist did not help her comprehend her medical records: "She discounted the meaning of it. She pretended that it wasn't going to hurt. Later I got hold of her correspondence with the hospital. After she saw me, she wrote to tell them they should be really proud and that I was doing great as my 'true sex, i.e., female' " (qtd. in Hegarty 2000: 120).

9 Dreger and Chase contacted one another after Dreger published an article on nineteenth-century views of hermaphroditism (Dreger 1995). They grew to become close friends and allies in the struggle to change the medical management of intersex, and for many years their work was decidedly collaborative. For Dreger's detailed account of the development of their collaboration, see Dreger 2004.

10 Interestingly, Chase was not present at the protest that put ISNA on the map. Instead, she worked very closely with the "transexual" activist Riki Wilchins and members of Transexual Menace to "get bodies" for the protest. For a personal account of the protest, see Valentine and Wilchins 1997.

11 The first push for changing the nomenclature that I heard came from Chase and Dreger at the NATFI meetings in 2000. A subsequent article in a medical journal laid out their arguments for changing the medical terminology surrounding intersexuality (Dreger et al. 2005). It was not until the publication of the DSD guidelines in 2005, however, that intersex adults began having an intense discussion about the new nomenclature, especially concerning the term *disorders of sex development*. Several individuals who had helped contribute to the handbooks argued the new term medicalized and pathologized intersex conditions and people (for one

argument against the DSD terminology, see www.intersexualite.org/Response–to–Intersex–Initiative.html). This prompted a response from ISNA about why it had adopted the term (see www.isna.org/node/1066). Several people who had supported ISNA in its beginnings distanced themselves from the group over this issue and asked that their objections be noted in the DSD guidelines. Dreger, who was instrumental in bringing about the change in nomenclature, later expressed reservations about the term on her Web site (see www.alicedreger.com/dsd.html).

12 Chase says she has spoken to roughly five hundred intersex adults. Statistics are not easy to come by, but a commonly cited figure is that one in two thousand individuals are born with an intersex condition or undergo genital surgery for atypical genitals (Blackless et al., 2000). If so, one might estimate there are one hundred thousand people with intersex diagnoses living in the United States today—so Chase has had contact with fewer than half of one percent of them.

13 Other barriers to retrospective and prospective studies include a range of diagnoses and physiologies that make comparisons among individuals and groups difficult; scattered surgical records among hospitals; insufficient pre-and post-surgical data (e.g., brief surgical reports); difficulty or inability to locate patients; patients' unwillingness to participate; and insufficient funding for the costs of follow-up.

14 This work identified several heuristics, or cognitive processes that simplify decision making, relevant to medicine. One is what it called "representativeness" errors, made when thinking is overly influenced by what is usually true. With this type of error, physicians assume commonality between objects that have a similar appearance, that is, they fail to consider possibilities that do not fit their templates of a disease and thus attribute symptoms to the wrong cause. While representativeness is often very useful in clinical practice because it can render complex problems manageable, it can also result in the neglect of relevant information. More recently, Croskerry has identified an affective heuristic, which is the tendency to make decisions on what we wish were true (Croskerry 2006). For an excellent review of cognitive heuristics in clinical decision making, see Croskerry 2000.

15 In children, cystoscopies are done under general anesthesia, which carries a risk, but they can also produce more urinary tract infections (UTIs) as a result of the procedure. Moreover, some girls find the procedure scary and even humiliating, especially as they approach puberty.

16 In addition, the focus of Schober's study, like that of the other two surgeons named, is to assess gender identity (not surgical outcome) in "genetic males" assigned female. The criticism that she is just "reading stuff" rather than practicing surgery is irrelevant because the project does not involve conducting surgery on patients but rather interviewing them about their gender identity. Finally, the surgeon accused Schober's work of weakness because it is retrospective rather than hands-on, but the nature of the study requires a retrospective approach: these children have already been gender-assigned, and she is trying to assess their

quality of life and acceptance of this assignment. A prospective study, of which there are very few, would require randomly assigning these children a gender and examining how they fare, a methodology that would obviously be unethical.

17 This trend has been contemporary with a proliferation of guidelines and protocols for the treatment of various conditions (Woolf et al. 1999). Some of these have been developed by experts from professional medical societies and organizations and from government panels, but some are produced commercially and purchased by health care organizations. As a result, conflicting guidelines may circulate.

18 Exclusionary practices were built into the structure of the meeting. The meeting was invitation only; the only non-clinicians invited to participate were Chase and another patient advocate. No parents were invited to participate. Moreover, prior to the meeting, attendees were assigned to one of six workgroups, each of which addressed a particular aspect of diagnosis and treatment. Advocates were not directly involved in the preparatory teams because, as one participant phrased it, "they are not academic experts." Within each work group, all participants were asked to respond to a series of questions. Advocates (and indeed most other participants) had no role in formulating or prioritizing the questions and the issues to be addressed, nor did they have any say concerning their work-group assignment. Those with clinical and research specialties had straightforward assignments, but advocates' interests do not fall neatly into clinical frameworks and the parsing of the relevant issues. From Chase's perspective, for example, the work group to which she was assigned dealt with noncontroversial issues of little importance to her, thus limiting her participation in the more contentious areas of surgery, outcome data, and acceptable evidence. The workgroup that dealt with the surgical management, for example, was composed entirely of surgeons and a senior pediatric endocrinologist who had trained with Lawson Wilkins. This group was charged with answering deeply contested issues such as whether surgery should be performed to reduce the clitoris and on whom and when vaginal surgery should be performed. Although one would expect surgeons to participate in formulating answers to these questions, relying on them exclusively represents a failure to bring diverse perspectives to the issue. As a result, those who settled questions regarding surgery included only those whose professional training would lead them to believe that the appropriate and even ethical response to atypical genitals includes surgery. This explains why the consensus statement failed to resolve the most controversial aspects of treatment—early genital surgery (Karkazis 2006). When the groups convened, the most important work had already been completed, all without any input from patient advocates, who were brought into the process at a late stage and even then to only limited participation.

Although most specialists had someone from their discipline to vet the final written document, advocates were not allowed to participate in the process and thus did not know if the document to which their names would be attached contained statements or recommendations they found objectionable. This was particularly important because advocates had already made concessions knowing many clinicians would not agree to any broader changes. As two of the only nonmedical participants, their power was already limited. Fear of being too adversarial and thus not being included in any future meetings circumscribed their participation even more. Participants sympathetic to their concerns helped to get their viewpoint heard, but the situation also created a dynamic in which scientific and medical experts were articulating their needs and protecting their interests.

Participants were unable to finalize a draft of the document prior to the end of the meeting, which left the organizers to complete this afterward. Only the heads of each work group received the document for review. In the end, well-meaning sympathizers shared the document with patient advocates. It excluded several points they believed had been agreed on in the meeting, and they wrote a letter expressing their desire to have them reinserted. Shortly after sending the letter, they received the final draft of the document which had already been submitted to several journals for publication. Some changes had been reinserted, but some had not. There was no dialogue as to how these decisions were made. One could argue that this is the nature of a consensus meeting, and indeed it is, but this sidesteps the stark power differentials between the medical experts and patient advocates, whose meaningful participation was severely curtailed.

REFERENCES

Aaronson, I. A. 1994. "Micropenis: Medical and Surgical Implications." *Journal of Urology* 152 (1): 4–14.

———. 2001. "Members, North American Task Force on Intersexuality." Letter to the members of the North American Task Force on Intersexuality. Copy in K. Karkazis's personal collection.

Adams, V., and S. L. Pigg, eds. 2005. *Sex in Development: Science, Sexuality and Morality in Global Perspective.* Durham, NC: Duke University Press.

Alaszewski, A. 2005. "A Person-Centered Approach to Communicating Risk." *PLoS Medicine* 2 (2): e41.

Alessandrescu, D., et al. 1996. "Neocolpopoiesis with Split-Thickness Skin Graft as a Surgical Treatment of Vaginal Agenesis: Retrospective Review of 201 Cases." *American Journal of Obstetrics and Gynecology* 175 (1): 131–38.

Alizai, N. K., et al. 1999. "Feminizing Genitoplasty for Congenital Adrenal Hyperplasia: What Happens at Puberty?" *Journal of Urology* 161 (5): 1588–91.

Allen, L. E., B. E. Hardy, and B. M. Churchill. 1982. "The Surgical Management of the Enlarged Clitoris." *Journal of Urology* 128 (2): 351–54.

American Academy of Pediatrics. 2000. "Evaluation of the Newborn with Developmental Anomalies of the External Genitalia." *Pediatrics* 106 (1): 138–42.

Angier, N. 1996. "Intersexual Healing: An Anomaly Finds a Group." *New York Times*, February 4.

———. 1997. "New Debate over Surgery on Genitals." *New York Times*, May 13.

———. 2000. "X+Y=Z: An Account of the John/Joan Case, in Which Doctors Tried to Turn a Baby Boy into a Girl." *New York Times Book Review*, February 20, 10.

Arney, W. R. 1982. *Power and the Profession of Obstetrics.* Chicago: University of Chicago Press.

Azziz, R., et al. 1986. "Congenital Adrenal Hyperplasia: Long-Term Results Following Vaginal Reconstruction." *Fertility and Sterility* 46 (6): 1011–14.

Bailez, M. M., et al. 1992. "Vaginal Reconstruction after Initial Construction of the External Genitalia in Girls with Salt-Wasting Adrenal Hyperplasia." *Journal of Urology* 148 (2): 680–84.

Bangsboll, S., et al. 1992. "Testicular Feminization Syndrome and Associated Gonadal Tumors in Denmark." *Acta Obstetricia et Gynecologica Scandinavica* 71 (1): 63–66.

Barry, E. 1996. "United States of Ambiguity." *Boston Phoenix*, November 22.

Baskin, L. S. 2003. "Working with Intersex." *University of California at San Francisco Pediatric Urology Newsletter*, 3.

Baskin, L. S., and M. B. Ebbers. 2006. "Hypospadias: Anatomy, Etiology, and Technique." *Journal of Pediatric Surgery* 41 (3): 463–72.

Baskin, L. S., et al. 1999. "Anatomical Studies of the Human Clitoris." *Journal of Urology* 162 (3): 1015–20.

Bayer, R. 1981. *Homosexuality and American Psychiatry: The Politics of Diagnosis*. New York: Basic Books.

Bellinger, M. F. 1993. "Subtotal De-epithelialization and Partial Concealment of the Glans Clitoris: A Modification to Improve the Cosmetic Results of Feminizing Genitoplasty." *Journal of Urology* 150 (2): 651–53.

Berenbaum, S. A., and J. M. Bailey. 2003. "Effects on Gender Identity of Prenatal Androgens and Genital Appearance: Evidence from Girls with Congenital Adrenal Hyperplasia." *Journal of Clinical Endocrinology and Metabolism* 88 (3): 1102–6.

Berenbaum, S. A., S. C. Duck, and K. Bryk. 2000. "Behavioral Effects of Prenatal Versus Postnatal Androgen Excess in Children with 21-Hydroxylase-Deficient Congenital Adrenal Hyperplasia." *Journal of Clinical Endocrinology and Metabolism* 85 (2): 727–33.

Berenbaum, S. A., and M. Hines. 1992. "Early Androgens Are Related to Childhood Sex-Typed Toy Preferences." *Psychological Science* 3: 203–6.

Berg, R. and G. Berg. 1983a. "Castration Complex: Evidence from Men Operated for Hypospadias." *Acta Psychiatrica Scandinavica* 68: 143–53.

———. 1983b. "Penile Malformation, Gender Identity, and Sexual Orientation." *Acta Psychiatrica Scandinavica* 68 (3): 154–66.

Berg, R., G. Berg, and J. Svensson. 1982. "Penile Malformation and Mental Health: A Controlled Psychiatric Study of Men Operated for Hypospadias in Childhood." *Acta Psychiatrica Scandinavica* 66 (5): 398–416.

Berg, R., J. Svensson, and G. Astrom. 1981. "Social and Sexual Adjustment of Men Operated for Hypospadias during Childhood: A Controlled Study." *Journal of Urology* 125 (3): 313–17.

Bin-Abbas, B., et al. 1999. "Congenital Hypogonadotropic Hypogonadism and Micropenis: Why Sex Reversal Is Not Indicated." *Journal of Pediatrics* 134 (5): 579–83.

"Biological Imperatives (the Sexes)." 1973. *Time*, January 8, 34.

Blackless, M., et al. 2000. "How Sexually Dimorphic Are We? Review and Synthesis." *American Journal of Human Biology* 12: 151–66.

Bleier, R. 1984. *Science and Gender: A Critique of Biology and Its Theories on Women*. New York: Pergamon.

Borruto, F. 1992. "Mayer-Rokitansky-Kuster Syndrome: Vecchietti's Personal Series." *Clinical and Experimental Obstetrics and Gynecology* 19 (4): 273–74.

Bosk, C. L. 1979. *Forgive and Remember: Managing Medical Failure*. Chicago: University of Chicago Press.

"Boy or Girl? What Happens When Doctors Choose Your Child's Sex?"1997. *Primetime*, ABC, September 3.

Bradley, S. J., et al. 1998. "Experiment of Nurture: Ablatio Penis at Two Months, Sex Reassignment at Seven Months, and a Psychosexual Follow-Up in Young Adulthood." *Pediatrics* 102 (1): e9.

Brown, P., et al. 2004. "Embodied Health Movements: New Approaches to Social Movements in Health." *Sociology of Health and Illness* 26 (1): 50–80.

Bryant, K. 2006. "Making Gender Identity Disorder of Childhood: Historical Lessons for Contemporary Debates." *Sexuality Research and Social Policy* 3 (3): 23–39.

——. 2007. "The Politics of Pathology and the Making of Gender Identity Disorder." PhD diss., University of California at Santa Barbara.

Buss, J. G., and R. A. Lee. 1989. "McIndoe Procedure for Vaginal Agenesis: Results and Complications." *Mayo Clinic Proceedings* 64 (7): 758–61.

Butler, J. 1990. *Gender Trouble: Feminism and the Subversion of Identity*. New York: Routledge.

Byne, W. 1994. "The Biological Evidence Challenged." *Scientific American*, May, 50–56.

——. 1995. "Science and Belief: Psychological Research on Sexual Orientation." *Journal of Homosexuality* 28 (3–4): 303–44.

Capel, B., and D. Coveney. 2004. "Frank Lillie's Freemartin: Illuminating the Pathway to Twenty-first-Century Reproductive Endocrinology." *Journal of Experimental Zoology* 301 (11): 853–56.

Cappon, D., C. Ezrin, and P. Lynes. 1959. "Psychosexual Identification (Psychogender) in the Intersexed." *Canadian Psychiatric Association Journal* 4 (2): 90–106.

Cawadias, A. P. 1943. *Hermaphroditos: The Human Intersex*. London: William Heinemann Medical Books.

Chalker, R. 2000. *The Clitoral Truth: The Secret World at Your Fingertips*. New York: Seven Stories.

Chase, C. 1993. "Intersexual Rights." *Sciences* 33 (4): 3.

——. 1994. " 'Corrective' Surgery Unnecessary." *Johns Hopkins Magazine*, February, 6–7.

——. 1997. "Spare the Knife, Study the Child." *Ob.Gyn. News* 32: 14–15.

——. 1998. "Hermaphrodites with Attitude: Mapping the Emergence of Intersex Political Activism." *GLQ: A Journal of Gay and Lesbian Studies* 4 (2): 189–211.

Chen K. L., J. J. Yoo, and A. Atala. 2006. "Total Corpora Replacement for Erectile Dysfunction." Abstract 1323 for the American Urological Association Annual Meeting Program, Atlanta, May 20–25, 2006.

Cheng, P. S., and J.-P. Chanoine. 2001. "Should the Definition of Micropenis Vary According to Ethnicity?" *Hormone Research* 55 (6): 278–81.

Chudacoff, R. M., et al. 1996. "Tissue Expansion Vaginoplasty for Treatment of Congenital Vaginal Agenesis." *Obstetrics and Gynecology* 87 (5): 865–68.

Clarke, A. 1998. *Disciplining Reproduction: Modernity, American Life Sciences, and "the Problems of Sex."* Berkeley: University of California Press.

Colapinto, J. 1997. "The True Story of John/Joan." *Rolling Stone*, December 11, 54–73, 92–97.

——. 2000. *As Nature Made Him: The Boy Who Was Raised as a Girl*. New York: HarperCollins.

Collier, J. L. 1973. "Male and Female Created He Them, or Did He?" *New York Times*, February 25.

Costa, E. M. F., et al. 1997. "Management of Ambiguous Genitalia in Pseudohermaphrodites: New Perspectives on Vaginal Dilation." *Fertility and Sterility* 67 (2): 229–32.

Coventry, M. 1998. "The Tyranny of the Esthetic: Surgery's Most Intimate Violation." *On the Issues: The Progressive Woman's Quarterly* 7 (3): 16–23, 60–61.

Cowley, G. 1997. "Gender Limbo." *Newsweek*, May 19, 64–67.

Creighton, S. 2001. "Surgery for Intersex." *Journal of the Royal Society of Medicine* 94 (5): 218–20.

——. 2004. "Long-Term Outcome of Feminization Surgery: The London Experience." *BJU International* 93, supplement 3: 44–46.

Creighton, S., and C. Minto. 2001a. "Managing Intersex." *British Medical Journal* 323 (7324): 1264–65.

Creighton, S., and C. Minto. 2001b. "Sexual Function in Adult Women with Complete Androgen Insensitivity Syndrome." *Journal of Pediatric and Adolescent Gynecology* 14 (3): 144–45.

Creighton, S., C. Minto, and S. Steele. 2001a. "Cosmetic and Anatomical Outcomes Following Feminising Childhood Surgery for Intersex Conditions." *Journal of Pediatric and Adolescent Gynecology* 14 (3): 142.

——. 2001b. "Objective Cosmetic and Anatomical Outcomes at Adolescence of Feminising Surgery for Ambiguous Genitalia Done in Childhood." *Lancet* 358 (9276): 124–25.

Creighton, S., C. Minto, and C. Woodhouse. 2001. "Long Term Sexual Function in Intersex Conditions with Ambiguous Genitalia." *Journal of Pediatric and Adolescent Gynecology* 14 (3): 141–42.

Croskerry, P. 2000. "The Cognitive Imperative: Thinking about How We Think." *Academic Emergency Medicine* 7 (11): 1223–31.

——. 2006. "The Affective Imperative: Coming to Terms with Our Emotions." *Academic Emergency Medicine* 14 (2): 184–85.

Crouch, N. S., et al. 2004. "Genital Sensation after Feminizing Genitoplasty for Congenital Adrenal Hyperplasia: A Pilot Study." *BJU International* 93 (1): 135–38.

Daaboul, J., and J. Frader. 2001. "Ethics and the Management of the Patient with Intersex: A Middle Way." *Journal of Pediatric Endocrinology and Metabolism* 14 (9): 1575–83.

Danish, R. K., et al. 1980. "Micropenis: II; Hypogonadotropic Hypogonadism." *Johns Hopkins Medical Journal* 146 (5): 177–84.

Daston, L., and K. Park. 1995. "The Hermaphrodite and the Orders of Nature: Sexual Ambiguity in Early Modern France." *GLQ: A Journal of Gay and Lesbian Studies* 1 (4): 419–38.

Dayner, J. E., P. A. Lee, and C. P. Houk. 2004. "Medical Treatment of Intersex: Parental Perspectives." *Journal of Urology* 172 (4): 1762–65.

De Jong, T. P., and T. Boemers. 1995. "Neonatal Management of Female Intersex by Clitorovaginoplasty." *Journal of Urology* 154 (2): 830–32.

Del Vecchio Good, M.-J. 2001. "The Biotechnical Embrace." *Culture, Medicine, and Psychiatry* 25 (4): 395–410.

Denny, K. 1999. "Evidence-Based Medicine and Medical Authority." *Journal of Medical Humanities* 20 (4): 247–63.

Denzin, N. K., and Y. S. Lincoln, eds. 2005. *The Sage Handbook of Qualitative Research.* 3rd. ed. Thousand Oaks, CA: Sage.

Dessens, A., F. Slijper, and S. Drop. 2005. "Gender Dysphoria and Gender Change in Chromosomal Females with Congenital Adrenal Hyperplasia." *Archives of Sexual Behavior* 32 (4): 389–97.

Dewhurst, C. J., and R. R. Gordon. 1969. *The Intersexual Disorders.* London: Ballière, Tindall, and Cassell.

Diamond, D. A. 2004. "Gender Assignment Survey of AAP Urology Section Fellows." Paper presented at the American Academy of Pediatrics Annual Meeting, San Francisco, October 11.

Diamond, M. n.d. AIS survey, version 3.7.1.

——. 1965. "A Critical Evaluation of the Ontogeny of Human Sexual Behavior." *Quarterly Review of Biology* 40 (2): 145–75.

——. 1982. "Sexual Identity: Monozygotic Twins Reared in Discordant Sex Roles and a BBC Follow-Up." *Archives of Sexual Behavior* 11 (2): 181–85.

——. 1999. "Pediatric Management of Ambiguous and Traumatized Genitalia." *Journal of Urology* 162 (3): 1021–28.

Diamond, M., and H. K. Sigmundson. 1997a. "Management of Intersexuality: Guidelines for Dealing with Persons with Ambiguous Genitalia." *Archives of Pediatrics and Adolescent Medicine* 151 (10): 1046–50.

——. 1997b. "Sex Reassignment at Birth: Long-Term Review and Clinical Implications." *Archives of Pediatrics and Adolescent Medicine* 151 (3): 298–304.

Di Benedetto, V., et al. 1997. "The Anterior Sagittal Transanorectal Approach: A Modified Approach to One-Stage Clitoral Vaginoplasty in Severely Masculinized Female Pseudohermaphrodites; Preliminary Results." *Journal of Urology* 157 (1): 330–32.

Dickinson, R. L. 1949. *Human Sex Anatomy: A Topographical Hand Atlas.* Baltimore: Williams and Wilkins.

Dietrich, M. R. 2003. "Richard Goldschmidt: Hopeful Monsters and Other 'Heresies.'" *Nature Reviews Genetics* 4 (1): 68–74.

Dittmann, R. W., et al. 1990. "Congenital Adrenal Hyperplasia: I; Gender-Related Behavior and Attitudes in Female Patients and Sisters." *Psychoneuroendocrinology* 15 (5–6): 401–20.

Doell, R., and H. E. Longino. 1988. "Sex Hormones and Human Behavior: A Critique of the Linear Model." *Journal of Homosexuality* 15 (3–4): 55–79.

Doll, D. C., et al. 1978. "Seminoma in a Twelve-Year-Old Male with 46 XY/45 XO Karyotype." *Cancer* 42 (4): 1823–25.

Donahoe, P. K, et al. 1991. "Clinical Management of Intersex Abnormalities." *Current Problems in Surgery* 28 (8): 513–79.

Dreger, A. D. 1995. "Doubtful Sex: The Fate of the Hermaphrodite in Victorian Medicine." *Victorian Studies* 38 (3): 336–70.

——. 1998a. *Hermaphrodites and the Medical Invention of Sex*. Cambridge: Harvard University Press.

——. 1998b. "When Medicine Goes Too Far in the Pursuit of Normality." *New York Times*, July 28.

——. 1999. *Intersex in the Age of Ethics*. Hagerstown, MD: University Publishing Group.

——. 2000. "Jarring Bodies: Thoughts on the Display of Unusual Anatomies." *Perspectives in Biology and Medicine* 43 (2): 161–72.

——. 2004. "Cultural History and Social Activism: Scholarship, Identities, and the Intersex Rights Movement." In *Locating Medical History: The Stories and Their Meanings*, ed. F. Huisman and J. H. Warner, 390–409. Baltimore: Johns Hopkins University Press. .

——, ed. 1998. "Intersexuality." Special issue, *Journal of Clinical Ethics* 9 (4).

Dreger, A. D., et al. 2005. "Changing the Nomenclature/Taxonomy for Intersex: A Scientific and Clinical Rationale." *Journal of Pediatric Endocrinology and Metabolism* 18 (8): 729–33.

Ehrenreich, B., and D. English. 1973. *For Her Own Good: 150 Years of the Experts' Advice to Women*. Garden City, NY: Anchor.

Ehrhardt, A., R. Epstein, and J. Money. 1968. "Fetal Androgens and Female Gender Identity in the Early Treated Adrenogenital Syndrome." *Johns Hopkins Medical Journal* 122 (3): 160–67.

Ehrhardt, A., K. Evers, and J. Money. 1968. "Influence of Androgen and Some Aspects of Sexually Dimorphic Behavior in Women with the Late-Treated Adrenogenital Syndrome." *Johns Hopkins Medical Journal* 123 (3): 115–22.

Ehrhardt, A., and J. Money. 1967. "Progestin-Induced Hermaphroditism: IQ and Psychosexual Identity in a Study of Ten Girls." *Journal of Sex Research* 3 (1): 83–100.

Elder, J. S. 1997. "Editorial Comment." *Journal of Urology* 158 (4): 1635.

Epstein, J. 1990. "Either/or—Neither/Both: Sexual Ambiguity and the Ideology of Gender." *Genders* 7: 99–142.

Epstein, S. 1996. *Impure Science: AIDS, Activism, and the Politics of Knowledge*. Berkeley: University of California Press.

Erichsen, J. E., M. Beck, and R. Johnson. 1895. *The Science and Art of Surgery: Being a Treatise on Surgical Injuries, Diseases, and Operations*. London: Longmans, Green.

Farber, M., and G. W. Mitchell. 1978. "Surgery for Congenital Absence of the Vagina." *Obstetrics and Gynecology* 51 (3): 364–67.

Farkas, A., B. Chertin, and I. Hadas-Halpren. 2001. "1-Stage Feminizing Genitoplasty: 8 Years of Experience with 49 Cases." *Journal of Urology* 165 (6): 2341–346.

Fausto-Sterling, A. 1985. *Myths of Gender: Biological Theories about Women and Men*. New York: Basic Books.

——. 1993. "The Five Sexes: Why Male and Female Are Not Enough." *Sciences* 33 (2): 20–25.

——. 1995a. "Gender, Race, and Nation: The Comparative Anatomy of 'Hottentot'

Women in Europe, 1815–1817." In *Deviant Bodies: Critical Perspectives on Difference in Science and Popular Culture*, ed. J. Terry and J. Urla, 19–48. Bloomington: Indiana University Press.

——. 1995b. "How to Build a Man." In *Constructing Masculinity*, ed. M. Berger, B. Wallis, and S. Watson, 127–34. New York: Routledge.

——. 2000. *Sexing the Body: Gender Politics and the Construction of Human Sexuality*. New York: Basic Books.

Fedele, L., et al. 1994. "Laparoscopic Creation of a Neovagina in Mayer-Rokitansky-Kuster-Hauser Syndrome by Modification of Vecchietti's Operation." *American Journal of Obstetrics and Gynecology* 171 (1): 268–69.

Feder, E. K. 2002. "Doctor's Orders: Parents and Intersexed Children." In *The Subject of Care: Feminist Perspectives on Dependency*, ed. E. F. Kittay and Feder, 294–320. Lanham, MD: Rowman and Littlefield.

——. 2007. *Family Bonds: Genealogies of Race and Gender*. New York: Oxford University Press.

Federation of Feminist Women's Health Centers. 1981. *A New View of a Woman's Body: A Fully Illustrated Guide*. New York: Simon and Schuster.

Fee, E. 1979. "Nineteenth-Century Craniology: The Study of the Female Skull." *Bulletin of the History of Medicine* 53: 415–33.

Feldman, K. W., and D. W. Smith. 1975. "Fetal Phallic Growth and Penile Standards for Newborn Male Infants." *Journal of Pediatrics* 86 (3): 395–98.

Fichtner, J., et al. 1995. "Analysis of Meatal Location in Five Hundred Men: Wide Variation Questions Need for Meatal Advancement in All Pediatric Anterior Hypospadias Cases." *Journal of Urology* 154 (2): 833–34.

Flack, C. E., M. A. Barraza, and P. S. Stevens. 1993. "Vaginoplasty: Combination Therapy Using Labia Minora Flaps and Lucite Dilators; Preliminary Report." *Journal of Urology* 150 (2): 654–56.

Flatau, E., et al. 1975. "Letter: Penile Size in the Newborn Infant." *Journal of Pediatrics* 87 (4): 663–64.

Fliegner, J. R. 1996. "Long-Term Satisfaction with Sheares Vaginoplasty for Congenital Absence of the Vagina." *Australian and New Zealand Journal of Obstetrics and Gynecology* 36 (2): 202–4.

Fonkalsrud, E. W., S. Kaplan, and B. Lippe. 1977. "Experience with Reduction Clitoroplasty for Clitoral Hypertrophy." *Annals of Surgery* 186 (2): 221–26.

Foucault, M. 1973. *The Birth of the Clinic: An Archaeology of Medical Perception*. Trans. A. M. S. Smith. New York: Pantheon.

——. 1978. *The History of Sexuality*. Vol. 1, *An Introduction*. Trans. Robert Hurley. New York: Pantheon.

——. 1980. "Introduction." In Herculine Barbin, *Herculine Barbin: Being the Recently Discovered Memoirs of a Nineteenth-Century Hermaphrodite*. Trans. Richard McDougall, vii–xvii. New York: Pantheon.

Frank, R. 1938. "The Formation of an Artificial Vagina without Operation." *American Journal of Obstetrics and Gynecology* 35 (6): 1053–55.

Franklin, S. 1991. "Fetal Fascinations: New Dimensions to the Medical-Scientific Construction of Fetal Personhood." In *Off Centre: Feminism and Cultural Studies*, ed. Franklin, C. Lury, and J. Stacey, 190–205. New York: HarperCollins.

——. 1993. "Essentialism, Which Essentialism? Some Implications of Reproductive and Genetic Technoscience." In *Issues in Biological Essentialism versus Social Constructionism in Gay and Lesbian Identities*, ed., 27–39. J. Dececco. London: Harrington Park Press.

Freedman, E. B. 1982. "Sexuality in Nineteenth-Century America: Behavior, Ideology, and Politics." *Reviews in American History* 10 (4): 196–215.

Freud, S. 1962. *Three Essays on the Theory of Sexuality*. Trans. James Strachey. New York: Basic. Originally published as *Drei Abhandlungen zur Sexualtheorie* (Frankfurt am Main: Fischer, 1905).

Frimberger, D., and J. Gearhart. 2005. "Ambiguous Genitalia and Intersex." *Urologia Internationalis* 75 (4): 291–97.

Gantt, P. A., et al. 1980. "A Clinical and Cytogenetic Study of Fifteen Patients with 45,X/46XY Gonadal Dysgenesis." *Fertility and Sterility* 34 (3): 216–21.

Garcia, J., and H. W. Jones Jr. 1977. "The Split Thickness Graft Technic for Vaginal Agenesis." *Obstetrics and Gynecology* 49 (3): 328–32.

Gearhart, J. P., A. Burnett, and J. Owen. 1995. "Measurement of Evoked Potentials during Feminizing Genitoplasty: Technique and Applications." *Journal of Urology* 153 (2): 486–87.

Gould, S. J. 1981. *The Mismeasure of Man*. New York: Norton.

Graziano, K., et al. 2002. "Vaginal Reconstruction for Ambiguous Genitalia and Congenital Absence of the Vagina: A 27-Year Experience." *Journal of Pediatric Surgery* 37 (7): 955–60.

Groneman, C. 1994. "Nymphomania: The Historical Construction of Female Sexuality." *Signs: Journal of Women in Culture and Society* 19 (2): 337–67.

Gross, R. E., J. Randolph, and J. F. Crigler. 1966. "Clitorectomy for Sexual Abnormalities: Indications and Technique." *Surgery* 59 (2): 300–308.

Gross, T. 1997. "Fresh Air Interviews Chase, Khoury, Neilson." *Fresh Air*, NPR, December 11.

Grosz, E. 1994. *Volatile Bodies: Toward a Corporeal Feminism*. Bloomington: Indiana University Press.

Grumbach, M. M., and F. A. Conte. 1998. "Disorders of Sex Differentiation." In *Williams Textbook of Endocrinology*, 9th ed., ed. J. D. Wilson et al., 1303–1425. Philadelphia: Saunders.

Grumbach, M. M., F. A. Conte, and I. A. Hughes. 2003. "Disorders of Sex Differentiation." In *Williams Textbook of Endocrinology*, 10th ed., ed. P. R. Larsen et al., 842–1002. Philadelphia: Saunders.

Guttman, C. 2000. "Task Force to Draft Guidelines on Intersex Disorders." *Urology Times*, September, 26.

Hall, C. M., et al. 2004. "Behavioral and Physical Masculinization Are Related to Genotype in Girls with Congenital Adrenal Hyperplasia." *Journal of Clinical Endocrinology and Metabolism* 89 (1): 419–24.

Hall, L. A. 2001. "Clitoris." In *The Oxford Companion to the Body*, ed. C. Blakemore and S. Jennett, 160–62. Oxford: Oxford University Press.

Hampson, J. G. 1955. "Hermaphroditic Genital Appearance, Rearing, and Eroticism in Hyperadrenocorticism." *Bulletin of the Johns Hopkins Hospital* 97 (4): 265–73.

Hampson, J. L., and J. G. Hampson. 1961. "The Ontogenesis of Sexual Behavior in Man." In *Sex and Internal Secretions*, 3rd. ed., ed. W. C. Young, 1401–32. Baltimore: Williams and Wilkins.

Hampson, J. L., J. G. Hampson, and J. Money. 1955. "The Syndrome of Gonadal Agenesis (Ovarian Agenesis) and Male Chromosomal Pattern in Girls and Women: Psychologic Studies." *Bulletin of the Johns Hopkins Hospital* 97 (3): 207–26.

Haraway, D. J. 1991. *Simians, Cyborgs, and Women: The Reinvention of Nature*. New York: Routledge.

Harding, S. G. 1991. *Whose Science? Whose Knowledge? Thinking from Women's Lives*. Ithaca: Cornell University Press.

Hausman, B. L. 1995. *Changing Sex: Transsexualism, Technology, and the Idea of Gender*. Durham, NC: Duke University Press.

——. 2000. "Do Boys Have to Be Boys? Gender, Narrativity, and the John/Joan Case." *National Women's Studies Association Journal* 12 (3): 114–38.

Heath, D., et al. 1999. "Nodes and Queries: Linking Locations in Networked Fields of Inquiry." *American Behavioral Scientist* 43 (3): 450–63.

Hecker, B. R., and S. McGuire. 1977. "Psychosocial Function in Women Treated for Vaginal Agenesis." *American Journal of Obstetrics and Gynecology* 129 (5): 543–47.

Hecker, W. C. 1982. "Timing and Technic of the Surgical Correction of Childhood Intersex Genitals Including the Procedure for Partial Vaginal Aplasia." *Z Kinderchir* 37 (3): 93–99.

Hegarty, P., and C. Chase. 2000. "Intersex Activism, Feminism, and Psychology: Opening a Dialogue on Theory, Research, and Clinical Practice." *Feminism and Psychology* 10 (1): 117–32.

Hendren, W. H., and A. Atala. 1995. "Repair of High Vagina in Girls with Severely Masculinized Anatomy from the Adrenogenital Syndrome." *Journal of Pediatric Surgery* 30 (1): 91–94.

Hendren, W. H., and J. D. Crawford. 1969. "Adrenogenital Syndrome: The Anatomy of the Anomaly and Its Repair: Some New Concepts." *Journal of Pediatric Surgery* 4 (1): 49–58.

Hendricks, M. 1993. "Is It a Boy or a Girl?" *Johns Hopkins Magazine*, November 10–16.

Hensle, T. W., and W. A. I. Kennedy. 1998. "Surgical Management of Intersexuality." In *Campbell's Urology*, 7th. ed., ed. P. Walsh et al., 2:2155–71. 3 vols. Philadelphia: Saunders.

Hensle, T. W., and E. A. Reiley. 1998. "Vaginal Replacement in Children and Young Adults." *Journal of Urology* 159 (3): 1035–38.

Hester, J. D. 2003. "Rhetoric of the Medical Management of Intersexed Children." *Genders Online Journal* 38. www.genders.org/g38/g38—hester.html (accessed November 13, 2007).

——. 2004. "Intersex(es) and Informed Consent: How Physicians' Rhetoric Constrains Choice." *Theoretical Medicine and Bioethics* 25 (1): 21–49.

Hirschauer, S. 1998. "Performing Sexes and Genders in Medical Practices." In *Differences in Medicine: Unraveling Practices, Techniques, and Bodies*, ed. M. Berg and A. Mol, 13–27. Durham, NC: Duke University Press.

Hitchcock, R. J. I., and P. S. Malone. 1994. "Colovaginoplasty in Infants and Children." *British Journal of Urology* 73 (2): 196–99.

Hojsgaard, A., and I. Villadsen. 1995. "McIndoe Procedure for Congenital Vaginal Agenesis: Complications and Results." *British Journal of Plastic Surgery* 48 (2): 97–102.

Holmes, M. 1994. "Medical Politics and Cultural Imperatives: Intersexuality beyond Pathology and Erasure." MA thesis, York University.

——. 2002. "Rethinking the Meaning and Management of Intersexuality." *Sexualities* 5 (2): 159–80.

Holt, L. E., J. Howland, and R. McIntosh. 1936. *Holt's Diseases of Infancy and Childhood: A Textbook for the Use of Students and Practitioners*, 10th ed. London: D. Appleton-Century.

Houk, C. P., et al. 2006. "Summary of Consensus Statement on Intersex Disorders and Their Management: International Intersex Consensus Conference." *Pediatrics* 118 (2): 753–57.

Hughes, I. A., et al. 2006. "Consensus Statement on Management of Intersex Disorders." *Archives of Disease in Childhood* 91 (7): 554–63.

"I Gotta Be Me." 1997. *Dateline*, NBC, June 17.

Ingersoll, F. M., and J. E. Finesinger. 1947. "A Case of Male Pseudohermaphroditism: The Importance of Psychiatry in the Surgery of This Condition." *Surgical Clinics of North America* 27: 1222.

Ingram, J. M. 1981. "The Bicycle Seat Stool in the Treatment of Vaginal Agenesis and Stenosis: A Preliminary Report." *American Journal of Obstetrics and Gynecology* 140 (8): 867–73.

Intersex Society of North America. 1995–96. *Hermaphrodites with Attitude* (newsletter), fall–winter issue.

Intons-Peterson, M. J., and M. Reddel. 1984. "What do People Ask about a Neonate?" *Developmental Psychology* 20 (3): 358–59.

Isaacs, D., and D. Fitzgerald. 1999. "Seven Alternatives to Evidence-Based Medicine." *British Medical Journal* 319 (7225): 1618.

"Is Early Vaginal Reconstruction Wrong for Some Intersex Girls?" 1997. *Urology Times*, Februrary, 10–12.

Joffe, C. E., T. A. Weitz, and C. L. Stacey. 2004. "Uneasy Allies: Pro-Choice Physicians, Feminist Health Activists, and the Struggle for Abortion Rights." *Sociology of Health and Illness* 26 (6): 775–96.

Johnson, N., A. Batchelor, and R. J. Lilford. 1991. "Experience with Tissue Expansion Vaginoplasty." *British Journal of Obstetrics and Gynaecology* 98 (6): 564–68.

Jones, H. W., Jr. 1974. "Surgical Construction of the Female Genitalia." *Clinics in Plastic Surgery* 1 (2): 255.

Jones, H. W., Jr., and W. W. Scott. 1958. *Hermaphroditism, Genital Anomalies, and Related Endocrine Disorders*. Baltimore: Williams and Wilkins.

Joseph, V. T. 1997. "Pudendal Thigh Flap Vaginoplasty in the Reconstruction of Genital Anomalies." *Journal of Pediatric Surgery* 32 (1): 62–65.

Kahneman, D., P. Slovic, and A. Tversky. 1982. *Judgment under Uncertainty: Heuristics and Biases*. Cambridge: Cambridge University Press.

Kahneman, D., and A. Tversky. 1972. "Subjective Probability: A Judgment of Representativeness." *Cognitive Psychology* 3 (3): 430–54.

Kandel, E. R., J. H. Schwartz, and T. M. Jessell, eds. 2000. *Principles of Neural Science*, 4th ed. New York: McGraw-Hill.

Karkazis, K. A. 2006. "Early Genital Surgery to Remain Controversial." *Pediatrics* 118 (2): 814–15.

Katz, J. 1995. *The Invention of Heterosexuality*. New York: Dutton.

Kenen, S. H. 1998. "Scientific Studies of Human Sexual Difference in Interwar America." PhD diss., University of California at Berkeley.

Kessler, S. J. 1990. "The Medical Construction of Gender: Case Management of Intersexual Infants." *Signs: Journal of Women in Culture and Society* 16 (1): 3–26.

——. 1998. *Lessons from the Intersexed*. New Brunswick, NJ: Rutgers University Press.

Kessler, S. J., and W. McKenna. 1978. *Gender: An Ethnomethodological Approach*. New York: John Wiley.

Kevles, D. 1986. *In the Name of Eugenics: Genetics and the Uses of Human Heredity*. Berkeley: University of California Press.

Kinsey, A. C., et al. 1953. *Sexual Behavior in the Human Female*. Philadelphia: Saunders.

Kleinman, A. 1988. *The Illness Narratives: Suffering, Healing, and the Human Condition*. New York: Basic Books.

Kolodny, R. C. 1979. *Textbook of Sexual Medicine*. New York, Little Brown.

Koyama, E., and L. Weasel. 2002. "From Social Construction to Social Justice: Transforming How We Teach about Intersexuality." *Women's Studies Quarterly* 30 (3–4): 169–78.

Kumar, H., et al. 1974. "Clitoroplasty: Experience during a Nineteen-Year Period." *Journal of Urology* 111 (1): 81–84.

Laing, R. D. 1960. *The Divided Self: An Existential Study in Sanity and Madness*. London: Tavistock.

Lakoff, A., and S. Collier. 2004. "Ethics and the Anthropology of Modern Reason." *Anthropological Theory* 4 (4): 419–34.

Laqueur, T. 1990. *Making Sex: Body and Gender from the Greeks to Freud*. Cambridge: Harvard University Press.

Lashley, M., et al. 2000. "Informed Proxy Consent: Communication between Pediatric Surgeons and Surrogates about Surgery." *Pediatrics* 105 (3): 591–97.

Lattimer, J. K. 1961. "Relocation and Recession of the Enlarged Clitoris with Preservation of the Glans: An Alternative to Amputation." *Journal of Urology* 86 (1): 113–16.

Lawson Wilkins Pediatric Endocrine Society/European Society for Paediatric Endocrinology (LWPES/ESPE) CAH Working Group. 2002. "Consensus Statement on

21-Hydroxylase Deficiency from the Lawson Wilkins Pediatric Endocrine Society and the European Society for Paediatric Endocrinology." *Journal of Clinical Endocrinology and Metabolism* 87 (9): 4048–53.

Leavitt, J. W. 1986. *Brought to Bed: Childbearing in America, 1750 to 1950*. New York: Oxford University Press.

Lee, P. A. 2001. "Should We Change Our Approach to Ambiguous Genitalia?" *Endocrinologist* 11 (2): 118–23.

Lee, P. A., et al. 1980a. "Micropenis: I; Criteria, Etiologies and Classification." *Johns Hopkins Medical Journal* 146 (4): 156–63.

Lee, P. A., et al. 1980b. "Micropenis: III; Primary Hypogonadism, Partial Androgen Insensitivity Syndrome, and Idiopathic Disorders." *Johns Hopkins Medical Journal* 147 (5): 175–81.

Lee, P. A., et al. 2006. "Consensus Statement on Management of Intersex Disorders: International Consensus Conference on Intersex." *Pediatrics* 118 (2): e488–500.

Leight, K. 2006. "Cares Letter to Consensus Meeting Receives International Support." *CARES Foundation Newsletter* 5 (1): 14–15.

Leveroni, C. L., and S. A. Berenbaum. 1998. "Early Androgen Effects on Interest in Infants: Evidence from Children with Congenital Adrenal Hyperplasia." *Developmental Neuropsychology* 14: 321–40.

Lightfoot-Klein, H. 1989. *Prisoners of Ritual*. New York: Harrington Park Press.

Lillie, F. R. 1916. "Theory of the Freemartin." *Science* 43: 611.

——. 1917. "The Free-Martin', a Study of the Action of Sex Hormones in Foetal Life of Cattle." *Journal of Experimental Zoology* 23: 371–452.

Lindenbaum, S., and M. Lock. 1993. *Knowledge, Power, and Practice: The Anthropology of Medicine and Everyday Life*. Berkeley: University of California Press.

Lloyd, J., et al. 2005. "Female Genital Appearance: 'Normality' Unfolds." *British Journal of Obstetrics and Gynaecology* 112 (5): 643–6.

Lobe, T. E., et al. 1987. "The Complications of Surgery for Intersex: Changing Patterns over Two Decades." *Journal of Pediatric Surgery* 22 (7): 651–52.

Low, Y., and J. M. Hutson. 2003. "Rules for Clinical Diagnosis in Babies with Ambiguous Genitalia." *Journal of Paediatrics and Child Health* 39 (6): 406–13.

Macedo, A., Jr., and M. Srougi. 2001. "Surgery to the External Genitalia." *Current Opinion in Urology* 11 (6): 585–90.

Mak, G. 2005. " 'So We Must Go Behind Even What the Microscope Can Reveal': The Hermaphrodite's 'Self' in Medical Discourse at the Start of the Twentieth Century." *GLQ: A Journal of Gay and Lesbian Studies* 11 (1): 65–94.

Marsh, J. L. 2006. "To Cut or Not to Cut? A Surgeon's Perspective on Surgically Shaping Children." In *Surgically Shaping Children: Technology, Ethics, and the Pursuit of Normality*, ed. E. Parens, 113–24. Baltimore: Johns Hopkins University Press.

Martin, E. 1987. *The Woman in the Body: A Cultural Analysis of Reproduction*. Boston: Beacon.

Martinez-Mora, J., et al. 1992. "Neovagina in Vaginal Agenesis: Surgical Methods and Long-Term Results." *Journal of Pediatric Surgery* 27 (1): 10–14.

Masters, W. H., and V. E. Johnson. 1966. *Human Sexual Response*. Boston: Little, Brown.

Matta, C. 2005. "Ambiguous Bodies and Deviant Sexualities: Hermaphrodites, Homosexuality, and Surgery in the United States, 1850–1904." *Perspectives in Biology and Medicine* 48 (4): 74–83.

May, S. K. 2003. "Hypospadias." PhD. diss., Maimonides University.

Mazur, T. 2005. "Gender Dysphoria and Gender Change in Androgen Insensitivity or Micropenis." *Archives of Sexual Behavior* 34 (4): 411–21.

McKenna, W., and S. J. Kessler. 2000. "Retrospective Response." *Feminism and Psychology* 10 (1): 66–72.

Meldrum, K. K., R. Mathews, and J. P. Gearhart. 2001. "Hugh Hampton Young: A Pioneer in Pediatric Urology." *Journal of Urology* 166 (4): 1415–17.

Melton, L. 2001. "New Perspectives on the Management of Intersex." *Lancet* 357 (9274): 2110.

Meyer-Bahlburg, H. 1999. "Gender Assignment and Reassignment in 46,XY Pseudohermaphroditism and Related Conditions." *Journal of Clinical Endocrinology and Metabolism* 84 (10): 3455–58.

——. 2000. "Psychobiological Models of Psychosexual Differentiation." *Zeitschrift fur Humanontogenetik* 3: 49–53.

——. 2003. "Rationale for Gender Assignment." *Endocrinologist* 13 (3): 224–26.

——. 2005a. "Gender Identity Outcome in Female-Raised 46,XY Persons with Penile Agenesis, Cloacal Exstrophy of the Bladder, or Penile Ablation." *Archives of Sexual Behavior* 34 (4): 423–38.

——. 2005b. "Intersex—Fact or Fiction? (Review of *Intersex and Identity* by Sharon Preves)." *Journal of Sex Research* 42 (2): 177–80.

——. 2005c. "Introduction: Gender Dysphoria and Gender Change in Persons with Intersexuality." *Archives of Sexual Behavior* 34 (4): 371–73.

Meyer-Bahlburg, H., et al. 1996. "Gender Change from Female to Male in Classical Congenital Adrenal Hyperplasia." *Hormones and Behavior* 30 (4): 319–22.

Meyer-Bahlburg, H., et al. 2001. "Does Genital Status at Birth Predict Gender-Related Behavior in Girls with Congenital Adrenal Hyperplasia?" Paper presented at the International Academy of Sex Research 27th Annual Meeting, Montreal, Quebec, Canada, July 11–14.

Meyer-Bahlburg, H., et al. 2004. "Prenatal Androgenization Affects Gender-Related Behavior But Not Gender Identity in Five-to-Twelve-Year-Old Girls with Congenital Adrenal Hyperplasia." *Archives of Sexual Behavior* 33 (2): 97–104.

Migeon, C. J., and A. B. Wisniewski. 1998. "Sexual Differentiation: From Genes to Gender." *Hormone Research* 50 (5): 245–51.

Miller, A. M., and C. S. Vance. 2004. "Sexuality, Human Rights, and Health." *Health and Human Rights* 7 (2): 5–15.

Millett, K. 1970. *Sexual Politics*. Garden City, NY: Doubleday.

Millman, M. 1976. *The Unkindest Cut: Life in the Backrooms of Medicine*. New York: Morrow.

Mininberg, D. T. 1982. "Phalloplasty in Congenital Adrenal Hyperplasia." *Journal of Urology* 128 (2): 355–56.

Minto, C. 2000. "Survey for Adult Women with CAH." Unpublished survey.

Minto, C., et al. 2003a. "The Effect of Clitoral Surgery on Sexual Outcome in Individuals Who Have Intersex Conditions with Ambiguous Genitalia: A Cross-Sectional Study." *Lancet* 361 (9365): 1252–57.

Minto, C., et al. 2003b. "Sexual Function in Women with Complete Androgen Insensitivity Syndrome." *Fertility and Sterility* 80 (1): 157–64.

Money, J. 1952. "Hermaphroditism: An Inquiry into the Nature of a Human Paradox." PhD diss., Harvard University.

———. 1955. "Hermaphroditism, Gender, and Precocity in Hyper-adrenocorticism: Psychologic Findings." *Bulletin of the Johns Hopkins Hospital* 96 (6): 253–64.

———. 1956. "Hermaphroditism: Recommendations Concerning Case Management." *Journal of Clinical Endocrinology and Metabolism* 16 (4): 547–56.

———. 1961. "Components of Erotocism in Man: II; The Orgasm and Genital Somesthesia." *Journal of Nervous and Mental Disease* 132: 289–97.

———. 1970. "Critique of Dr. Zuger's Manuscript." *Psychosomatic Medicine* 32 (5): 463–67.

———. 1974. "Psychologic Consideration of Sex Assignment in Intersexuality." *Clinics in Plastic Surgery* 1 (2): 215–22.

———. 1975. "Ablatio Penis: Normal Male Infant Sex-Reassigned as a Girl." *Archives of Sexual Behavior* 4 (1): 65–71.

———. 1986. *Venuses Penuses: Sexology, Sexosophy, and Exigency Theory.* Buffalo: Prometheus.

———. 1994. *Sex Errors of the Body and Related Syndromes: A Guide to Counseling Children, Adolescents, and Their Families.* Baltimore: Paul H. Brookes.

———. 1998a. "Ablatio Penis: Nature/Nurture Redux." In *Sin, Science, and the Sex Police: Essays on Sexology and Sexosophy*, ed. Money, 297–326. Amherst, NY: Prometheus.

———. 1998b. "Case Consultation: Ablatio Penis." *Medicine and Law* 17 (1): 113–23.

———. 1998c. "Medicine and Law: Case Consultation, Ablatio Penis." In *Sin, Science, and the Sex Police: Essays on Sexology and Sexosophy*, ed. Money, 286–96. Amherst, NY: Prometheus.

———. 2002. *A First Person History of Pediatric Psychoendocrinology.* New York: Kluwer Academic/Plenum Publishers.

Money, J., and A. A. Ehrhardt. 1972. *Man and Woman, Boy and Girl: The Differentiation and Dimorphism of Gender Identity from Conception to Maturity.* Baltimore: Johns Hopkins University Press.

Money, J., J. G. Hampson, and J. L. Hampson. 1955a. "An Examination of Some Basic Sexual Concepts: The Evidence of Human Hermaphroditism." *Bulletin of the Johns Hopkins Hospital* 97 (4): 301–19.

———. 1955b. "Hermaphroditism: Recommendations Concerning Assignment of Sex, Change of Sex, and Psychologic Management." *Bulletin of the Johns Hopkins Hospital* 97 (4): 284–300.

———. 1956. "Sexual Incongruities and Psychopathology: The Evidence of Human Hermaphroditism." *Bulletin of the Johns Hopkins Hospital* 98 (1): 43–57.

———. 1957. "Imprinting and the Establishment of Gender Role." *Archives of Neurology and Psychiatry* 77 (3): 333–36.

Money, J., G. K. Lehne, and F. Pierre-Jerome. 1985. "Micropenis: Gender, Erotosexual Coping Strategy, and Behavioral Health in Nine Pediatric Cases Followed to Adulthood." *Comprehensive Psychiatry* 26 (1): 29–42.

Money, J., R. Potter, and C. Stoll. 1969. "Sex Reannouncement in Hereditary Sex Deformity: Psychology and Sociology of Habilitation." *Social Science and Medicine* 3 (2): 207–16.

Money, J., and P. Tucker. 1975. *Sexual Signatures: On Being a Man or Woman*. Boston: Little, Brown.

Moore, K. L., and M. L. Barr. 1955. "Smears from the Oral Mucosa in the Detection of Chromosomal Sex." *Lancet* 269 (6880): 57–58.

Moore, K. L., M. A. Graham, and M. L. Barr. 1953. "The Detection of Chromosomal Sex in Hermaphrodites from a Skin Biopsy." *Surgery, Gynecology, and Obstetrics* 96 (6): 641–48.

Moore, L. J., and A. E. Clarke. 1995. "Genital Conventions and Transgressions: Graphic Representations in Anatomy Texts, c1900–1991." *Feminist Studies* 22 (1): 255–301.

Morello-Frosch, R., et al. 2006. "Embodied Health Movements: Responses to a 'Scientized' World." In *The New Political Sociology of Science: Institutions, Networks, and Power*, ed. S. Frickel and K. Moore, 244–71. Madison: University of Wisconsin Press.

Moreno, A., and J. Goodwin. 1998. "Am I a Woman or a Man?" *Mademoiselle*, March, 178–81, 208.

Morgen, S. 2002. *Into Our Own Hands: The Women's Health Movement in the United States, 1969–1990*. New Brunswick, NJ: Rutgers University Press.

Morland, I. 2004. "Thinking with the Phallus." *Psychologist* 17 (8): 448–50.

Morris, R. 1995. "All Made Up: Performance Theory and the New Anthropology of Sex and Gender." *Annual Review of Anthropology* 24: 567–92.

Mouriquand, P. D. 2004. "Possible Determinants of Sexual Identity: How to Make the Least Bad Choice in Children with Ambiguous Genitalia." *BJU International* 93, supplement 3: 1–2.

Mouriquand, P. D., and P. Y. Mure. 2004. "Current Concepts in Hypospadiology." *BJU International* 93, supplement 3: 26–34.

Mukherjee, B., et al. 2007. "Psychopathology, Psychosocial, Gender and Cognitive Outcomes in Patients with Cloacal Exstrophy." *Journal of Urology* 178(2): 630–35.

Mulaikal, R. M., C. J. Migeon, and J. A. Rock. 1987. "Fertility Rates in Female Patients with Congenital Adrenal Hyperplasia Due to 21-Hydroxylase Deficiency." *New England Journal of Medicine* 316 (4): 178–82.

Muller, J., et al. 1999. "Management of Males with 45,X/46,XY Gonadal Dysgenesis." *Hormone Research* 52 (1): 11–14.

Muller, U., et al. 1983. "Correlation between Testicular Tissue and H-Y Phenotype in Intersex Patients." *Clinical Genetics* 23 (1): 49–57.

Mureau, M. A., et al. 1995a. "Genital Perception of Children, Adolescents, and Adults Operated on for Hypospadias: A Comparative Study." *Journal of Sex Research* 32 (4): 289–98.

Mureau, M. A., et al. 1995b. "Psychosexual Adjustment of Children and Adolescents after Different Types of Hypospadias Surgery: A Norm-Related Study." *Journal of Urology* 154 (5): 1902–7.

Mureau, M. A., et al. 1995c. "Psychosexual Adjustment of Men Who Underwent Hypospadias Repair: A Norm-Related Study." *Journal of Urology* 154 (4): 1351–55.

Mureau, M. A., et al. 1996. "Satisfaction with Penile Appearance after Hypospadias Surgery: The Patient and Surgeon View." *Journal of Urology* 155 (2): 703–6.

Mureau, M. A., et al. 1997. "Psychosocial Functioning of Children, Adolescents, and Adults Following Hypospadias Surgery: A Comparative Study." *Journal of Pediatric Psychology* 22 (3): 371–87.

Mykhalovskiy, E., and L. Weir. 2004. "The Problem of Evidence-Based Medicine: Directions for Social Science." *Social Science and Medicine* 59 (5): 1059–69.

Nelkin, D., and M. S. Lindee. 1995. *DNA Mystique: The Gene as a Cultural Icon*. New York: Freedman.

Newman, K., J. Randolph, and K. Anderson. 1992. "The Surgical Management of Infants and Children with Ambiguous Genitalia: Lessons Learned from Twenty-Five Years." *Annals of Surgery* 215 (6): 644–53.

Newman, K., J. Randolph, and S. Parson. 1992. "Functional Results in Young Women Having Clitoral Reconstruction as Infants." *Journal of Pediatric Surgery* 27 (2): 180–84.

Nihoul-Fékété, C. 1981. "Feminizing Genitoplasty in the Intersex Child." *Pediatric and Adolescent Endocrinology* 8: 247–60.

Nonomura, M., et al. 1991. "Does a Gonadotropin-Releasing Hormone Analogue Prevent Cisplatin-Induced Spermatogenic Impairment? An Experimental Study in the Mouse." *Urological Research* 19 (2): 135–40.

Oberfield, S. E., et al. 1989. "Clitoral Size in Full-Term Infants." *American Journal of Perinatology* 6 (4): 453–54.

Ochoa, B. 1998. "Trauma of the External Genitalia in Children: Amputation of the Penis and Emasculation." *Journal of Urology* 160 (3): 1116–19.

O'Connell, H. E., et al. 1998. "Anatomical Relationship between Urethra and Clitoris." *Journal of Urology* 159 (6): 1892–97.

Odem, M. E. 1995. *Delinquent Daughters: Protecting and Policing Adolescent Female Sexuality in the United States, 1885–1920*. Chapel Hill: University of North Carolina Press.

Oesterling, J. E., J. P. Gearhart, and R. D. Jeffs. 1987. "A Unified Approach to Early Reconstructive Surgery of the Child with Ambiguous Genitalia." *Journal of Urology* 138 (4): 1079–84.

Oudshoorn, N. 1994. *Beyond the Natural Body: An Archaeology of Sex Hormones*. London: Routledge.

Paparel, P., et al. 2001. "Approche actuelle de l'hypospade chez l'enfant" [Current approach to hypospadias in children]. *Progrès en Urologie* 11 (4): 741–51.

Parker, M. 2002. "Whither Our Art? Clinical Wisdom and Evidence-Based Medicine." *Medicine, Health Care, and Philosophy* 5: 273–80.

Phoenix, C., and R. Goy. 1959. "Organizing Action of Prenatally Administered Testosterone Proprionate on the Tissues Mediating Mating Behavior in the Female Guinea Pig." *Endocrinology* 65: 369–82.

Phornphutkul, C., A. Fausto-Sterling, and P. A. Gruppuso. 2000. "Gender Self-Reassignment in an XY Adolescent Female Born with Ambiguous Genitalia." *Pediatrics* 106 (1): 135–37.

Pinter, A., and G. Kosztolanyi. 1990. "Surgical Management of Neonates and Children with Ambiguous Genitalia." *Acta Paediatrica Hungarica* 30 (1): 111–21.

Prader, A. 1958. "Vollkomen Manliche aubere Genitalentwicklung und Salverlust-syndrom bei Madchen mit Kongelitalem adronogenitalem Syndrome" [Perfect male external genital development and salt-loss syndrome in girls with congenital andrenogenital syndrome]. *Helvetica Paediatrica Acta* 13 (1): 5–14.

Premawardhana, L. D. K. E., et al. 1997. "Longer Term Outcomes in Females with Congenital Adrenal Hyperplasia (CAH): The Cardiff Experience." *Clinical Endocrinology* 46 (3): 327–32.

Preves, S. E. 2003. *Intersex and Identity: The Contested Self.* New Brunswick, NJ: Rutgers University Press.

——. 2004. "Out of the OR and into the Streets: Exploring the Impact of Intersex Media Activism." *Research in Political Sociology* 13: 179–223.

Purves, T., et al. 2007. "Vaginal Reconstruction for Patients with Complete Androgen Insensitivity: Challenging the Dogma." Paper presented at the American Academy of Pediatrics Annual Meeting, San Francisco, October 27–30.

Rajfer, J., R. M. Ehrlich, and W. E. Goodwin. 1982. "Reduction Clitoroplasty via Ventral Approach." *Journal of Urology* 128 (2): 341–43.

Ramani, P., C. K. Yeung, and S. S. Habeebu. 1993. "Testicular Intratubular Germ Cell Neoplasia in Children and Adolescents with Intersex." *American Journal of Surgical Pathology* 17 (11): 1124–33.

Randolph, J., and W. Hung. 1970. "Reduction Clitoroplasty in Females with Hypertrophied Clitoris." *Journal of Pediatric Surgery* 5 (2): 224–30.

Randolph, J., W. Hung, and M.C. Rathlev. 1981. "Clitoroplasty for Females Born with Ambiguous Genitalia: A Long-Term Study of Thirty-Seven Patients." *Journal of Pediatric Surgery* 16 (6): 882–87.

Rapp, R. 1999. *Testing Women, Testing the Fetus: The Social Impact of Amniocentesis in America.* New York, Routledge.

Rapp, R., D. Heath, and K. Taussig. 2001. "Genealogical Dis-ease: Where Hereditary Abnormality, Biomedical Explanation and Family Responsibility Meet." In *Relative Values: Reconfiguring Kinship Studies*, ed. S. Franklin and S. McKinnon. Durham, NC: Duke University Press.

Redick, A. 2004. "American History XY: The Medical Treatment of Intersex, 1916–1955." PhD diss., New York University.

Reilly, J. M., and C. R. J. Woodhouse 1989. "Small Penis and the Male Sexual Role." *Journal of Urology* 142 (2): 569–71.

Reiner, W. G. 1996. "Case Study: Sex Reassignment in a Teenage Girl." *Journal of the American Academy of Child and Adolescent Psychiatry* 35 (6): 799–803.

——. 1997. "Sex Assignment in the Neonate with Intersex or Inadequate Genitalia." *Archives of Pediatrics and Adolescent Medicine* 151 (10): 1044–45.

——. 1999. "Assignment of Sex in Neonates with Ambiguous Genitalia." *Current Opinions in Pediatrics* 11 (4): 363–65.

——. 2000a. "Ambiguity and Microphallus: Gender Assignment and Reassignment." Paper presented at the Midwest Pediatric Endocrine Society conference, Chicago, Illinois, September 9.

——. 2000b. "Philosophy and Ethics of Gender Identity." Paper presented for the John Duckett Lectureship in Pediatric Urology, Ann Arbor, University of Michigan, July 21.

——. 2004. "Psychosexual Development in Genetic Males Assigned Female: The Cloacal Exstrophy Experience." *Child and Adolescent Psychiatric Clinics of North America* 13 (3): 657–74.

——. 2005. "Gender Identity and Sex-of-Rearing in Children with Disorders of Sexual Differentiation." *Journal of Pediatric Endocrinology and Metabolism* 18 (6): 549–53.

Reiner, W. G., and J. P. Gearhart. 2004. "Discordant Sexual Identity in Some Genetic Males with Cloacal Exstrophy Assigned to Female Sex at Birth." *New England Journal of Medicine* 350 (4): 333–41.

Reiner, W. G., J. P. Gearhart, and R. Jeffs. 1999. "Psychosexual Dysfunction in Males with Genital Anomalies: Late Adolescence, Tanner Stages IV to VI." *Journal of the American Academy of Child and Adolescent Psychiatry* 38 (7): 865–72.

Reiner, W. G., and B. P. Kropp. 2004. "A Seven-Year Experience of Genetic Males with Severe Phallic Inadequacy Assigned Female." *Journal of Urology* 172 (6): 2395–98.

Reis, E. 2005. "Impossible Hermaphrodites: Intersex in America, 1620–1960." *Journal of American History* 92 (2):411–41.

Riley, W. J., and A. Rosenbloom. 1980. "Clitoral Size in Infancy." *Journal of Pediatrics* 96 (5): 918–19.

Rink, R. C., and M. C. Adams. 1998. "Feminizing Genitoplasty: State of the Art." *World Journal of Urology* 16 (3): 212–18.

Robboy, S. J., et al. 1982. "Dysgenesis of Testicular and Streak Gonads in the Syndrome of Mixed Gonadal Dysgenesis: Perspective Derived from a Clinicopathologic Analysis of Twenty-One Cases." *Human Pathology* 13 (8): 700–716.

Roberts, C. 2000. "Biological Behavior? Hormones, Psychology, and Sex." *National Womens Studies Association Journal* 12 (3): 1–20.

Robertson, I. 1977. *Sociology*. New York: Worth.

Rock, J. A., et al. 1983. "Success Following Vaginal Creation for Müllerian Agenesis." *Fertility and Sterility* 39 (6): 809–13.

Rothman, B. K. 1991. *In Labor: Women and Power in the Birthplace*. New York: Norton.

Rothman, D. J. 1991. *Strangers at the Bedside: A History of How Law and Bioethics Transformed Medical Decision Making*. New York: Basic Books.

Rubin, G. 1975. "The Traffic in Women: Notes on the Political Economy of Sex." In *Toward an Anthropology of Women*, ed. R. R. Reiter, 157–210. New York: Monthly Review Press.

——. 1992. "Thinking Sex: Notes for a Radical Theory of the Politics of Sexuality." In *Pleasure and Danger: Exploring Female Sexuality*, ed. C. S. Vance, 267–319. London: Pandora.

Russett, C. E. 1989. *Sexual Science: The Victorian Construction of Womanhood*. Cambridge: Harvard University Press.

Sackett, D. L., et al. 1996. "Evidence Based Medicine: What It Is and What It Isn't." *British Medical Journal* 312 (7023): 71–72.

Sakurai, H., et al. 2000. "The Use of Free Jejunal Autograft for the Treatment of Vaginal Agenesis: Surgical Methods and Long-Term Results." *British Journal of Plastic Surgery* 53 (4): 319–23.

Sandberg, D. E., et al. 2004. "Intersexuality: A Survey of Clinical Practice." *Pediatric Research (Abstract Issue APS-SPR)* 55 (4): abstract 869.

Sankar, P. 2004. "Communication and Miscommunication in Informed Consent to Research." *Medical Anthropology Quarterly* 18 (4): 429–46.

Sargent, A. G., ed. 1977. *Beyond Sex Roles*. St. Paul: West Publishing.

Sax, L. 2002. "How Common Is Intersex? A Response to Anne Fausto-Sterling." *Journal of Sex Research* 39 (3): 174–78.

——. 2005. *Why Gender Matters: What Parents and Teachers Need to Know about the Emerging Science of Sex Differences*. New York: Doubleday.

Schattner, A., and R. H. Fletcher. 2003. "Research Evidence and the Individual Patient." *QJM: Monthly Journal of the Association of Physicians* 96 (1): 1–5.

Scheck, A. 1997. "Attitudes Changing toward Intersex Surgery, But for the Better?" *Urology Times*, August, 44–45.

Schiebinger, L. L. 1989. *The Mind Has No Sex? Women in the Origins of Modern Science*. Cambridge: Harvard University Press.

——. 1993. *Nature's Body: Gender in the Making of Modern Science*. Boston: Beacon.

Schober, J. M. 1998. "Feminizing Genitoplasty for Intersex." In *Pediatric Surgery and Urology: Long Term Outcomes*, ed. M. D. Stringer et al., 549–58. London: Saunders.

——. 1999. "Quality-of-Life Studies in Patients with Ambiguous Genitalia." *World Journal of Urology* 17 (4): 249–52.

——. 2001. "Sexual Behaviors, Sexual Orientation, and Gender Identity in Adult Intersexuals: A Pilot Study." *Journal of Urology* 165 (6): 2350–53.

Schober, J. M., et al. 2002. "The Ultimate Challenge of Cloacal Exstrophy." *Journal of Urology* 167 (1): 300–304.

Schonfeld, W. A., and G. W. Beebe. 1942. "Normal Growth and Variation in the Male Genitalia from Birth to Maturity." *Journal of Urology* 48: 759–77.

Schützmann, K., et al. 2007. "Psychological Distress, Self-Harming Behavior, and Suicidal Tendencies in Adults with Disorders of Sex Development." *Archives of Sexual Behavior*, doi:10.1007/s10508-007-9241-9. http://www.springerlink.com/content/x86156w915435611r/ (electronically published October 18).

Scull, A., and D. Favreau. 1986. "The Clitoridectomy Craze." *Social Research* 53 (2): 243–60.

Sharp, R., et al. 1987. "Neonatal Genital Reconstruction." *Journal of Pediatric Surgery* 22 (2): 168–71.

Sheares, B. 1960. "Congenital Atresia of the Vagina: A New Technique for Tunnelling the Space between the Bladder and the Rectum and Construction of a New Vagina by the Modified Wharton Technique." *Journal of Obstetrics and Gynaecology of the British Empire* 67: 24–31.

Sheehan, E. A. 1997. "Victorian Clitoridectomy: Isaac Baker Brown and His Harmless Operative Procedure." In *The Gender/Sexuality Reader: Culture, History, Political Economy*, ed. R. N. Lancaster and M. di Leonardo, 325–34. New York: Routledge.

Sheldon, C. A., and J. W. Duckett. 1987. "Hypospadias." *Pediatric Clinics of North America* 34 (5): 1259–72.

Sherfey, M. J. 1972. *The Nature and Evolution of Female Sexuality*. New York: Random House.

Silverman, V., and S. Stryker, dirs. 2005. *Screaming Queens: The Riot at Compton's Cafeteria*. Frameline.

Slijper, F. M. 2003. "Clitoral Surgery and Sexual Outcome in Intersex Conditions." *Lancet* 361 (9365): 1236–37.

Slijper, F. M., et al. 1998. "Long-Term Psychological Evaluation of Intersex Children." *Archives of Sexual Behavior* 27 (2): 125–44.

Smith, E. D. 1997. "The History of Hypospadias." *Pediatric Surgery International* 12 (2–3): 81–85.

Solursh, D. S., et al. 2003. "The Human Sexuality Education of Physicians in North American Medical Schools." *International Journal of Impotence Research* 15, supplement 5: 41–45.

Sotiropoulos, A., et al. 1976. "Long-Term Assessment of Genital Reconstruction in Female Pseudohermaphrodites." *Journal of Urology* 115 (5): 599–601.

Starr, P. 1982. *The Social Transformation of American Medicine*. New York: Basic Books.

Stecker, J. F., et al. 1981. "Hypospadias Cripples." *Urologic Clinics of North America* 8 (3): 539–44.

Stoller, R. J. 1964. "A Contribution to the Study of Gender Identity." *Journal of the American Medical Association* 45: 220–26.

Sullivan, F. M., and R. J. MacNaughton. 1996. "Evidence in Consultations: Interpreted and Individualised." *Lancet* 348 (9032): 941–43.

Svensson, H., et al. 1997. "Staged Reconstruction of Hypospadias with Chordee: Outcome and Costs." *Scandanavian Journal of Plastic and Reconstructive Surgery and Hand Surgery* 31 (1): 51–55.

Terry, J., and J. Urla, eds. 1995. *Deviant Bodies: Critical Perspectives on Difference in Science and Popular Culture*. Bloomington: Indiana University Press.

Timmermans, S., and A. Angell. 2001. "Evidence-Based Medicine, Clinical Uncertainty, and Learning to Doctor." *Journal of Health and Social Behavior* 42: 342–59.

Toomey, C. 2001. "The Worst of Both Worlds." *Sunday Times Magazine*, October 28, 34–40.

Tovar, J. A. 2003. "Clitoral Surgery and Sexual Outcome in Intersex Individuals." *Lancet* 362 (9379): 247–48.

Tuana, N. 1996. "Fleshing Gender, Sexing the Body: Refiguring the Sex/Gender Distinction." *Southern Journal of Philosophy* 35: 53–73.

Turner, V. W. 1974. *Dramas, Fields, and Metaphors: Symbolic Action in Human Society*. Ithaca: Cornell University Press.

Urla, J., and A. Swedlund. 1995. "The Anthropometry of Barbie." In *Deviant Bodies: Critical Perspectives on Difference in Science and Popular Culture*, ed. J. Terry and Urla, 277–313. Bloomington: Indiana University Press.

Valentine, D. 2000. " 'I Know What I Am': The Category 'Transgender' in the Construction of Contemporary U.S. American Conceptions of Gender and Sexuality." PhD diss., New York University.

Valentine, D., and R. A. Wilchins. 1997. "One Percent on the Burn Chart: Gender, Genitals, and Hermaphrodites with Attitude." *Social Text* 15 (3–4): 215–22.

Vance, C. S. 1991. "Anthropology Rediscovers Sexuality: A Theoretical Comment." *Social Science and Medicine* 33 (8): 875–84.

Van Der Kamp, H. J., et al. 1992. "Evaluation of Young Women with Congenital Adrenal Hyperplasia: A Pilot Study." *Hormone Research* 37, supplement 3: 45–49.

Van Howe, R. S., and C. J. Cold. 1997. "Sex Reassignment at Birth: Long-Term Review and Clinical Implications." *Archives of Pediatrics and Adolescent Medicine* 151 (10): 1062.

Van Seters, A. P., and A. K. Slob. 1988. "Mutually Gratifying Heterosexual Relationship with Micropenis of Husband." *Journal of Sex and Marital Therapy* 14 (2): 98–107.

Van Wyk, J., and A. Calikoglu. 1999. "Should Boys with Micropenis Be Raised as Girls?" *Journal of Pediatrics* 134 (5): 537–38.

Vecchietti, G. 1965. "Neovagina nella sindrome di Rokitansky-Kuster-Hauser" [Creation of an artificial vagina in Rokitansky-Kuster-Hauser syndrome]. *Attualità di Ostetricia e Ginecologia* 11 (2): 131–47.

——. 1979. "Le Neo-vagin dans le syndrome de Rokitansky-Kuster-Hauser" [The Neovagina in the Robitansky-Kuster-Hauser syndrome]. *Revue Médicale de la Suisse Romande* 99 (9): 593–601.

Verghese, A. 2006. "AIDS at 25: An Epidemic of Caring." *New York Times*, June 4.

Verkauf, B. S., J. von Thron, and W. O'Brien. 1992. "Clitoral Size in Normal Women." *Obstetrics and Gynecology* 80 (1): 41–44.

Veronikis, D. K., G. B. Mcclure, and D. H. Nichols. 1997. "The Vecchietti Operation for Constructing a Neovagina: Indications, Instrumentation, and Techniques." *Obstetrics and Gynecology* 90 (2): 301–4.

Walker, M. G. 1998. "The Psychological Experience of Living with Hypospadias through Verbal Descriptions and Drawing." PhD diss., Union Institute.

Wallace, T. M., and H. S. Levin. 1990. "Mixed Gonadal Dysgenesis: A Review of Fifteen Patients Reporting Single Cases of Malignant Intratubular Germ Cell Neoplasia of the Testis, Endometrial Adenocarcinoma, and a Complex Vascular Anomaly." *Archives of Pathology and Laboratory Medicine* 114 (7): 679–88.

Walsh, P., et al., eds. 1992. *Campbell's Urology*. 6th ed. Philadelphia: Saunders.

Warnke, G. 2001. "Intersexuality and the Categories of Sex." *Hypatia* 16 (3): 126–37.

Weeks, J. 1986. *Sexuality*. London, Tavistock.

Wesley, J. R., and A. G. Coran. 1992. "Intestinal Vaginoplasty for Congenital Absence of the Vagina." *Journal of Pediatric Surgery* 27 (7): 885–89.

White, R. 1997. "Intersexuals." *All Things Considered*, NPR, November 28.

Wilkins, L. 1957. "Abnormal Sex Differentiation: Hermaphroditism and Gonadal Dysgenesis." In *The Diagnosis and Treatment of Endocrine Disorders in Childhood and Adolescence*, 2nd ed., 258–91. Springfield, IL: Thomas.

——. 1965. *The Diagnosis and Treatment of Endocrine Disorders in Childhood and Adolescence*, 3rd ed. Springfield, IL: Thomas.

Wilkins, L., et al. 1951. "Treatment of Congenital Adrenal Hyperplasia with Cortisone." *Journal of Clinical Endocrinology and Metabolism* 11 (1): 1–25.

Williams, E. 1964. "Congenital Absence of the Vagina: A Simple Operation for Its Relief." *Journal of Obstetrics and Gynaecology of the British Commonwealth* 71: 511–16.

Wiser, W. L., and G. W. Bates. 1984. "Management of Agenesis of the Vagina." *Surgery, Gynecology, and Obstetrics* 159 (2): 108–12.

Wisniewski, A. B., et al. 2000. "Complete Androgen Insensitivity Syndrome: Long-Term Medical, Surgical, and Psychosexual Outcome." *Journal of Clinical Endocrinology and Metabolism* 85 (8): 2664–69.

Wollett, A., D. White, and L. Lyon. 1982. "Observations of Fathers at Birth." In *Fathers: Psychological Perspectives*, ed. N. Beail and J. McGuire, 71–91. London: Junction Books.

Woolf, S. H., et al. 1999. "Clinical Guidelines: Potential Benefits, Limitations, and Harms of Clinical Guidelines." *British Medical Journal* 318 (7182): 527–30.

Young, H. H. 1937. *Genital Abnormalities, Hermaphroditism, and Related Adrenal Diseases*. Baltimore: Williams and Wilkins.

Young, H. H., and D. M. Davis. 1926. *Young's Practice of Urology, Based on a Study of 12,500 Cases*. Philadelphia: Saunders.

Young, I. M. 1990. *Throwing Like a Girl and Other Essays in Feminist Philosophy and Social Theory*. Bloomington: Indiana University Press.

Young, R. 2000. "Sexing the Brain: Measurement and Meaning in Biological Research on Human Sexuality." PhD diss., Columbia University.

Zderic, S. A., et al., eds. 2002. *Pediatric Gender Assignment: A Critical Reappraisal*. New York: Kluwer Academic/Plenum Publishers.

Zucker, K. J. 1996. "Commentary on Diamond's 'Prenatal Predisposition and the Clinical Management of Some Pediatric Conditions.'" *Journal of Sex and Marital Therapy* 22 (5): 148–60.

——. 1999. "Intersexuality and Gender Identity Differentiation." *Annual Review of Sex Research* 10: 1–69.

——. 2002. "Intersexuality and Gender Identity Differentiation." *Journal of Pediatric and Adolescent Gynecology* 15 (1): 3–13.

Zuger, B. 1970. "Gender Role Determination: A Critical Review of the Evidence from Hermaphroditism." *Psychosomatic Medicine* 32 (5): 449–67.

INDEX

KATRINA KARKAZIS

is a senior research scholar at the Stanford Center for Biomedical Ethics at Stanford University.

Library of Congress Cataloging-in-Publication Data
Karkazis, Katrina, 1970–
Fixing sex : intersex, medical authority, and lived experience / Katrina Karkazis.
p. ; cm.
Includes bibliographical references and index.
ISBN 978-0-8223-4302-8 (cloth : alk. paper)
ISBN 978-0-8223-4318-9 (pbk. : alk. paper)
1. Intersexuality. 2. Gender identity.
3. Intersexuality—Social aspects.
4. Intersex children. I. Title.
[DNLM: 1. Hermaphroditism—psychology.
2. Gender Identity. 3. Hermaphroditism—surgery.
WJ 712 K18f 2008]
QP267.K37 2008
616.6′94—dc22 2008013531